ADVANCED MATHS

A2 CORE
FOR AQA

John Wood
Rosemary Emanuel
Janet Crawshaw

PEARSON

Longman

Pearson Education Limited
Edinburgh Gate
Harlow
Essex
CM20 2JE
England
www.longman.co.uk

First published 2005
ISBN 0 582 84240 9

Design by Ken Vail Graphic Design

Cover design by Raven Design

Index by Indexing Specialists (UK) Ltd

Typeset by Techset, Gateshead

Printed in the U.K. by Scotprint, Haddington

The publisher's policy is to use paper manufactured from sustainable forests.

Live Learning, Live Authoring and Live Player are all trademarks of Live Learning LTD.

The Publisher wishes to draw attention to the Single-User Licence Agreement situated at the back of the book. Please read this agreement carefully before installing and using the CD-ROM.

AQA examination questions are reproduced by permission of the Assessment and Qualifications Alliance. All such questions have a reference in the margin. AQA can accept no responsibility whatsoever for accuracy of any solutions or answers to these questions. Please note that the following AQA (AEB) questions used are **not** from the live examinations for the current specification (Book: Q1, p. 98; Q9, p. 100; Q1, p. 316; Q12, p. 318; Q15, p. 318; Q3, p. 387; Q10, p. 389; Q11, p. 390; CD-ROM Extension Exam Questions: Q7; Q13, Q25, Q34). For GCSE Advanced Level subjects, new specifications were introduced in 2001.

On the CD-ROM we are grateful for permission from *Edexcel* to reproduce past exam questions. All such questions have a reference in the margin. *Edexcel* can accept no responsibility whatsoever for accuracy of any solutions or answers to these questions.

On the CD-ROM we are grateful for permission from *OCR* to reproduce past exam questions. All such questions have a reference in the margin. Oxford Cambridge and RSA Examinations is the copyright holder of all examination questions referenced '*OCR*' '*UCLES*' and '*Cambridge*'. *OCR* can accept no responsibility whatsoever for accuracy of any solutions or answers to these questions.

Every effort has been made to ensure that the structure and level of sample question papers matches the current specification requirements and that solutions are accurate. However, the publisher can accept no responsibility whatsoever for accuracy of any solutions or answers to these questions. Any such solutions or answers may not necessarily constitute all possible solutions.

Contents

Introduction

A2 Core for AQA is written to support the AQA A2 Mathematics specification (teaching from September 2004). Chapters 1–8 cover material from Unit C3 and Chapters 9–15 cover Unit C4. The accompanying text *AS Core for AQA* covers units C1 and C2. The authors hope that, in the course of preparing students for their A level examinations, this book will allow students to experience the intellectual excitement of advanced mathematics and to appreciate some of its powerful uses.

The book is designed to support all students, including those who need extra guidance and those who want to be challenged. It can be used by students and teachers in class, but the book and CD-ROM also provide a range of features for students working independently.

The book

Each chapter starts with a list of topics, from the AQA specification, to be covered in that chapter. The text has a full explanation of each topic, numerous worked examples and core exercises. The worked examples, covering all types of questions, are particularly useful for students who have to work on their own and for revision. They are presented in a user-friendly format, with solutions on the left of the page, and helpful comments – as might be made by a teacher – on the right.

The exercises in the book are carefully graded to give comprehensive coverage of the AQA specification. Students who need work of a more challenging nature, or are preparing for the Advanced Extension Awards, will find *Extension material*, containing harder questions on the specification and extra material beyond the specification, on the student CD-ROM.

At the end of each module is a *sample examination paper*.

Key points at the end of each chapter summarise the important results of the chapter.

Review exercises at the end of each chapter test synoptic understanding.

Extension material or *extension questions* in the book are marked ■ or with a blue band in the margin.

Past AQA examination questions are included in four *exam practice exercises*, each covering the topics of a number of chapters.

The *glossary* of mathematical terms and a *list of notation* are available for reference.

The student CD-ROM

The accompanying student CD-ROM is found in the back of this book. When additional support for a topic is available, a reference is given with a CD-ROM logo ⊘ next to it.

The student CD-ROM contains:

■ *'Live-authored' solutions* for the C3 and C4 sample examination paper. These talk students through model solutions to the exam questions as well as giving hints on exam technique. Details on how to install the *Live Player*® and access these solutions are included on the next page.

For most chapters, the CD-ROM also contains:

■ *Extension material* which supplements the usual course of study and extends topics beyond A level
■ *Extension exercises* to offer a challenge and help prepare students for Advanced Extension Awards papers
■ *Test yourself exercises* consisting of multiple choice questions for revision.

Answers are given either exactly or approximately. The usual approximation is to three significant figures or to one decimal place for angles measured in degrees. To discourage the rounding of numbers part way through a numerical calculation, numbers read from a calculator are written in the form 56.789 . . . The dots indicate that the first few digits of the unrounded display on the calculator have been written down.

The CD-ROM also contains a Miscellany of questions that provide a challenge and call on many of the skills and techniques presented in this course.

Acknowledgements

The authors are grateful to all those who have read the manuscript and made helpful suggestions and contributions. We are particularly indebted to John Backhouse, John Barrett, Philip Cooper, Andrew Davis, Jane Dyer, Frankie Elston, John Emanuel, Tony Fisher, John Gillard, Sheila Hill, David Hodgson, Peter Horril, Peter Houldsworth, Lyn Imeson, Barbara Morris, Laurence Pateman, Ian Potts, George Ross, Michael Ross, Rosemary Smith, John Spencer, Gareth West, Rhona West, Johanne Wood and Ben Yudkin.

We thank the excellent team at Longman for turning the manuscript into a book. The commitment to excellence by Longman has been a constant encouragement.

Using the CD-ROM

To use the CD-ROM, start the computer and insert the *A2 Core for AQA* CD-ROM. The program will automatically run and a menu will appear. You can choose the file you wish to view from the menu screen by clicking on the link for that file.

When the file appears, you can use Acrobat's back and forward buttons (or previous page and next page buttons) to move between pages.

If you wish to see and hear a worked solution to a question in one of the exam papers, click on the 'LA' icon that appears in the margin of each question paper. The first time you click on an 'LA' icon you will be asked if you want to install *Live Player*. Click 'Yes' to install and follow the instructions. The installation may take several minutes. After installation, the *Live Player* window will start.

The first time you click on an 'LA' icon a warning message will appear. Tick the option 'do not show this message again' and click 'open' to load the *Live Player* and play the worked solution.

The solution may be paused, replayed or forwarded using the control buttons at the bottom of the *Live Player* window.

Some questions have long solutions, so the live authored solution is divided into parts. These questions will have more than one 'LA' icon alongside. When one part of the solution ends simply click on the next icon to continue the solution. In some cases a Live Authored solution will refer to working from a previous part of a question.

If the software does not automatically run on your PC:

1. Select the My Computer icon on the Windows desktop
2. Select the CD-ROM drive icon
3. Select Open
4. Select a2_core.exe

Hardware requirements

Operating system: Windows 95(OS R2), 98, ME, 2000, NT or XP.

Pentium 100 (IBM Compatible PC) or equivalent PC

32 MB RAM or Higher

16 bit graphic card

4 speed CD-ROM drive (minimum 16× recommended)

SVGA colour monitor and 800 × 600 resolution

Sound card

At least 100 MB free hard disk space

Notation

Miscellaneous symbols

$=$	equals *or* is equal to
$\equiv$	is identical to
$>$	is greater than
$\geqslant$	is greater than or equal to
$<$	is less than
$\leqslant$	is less than or equal to
$\therefore$	therefore
$p \Rightarrow q$	p implies q; if p then q
$p \Leftarrow q$	p is implied by q; if q then p
$p \Leftrightarrow q$	p implies and is implied by q; p if, and only if, q
∞	infinity

Set notation

$\in$	is an element of
$\notin$	is not an element of
$\mathbb{N}$	the set of natural numbers
$\mathbb{Z}$	the set of integers
$\mathbb{Z}^+$	the set of positive integers
$\mathbb{Q}$	the set of rational numbers
$\mathbb{R}$	the set of real numbers
(a, b)	the open interval $\{x \in \mathbb{R}: a < x < b\}$
$[a, b]$	the closed interval $\{x \in \mathbb{R}: a \leqslant x \leqslant b\}$
$(a, b]$	the interval $\{x \in \mathbb{R}: a < x \leqslant b\}$

Ranges on a number line

Ranges can be illustrated graphically

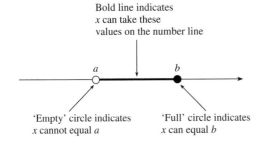

Bold line indicates x can take these values on the number line

'Empty' circle indicates x cannot equal a

'Full' circle indicates x can equal b

Functions

$f(x)$	the value of the function f at x
$f:x \mapsto y$	the function f maps the element x to the element y
f^{-1}	the inverse function of the function f
gf	the composite function of f and g defined by $gf(x) = g(f(x))$

$\lim_{x \to a} f(x)$	the limit of $f(x)$ as x tends to a
$\delta x, \Delta x$	an increment of x
$\dfrac{dy}{dx}$	the derivative of y with respect to x
$\dfrac{d^n y}{dx^n}$	the nth derivative of y with respect to x
$f'(x), f''(x), \ldots, f^{(n)}(x)$	the first, second, $\ldots$, nth derivatives of $f(x)$ with respect to x
$\int y \, dx$	the indefinite integral of y with respect to x
$\int_a^b y \, dx$	the definite integral of y with respect to x between the limits $x = a$ and $x = b$

Operations

$\displaystyle\sum_{i=1}^{n} a_i$	$a_1 + a_2 + \cdots + a_n$
$\sqrt{a}$ or $\sqrt{a}$	the positive square root of a
$\lvert a \rvert$	the modulus of a
$n!$	n factorial
$\dbinom{n}{r}$	the binomial coefficient $\dfrac{n!}{r!(n-r)!}$ for $n \in \mathbb{Z}^+$

Exponential and logarithmic functions

e	base of natural logarithms
$e^x, \exp x$	exponential function of x
$\log_a x$	logarithm to the base a of x
$\ln x, \log_e x$	natural logarithm of x
$\lg x, \log_{10} x$	logarithm of x to base 10

Vectors

$\mathbf{a}$	the vector $\mathbf{a}$
$\overrightarrow{AB}$	the vector represented in magnitude and direction by the directed line segment AB
$\hat{\mathbf{a}}$	a unit vector in the direction of $\mathbf{a}$
$\mathbf{i}, \mathbf{j}, \mathbf{k}$	unit vectors in the directions of the Cartesian coordinate axes
$\lvert \mathbf{a} \rvert, a$	the magnitude of $\mathbf{a}$
$\lvert \overrightarrow{AB} \rvert, AB$	the magnitude of $\overrightarrow{AB}$
$\mathbf{a} . \mathbf{b}$	the scalar product of $\mathbf{a}$ and $\mathbf{b}$

1 Functions

Earlier in the course you dealt with the relationships between two algebraic expressions. This chapter considers the relationships between two sets of numbers and introduces the concept of a function. Functions allow us to investigate relationships between different quantities and how a change in one quantity affects the other. They are used in a wide range of situations from modelling the economy to describing the motion of a spacecraft.

After working through this chapter you should

■ *be familiar with the concept of a function as a one–one or many–one mapping*

■ *understand the terms domain and range*

■ *be able to find the range of a function when given its domain*

■ *be able to form composite functions*

■ *be able to find an inverse function when it exists*

■ *be able to use the graph of a function to sketch the graph of the inverse function*

■ *be familiar with the modulus function* f(x) = |x|, *and its graph*

■ *be able to apply a combination of transformations to sketch the graph of a function.*

1.1 Mappings

This chapter looks at many relationships between two sets of numbers.

Such a relationship or **mapping** can be expressed as a rule or by an equation or in a table. The relationship or mapping can be illustrated by a diagram, such as a graph, or, by ordered pairs.

Consider the mapping between the elements of set $A = \{1, 2, 3, 4, 5\}$ and of set $B = \{3, 5, 7, 9, 11\}$.

The set A (the 'input' set) is called the **domain**.

The set B (the 'output' set) is called the **range**.

Each member of the domain maps to a member of the range.

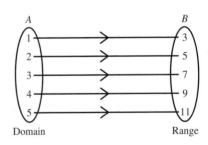

Each member of A is doubled and one is added to obtain the corresponding member of B.

If the elements of the domain are represented by x and of the range by y, the relationship is $y = 2x + 1$.

This can be expressed in function notation as

$$f(x) = 2x + 1$$

> Read as 'f of x equals 2x + 1'.

$$\text{or} \quad f : x \rightarrow 2x + 1$$

> Read as 'f maps x to 2x + 1'.

Domain $= \{1, 2, 3, 4, 5\}$

Range $= \{3, 5, 7, 9, 11\}$

Each element of the domain has an **image** in the range. For example, the image of 2 is 5. This can be expressed as

$\quad f(2) = 5$

or $\quad f : 2 \rightarrow 5$

The mapping can be expressed as a set of ordered pairs.

$\quad\quad$ (1, 3) $\quad$ (2, 5) $\quad$ (3, 7) $\quad$ (4, 9) $\quad$ (5, 11)

Member of domain $\quad$ Member of range

The mapping can be illustrated by plotting these ordered pairs as points.

Values on the *x*-axis represent the domain.

Values on the *y*-axis represent the range.

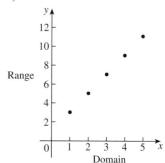

The value of *y* depends on the value of *x* so *y* is called the **dependent variable** and *x* the **independent variable**.

There are several ways in which sets, such as the domain or range, can be defined. In this chapter *x* is used for the domain and *y* or f(*x*) for the range.

Notation	Contents
$x \in \mathbb{R}$	*x* belongs to the set of real numbers $\mathbb{R}$.
$\{x : x \in \mathbb{R}, x \neq 0\}$	The set of values *x* such that *x* belongs to $\mathbb{R}$ and $x \neq 0$. NB read : as 'such that'
$x \neq 5$	Assume *x* is a real number. This would imply $x \in \mathbb{R}, x \neq 5$.
$y \in \mathbb{R}, 1 < y \leqslant 2$	*y* is a real number greater than 1 and less than or equal to 2.

Relations between *x* and *y* can be illustrated as graphs and divided into four types: **one–one**, **many–one**, **one–many** and **many–many**.

One–one relationships

For each value of *x*, there is a unique value of *y*, and vice versa.

A horizontal line cuts the graph, at most, once.

A vertical line cuts the graph, at most, once.

The relationship is **one–one**.

y is a function of *x*.

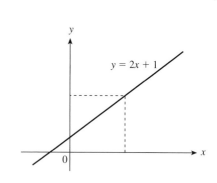

Many–one relationships

Many values of x (i.e. more than one) map to one value of y.

A horizontal line may cut the graph more than once.

A vertical line cuts the graph, at most, once.

The relationship is **many–one**.

y is a function of x.

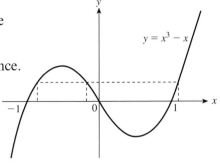

One–many relationships

One value of x maps to many (i.e. more than one) values of y.

A horizontal line cuts the graph, at most, once.

A vertical line may cut the graph more than once.

The relationship is **one–many**.

y is *not* a function of x (but x is a function of y).

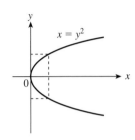

Many–many relationships

Many (i.e. more than one) values of x map to many (i.e. more than one) values of y.

For example, on $x^2 + y^2 = 9$, when $x = 1$, $y = \pm\sqrt{8}$, and when $y = 2$, $x = \pm\sqrt{5}$.

A horizontal line may cut the graph more than once.

A vertical line may cut the graph more than once.

The relationship is **many–many**.

y is *not* a function of x.

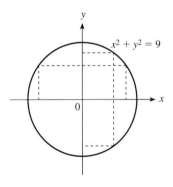

Only the first two types – one–one and many–one – are called **functions**. The others are included for completeness.

The essential feature of a function is that, for each member of the domain, there is one and *only* one member of the range.

Relationships like $f(x) = \pm\sqrt{x}$, where for a given value of x there is more than one value of $f(x)$ are not functions.

> **To be a function, a relationship must be one–one or many–one;**
> **it must be uniquely defined for all values of the domain.**

The possible values of x (the domain if dealing with a function) and the possible values of y (the range if dealing with a function) can be determined from the graph of the function.

Example 1 Find the range of given functions, and determine whether the functions are one–one or many–one.

When the elements of the domain are *listed*, substitute each in turn to find the elements of the range. In other cases, a sketch of the graph can be helpful in finding the range.

a $f(x) = 1 - 3x$. Domain: $\{0, 1, 2, 3, 4\}$.

The domain has five members 0, 1, 2, 3, 4.

$f(0) = 1$; $f(1) = -2$; $f(2) = -5$

$f(3) = -8$; $f(4) = -11$

Substituting each in turn gives the members of the range.

So, range is $\{-11, -8, -5, -2, 1\}$.

Each member of the domain maps to a different member of the range.

The order in the curly brackets need not relate to the order of the members of the domain. However, they should be in some order. Here they are given in ascending order.

∴   the function is one–one.

b $f(x) = 2x^2 - 1$. Domain: $x \in \mathbb{R}$.

Sketch the graph of $y = 2x^2 - 1$.

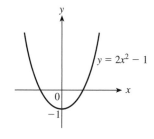

The domain (x-values) is all the real numbers.

From the sketch, the range is $y \in \mathbb{R}$, $y \geqslant -1$.

More than one value of x maps to a value of y.

For example, $f(1) = 1$ and $f(-1) = 1$.

Note: a horizontal line can cut more than once.

∴   the function is many–one.

c $f(x) = x^2 - 4$. Domain: $-1 \leqslant x \leqslant 3$.

Sketch the graph of $y = x^2 - 4$.

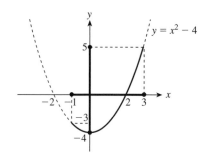

Note: Considering only the end values of the domain does not give the correct range.
$f(-1) = -3$ and $f(3) = 5$.
The range is not between -3 and 5 because, for example, $f(0) = -4$.

From the sketch, for $-1 \leqslant x \leqslant 3$, y can take values between -4 and 5.

So the range is $-4 \leqslant y \leqslant 5$.

The function is many–one.

d $f(x) = \dfrac{1}{x}$. Domain: $x \in \mathbb{R}$, $x \neq 0$.

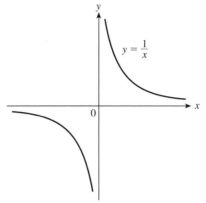

From the sketch, range is $y \in \mathbb{R}$, $y \neq 0$.

The function is one–one.

When a curve becomes closer and closer to a line but never reaches it, the line is called an **asymptote**. For $f(x) = \frac{1}{x}$ the x- and y-axes are both asymptotes. An asymptote which is not the x- or y-axis is shown with a dashed line. See Example 19 on page 29.

e $f(x) = x^4$. Domain: $1 < x \leqslant 2$.

Assume $x \in \mathbb{R}$.

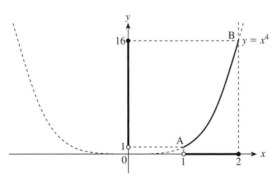

Range is $y \in \mathbb{R}$, $1 < y \leqslant 16$.

For $1 < x \leqslant 2$, the function is one–one.

f $f(x) = \begin{cases} 3x & 0 < x < 2 \\ x + 4 & 2 \leqslant x < 6 \end{cases}$.

Domain: $0 < x < 6$.

Range is $0 < y < 10$.

The function is one–one.

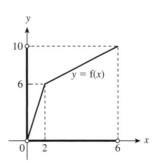

> This is the reciprocal function.

> The function is not defined for $x = 0$ because division by zero is not defined.

> Sketch the graph.

> The domain (x values) is all the real numbers excluding zero. The range (y values) is also all the real numbers excluding zero.

> A horizontal line can cut at most once.

> $x = 1$ is excluded from the domain so $y = f(1) = 1$ cannot be included in the range.
> $x = 2$ is included in the domain so $y = f(2) = 16$ is included in the range.

> Although $f(x) = x^4$ is a many–one function, for the domain given, the function is one–one. Between A and B a horizontal line cuts only once.

> Sketch the graph.

> $0 < x < 6 \Rightarrow 0 < y < 10$

> A horizontal line cuts at most once so the function is one–one.

Example 2

A function is not defined if it does not have a domain. However, in this example, the range is given and is used working backwards to find the domain. This is a useful exercise to understand the concepts involved.

Find the largest possible domain of given functions and determine whether the functions are one–one or many–one.

a $f(x) = x^2$. Range: $\{9, 16, 25\}$.

$x^2 = 9 \Rightarrow x = \pm 3$

$x^2 = 16 \Rightarrow x = \pm 4$

$x^2 = 25 \Rightarrow x = \pm 5$

Domain is $\{\pm 3, \pm 4, \pm 5\}$.

> The range has three members: 9, 16, 25. Putting x^2 equal to each in turn and solving for x will give the corresponding values of the domain.

> When the members of the range are listed, use this method to find the corresponding members of the domain.

Since more than one member of the domain maps to one member of the range the function is many–one.

b $f(x) = x^3$. Range: $-8 < y < 1$.

$x^3 = -8 \Rightarrow x = -2$

$x^3 = 1 \Rightarrow x = 1$

> Calculate the values at the end of the range.

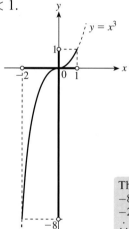

> Sketch the graph.

> The range (y-values) is $-8 < y < 1$.
> -8 and 1 are excluded from the range, so -2 and 1 will be excluded from the domain.
> $\therefore$ The domain (x-values) is $-2 < x < 1$.

From the sketch, the domain is $-2 < x < 1$.

The function is one–one.

> A horizontal line cuts at most once
> $\therefore$ one–one.

c $f(x) = 7x - 1$. Range: $6 \leqslant f(x) \leqslant 20$.

$7x - 1 = 6 \Rightarrow x = 1$

$7x - 1 = 20 \Rightarrow x = 3$

> Find the values of x corresponding to $f(x) = 6$ and $f(x) = 20$.

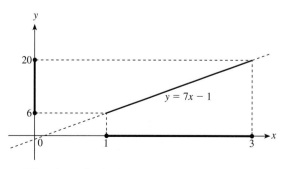

> Sketch the graph.

> The range is $6 \leqslant f(x) \leqslant 20$.
> The domain (x-values) is $1 \leqslant x \leqslant 3$.

The domain is $1 \leqslant x \leqslant 3$.

The function is one–one.

> A horizontal line cuts at most once
> $\therefore$ one–one.

d $f(x) = \dfrac{7}{(x-1)^2}$. Range: $f(x) \in \mathbb{R}$, $f(x) > 7$.

$f(x) = 7 \Rightarrow \dfrac{7}{(x-1)^2} = 7$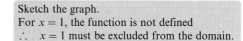

That is, $(x-1)^2 = 1$

Take square root of both sides remembering $\pm$ sign.

$x - 1 = \pm 1$

$\therefore \quad x = 0$ or $x = 2$.

Sketch the graph.
For $x = 1$, the function is not defined
$\therefore \quad x = 1$ must be excluded from the domain.

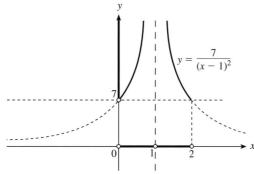

From the sketch the domain is $0 < x < 2$, $x \in \mathbb{R}$, $x \neq 1$.

The domain could be written $0 < x < 1$ and $1 < x < 2$.

The function is many–one.

A horizontal line cuts more than once $\therefore$ many–one.

Example 3   The function f is defined by $f(x) = x^2 - 2x - 8$, $x \in \mathbb{R}$.

a Find $f(-2)$.

b Solve $f(x) = 5x$.

c By completing the square, find the range of f.

Solution   **a** $f(-2) = (-2)^2 - 2 \times (-2) - 8$

$= 4 + 4 - 8$

$= 0$

b $f(x) = 5x \Rightarrow x^2 - 2x - 8 = 5x$

$x^2 - 7x - 8 = 0$

$(x - 8)(x + 1) = 0$

$x = 8$ or $x = -1$

c $f(x) = x^2 - 2x - 8$

$= (x - 1)^2 - 9$

Complete the square.

To find the range, either sketch the graph, or notice that $(x - 1)^2 \geqslant 0$
$\therefore \quad (x - 1)^2 - 9 \geqslant -9$.
For the sketch, $f(x) = x^2$ has been translated $\begin{pmatrix} 1 \\ -9 \end{pmatrix}$.
Only values of $y \geqslant -9$ are in the range.

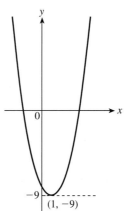

The range is $y \in \mathbb{R}$, $y \geqslant -9$.

1 By sketching their graphs, or otherwise, find the range of these functions, and state whether each function is one–one or many–one.

    a $f : x \rightarrow 2x$ Domain: $\{0, 2, 4, 8\}$

    b $f : x \rightarrow x^2 + 1$ Domain: $x \in \mathbb{R}$

    c $f : x \rightarrow 3x - 1$ Domain: $x \in \mathbb{R},\ -2 < x < 2$

    d $f : x \rightarrow \dfrac{1}{x}$ Domain: $x \in \mathbb{R},\ x \neq 0$

    e $f : x \rightarrow 2x^3$ Domain: $x \in \mathbb{R},\ -1 \leqslant x \leqslant 4$

    f $f : x \rightarrow \begin{cases} \sqrt{x} & 0 \leqslant x < 1 \\ 2 - x & 1 \leqslant x \leqslant 4 \end{cases}$ Domain: $x \in \mathbb{R},\ 0 \leqslant x \leqslant 4$

    g $f : x \rightarrow \dfrac{1}{(1 + x)^2}$ Domain: $x \in \mathbb{R},\ x \neq -1$

2 Find the largest possible domains of these functions, and state whether each function is one–one or many–one.

    a $f(x) = 2x - 1$ Range: $\{5, 10, 15, 20\}$

    b $f(x) = \sqrt{x}$ Range: $y \in \mathbb{R},\ 2 < y < 7$

    c $f(x) = 4 - x^2$ Range: $y \in \mathbb{R},\ y < 0$

    d $f(x) = \dfrac{1}{x - 6}$ Range: $y \in \mathbb{R},\ \frac{1}{6} \leqslant y \leqslant 1$

    e $f(x) = x^4$ Range: $y \in \mathbb{R},\ 1 < y \leqslant 81$

3 The function f is defined by $f(x) = x^2 - 2x + 9$, $x \in \mathbb{R}$.

    a Express $f(x)$ in the form $f(x) = (x - a)^2 + b$.

    b Hence, or otherwise, sketch $y = f(x)$ and state the range of the function.

4 Given that $f(x) = x^2 + 3x - 4$, $x \in \mathbb{R}$, sketch $y = f(x)$ and hence

    a find the range of f

    b state the solution of $f(x) = 0$

5 **a** Given that $6 + 2x - x^2 \equiv p - (q - x)^2$, find p and q.

    b Hence, or otherwise, find the range of the function, f, where
        $f : x \rightarrow 6 + 2x - x^2$, $x \in \mathbb{R}$.

6 These functions have domain $x \in \mathbb{R}$. Find, by completing the square, or otherwise, the range of each function.

    a $f(x) = x^2 + 4x - 7$ **b** $f(x) = 2x^2 - 6x + 1$

    c $f(x) = x^2 - 5x + 5$ **d** $f(x) = 6 - 3x - x^2$

7 **a** The function f is defined by $f(x) = (x - 2)^2 + 5$ for the domain $x \in \mathbb{R}$, $1 \leqslant x \leqslant 5$. Sketch the curve with equation $y = f(x)$, marking the vertex of the parabola.

    b Show, with reference to the sketch or otherwise, that the range of f for the given domain is $y \in \mathbb{R}$, $5 \leqslant y \leqslant 14$.

    c Solve $f(x) = 9$.

8 The function f is defined for $x \in \mathbb{R}$, $x > 0$ as $f : x \to 4 + \dfrac{5}{x}$

    a Sketch the function for $x > 0$.

    b State the range of the function.

    c Solve $f(x) = x$.

9 The function f is defined as $f(x) = \dfrac{2x + 3}{x - 1}$ with $x \neq 1$.

    a Express $f(x)$ in the form

$$f(x) = a + \frac{b}{x - 1}$$

    where a and b are constants.

    b The diagram shows a sketch of $y = f(x)$. State the coordinates of the points A, B and C and the equations of the asymptotes.

    c State the range of f.

    d Solve $f(x) = x + 1$, leaving surds in the answer.

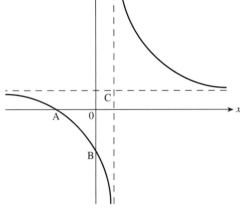

10 To be a function, a relationship must be one–one or many–one, and it must be uniquely defined for all values of the domain. Explain why these are *not* functions.

    a $x \to \begin{cases} 3 + x & 0 \leqslant x \leqslant 2 \\ 2x & 2 \leqslant x \leqslant 4 \end{cases}$

    b $x \to \pm\sqrt{x}$ for $x > 0$

    c $x \to y$ for the graph of y shown here.

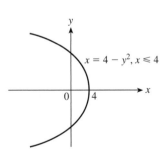

    d $x \to \dfrac{3}{x - 2}$ for $x \in \mathbb{R}$.

1.2 Composite functions

When two or more functions are combined, so that the output from the first function becomes the input to the second function, the result is called a **composite function.**

Consider the functions f and g defined by $f(x) = 2x + 1$ with domain $\{1, 2, 3, 4, 5\}$ and $g(x) = x^2$ with domain the range of f.

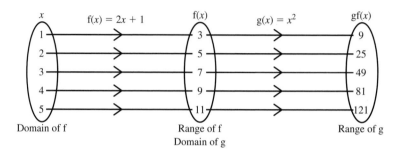

Here f is applied first, followed by g.
The function f is applied to x, and g is applied to $f(x)$. This is written

$$gf(x)$$

2nd function applied   1st function applied

The effect of f is to double the number and add 1.

The effect of g is to square the number.

The effect of f then g is to double the number, add 1 and then square.

The combined function of gf can be written $gf(x) = (2x + 1)^2$

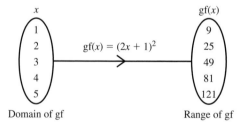

The following diagrams illustrate the process of finding composite functions. Once the process is understood it can be carried out mentally. The final algebraic expression may need to be simplified. If several functions are combined they are applied from the right.

$$fgh(x)$$

3rd   2nd   1st

Consider the functions f, g and h defined by $f(x) = 2x + 1$, $g(x) = x^2$ and $h(x) = \dfrac{1}{x}$, for all real values of x, $x \neq 0$.

To find fg. Do g first, then f.

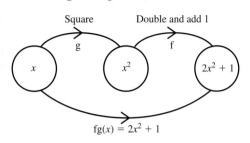

$$fg(x) = 2x^2 + 1$$

To find gh. Do h first, then g.

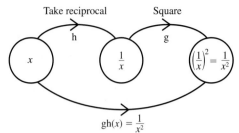

$$gh(x) = \frac{1}{x^2}$$

To find gfh. Do h, then f, then g.

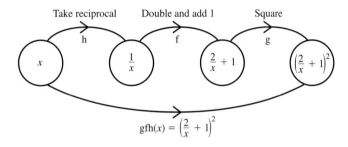

$$gfh(x) = \left(\frac{2}{x} + 1\right)^2$$

Example 4 The functions f, g and h are defined for $x \in \mathbb{R}$, $x \neq 0$ by $f(x) = 2x + 1$, $g(x) = x^2$ and $h(x) = \frac{1}{x}$. Find $fg(x)$ and $gh(x)$, $f^2(x)$, $h^2(x)$ and $fgh(x)$.

a $fg(x) = 2x^2 + 1$

> Do g first so x becomes x^2. Substitute x^2 in f.

> Or $fg(x) = f(g(x))$
> $= f(x^2)$
> $= 2x^2 + 1$

b $gh(x) = \left(\frac{1}{x}\right)^2$

$$= \frac{1}{x^2}$$

> Do h first so x becomes $\frac{1}{x}$. Substitute $\frac{1}{x}$ in g.

> Or $gh(x) = g(h(x))$
> $= g\left(\frac{1}{x}\right)$
> $= \left(\frac{1}{x}\right)^2$

c $f^2(x) = 2(2x + 1) + 1$
$$= 4x + 3$$

> $f^2(x)$ means $ff(x)$. f changes x to $2x + 1$
> Substitute $2x + 1$ in f.

d $h^2(x) = \dfrac{1}{\frac{1}{x}}$

$$= x$$

> h changes x to $\frac{1}{x}$. Substitute $\frac{1}{x}$ in h.

e $fgh(x) = 2 \times \dfrac{1}{x^2} + 1$

$$= \frac{2}{x^2} + 1$$

> h changes x to $\frac{1}{x}$. Substitute $\frac{1}{x}$ in g giving $\left(\frac{1}{x}\right)^2$.

> Then substitute $\frac{1}{x^2}$ in f giving $2 \times \frac{1}{x^2} + 1$.

Example 5 $f(x) = x^2$ $g(x) = 3x + 1$

    a   Show that, in general, $fg(x) \neq gf(x)$.

    b   Find the values of x such that $fg(x) = gf(x)$.

Solution   **a**
$$fg(x) = (3x + 1)^2$$
$$= 9x^2 + 6x + 1$$

> Do g first so x becomes $(3x + 1)$.
> Substitute $(3x + 1)$ in f.

$$gf(x) = 3x^2 + 1$$
$$\therefore \quad fg(x) \neq gf(x)$$

> Do f first so x becomes x^2.
> Substitute x^2 in g.

    b
$$fg(x) = gf(x)$$

> Put the two expressions equal to each other.

$$\text{So} \quad 9x^2 + 6x + 1 = 3x^2 + 1$$

> Tidy up.

$$6x^2 + 6x = 0$$

> Divide both sides by 6 and factorise.

$$x(x + 1) = 0$$

> Solve the quadratic equation.

$$\Rightarrow \qquad x = 0 \quad \text{or} \quad x = -1$$

So, the values of x such that $fg(x) = gf(x)$ are $x = -1$ and $x = 0$.

> ➤ | **The composite function, fg, can be formed only if the range of g is a subset of (i.e. is contained within) the domain of f.**

Example 6   *This example illustrates problems which may arise from incompatible domains and ranges.*

Given that $f: x \rightarrow 2x + 1$, domain $x \in \mathbb{R}$, $0 < x < 10$, and $g: x \rightarrow x^2$, domain $x \in \mathbb{R}$, $0 < x < 3$, can fg and gf be formed?

$f: x \rightarrow 2x + 1, \; x \in \mathbb{R}, \; 0 < x < 10$ $g: x \rightarrow x^2, \; x \in \mathbb{R}, \; 0 < x < 3$

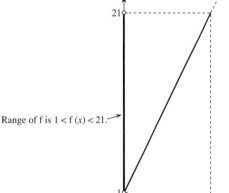

Range of f is $1 < f(x) < 21$.

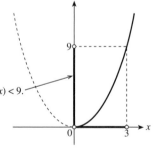

Range of g is $0 < g(x) < 9$.

Consider fg. Applying g first will give the range $0 < g(x) < 9$.
This range falls within domain of f so f can be applied. Hence fg can be formed.

Consider gf. Applying f first will give range $1 < f(x) < 21$.
The domain of g is $0 < x < 3$ so g can only be applied to the range of f which falls between 0 and 3, i.e. $1 < x < 3$.

So gf can only be formed if the domain of f is restricted to $0 < x < 1$.

Exercise 1B

In this exercise, the domain is $\mathbb{R}$, unless stated otherwise.

1 Given that the functions f and g are defined by $f(x) = 3x + 4$ and $g(x) = x^2$ find

    a $g(2)$ **b** $f(g(2))$ **c** $fg(x)$ **d** $fg(2)$

2 For each of these pairs of functions, find $fg(x)$ and $gf(x)$.

    a $f(x) = x + 1$, $g(x) = x + 2$ **b** $f(x) = 2x - 1$, $g(x) = 3x + 2$

    c $f(x) = x + 2$, $g(x) = x - 2$ **d** $f(x) = 2x + 1$, $g(x) = 4 - 3x$

    e $f(x) = x + 1$, $g(x) = x^2$ **f** $f(x) = \dfrac{1}{x}$, $x \neq 0$, $g(x) = 2x$

    g $f(x) = \dfrac{1}{x}$, $x \neq 0$, $g(x) = x + 1$ **h** $f(x) = (x + 1)(x - 2)$, $g(x) = 2x$

    (Notice how rarely fg and gf are the same.)

3 Given that the functions f, g and h are defined by $f: x \to x^2$, $g: x \to \dfrac{1}{x}$ and

    $h: x \to 3x - 1$ find expressions for these functions, expressing them in the form $fg: x \to \ldots$

    (Note $hgf(x) = h(gf(x))$ so the answer to part **b** can be used in part **g** for example).

    a fg   **b** gf   **c** gh   **d** ff

    e gg   **f** hh   **g** hgf   **h** fgh

    i fhg   **j** hfg   **k** ghf   **l** gfh

4 **a** Given that $f(x) = 3x + 1$ and $g(x) = 5x + k$, find k if $fg(x) = gf(x)$.

    b Check the answer to part **a** by showing that $fg(3) = gf(3)$.

5 The functions f, g and h are defined by

    $f(x) = x^2 - 1$, $g(x) = 3x + 2$ and $h(x) = \dfrac{1}{x}$.

    Solve these equations.

    a $fg(x) = 15$ **b** $gh(x) = -4$ **c** $hg(x) = x$ **d** $gg(x) = h(x)$

6 Given $f(x) = 2x + 7$, $g(x) = x^2$ and $h(x) = x^3$, find, stating the range, an expression for

    a $fg(x)$ **b** $fh(x)$ **c** $gf(x)$ **d** $hf(x)$

7 Suppose $f: x \to x^2 - 3$ and $g: x \to x + 4$.

    a Express fg and gf in the form $fg: x \to \ldots$ and $gf: x \to \ldots$

    b State the ranges of fg and of gf.

    c Find k such that $fg(k) = gf(k)$.

    d Find l leaving surds in the answer, such that $fg(l) = gf(3l)$.

8 **a** Given that $f(x) = x + 5$ find $ff(x)$, $fff(x)$ and $f^n(x)$, where $f^n(x)$ indicates that the function f is applied n times.

    b Find the function $g(x)$ such that $g^n(x) = 2^n x$

    c Find the function $h(x)$ such that $h^n(x) = \dfrac{x}{a^n}$

9 Suppose $f: x \rightarrow 2x - 1$, $g: x \rightarrow 3x^2 + 2$ and $h: x \rightarrow ax + b$ where a and b are positive constants.

a Show that $fg: x \rightarrow 6x^2 + 3$ and find the function gf.

b Find a and b such that $fgh: x \rightarrow 6x^2 + 12x + 9$.

c Find the values of x for which $fgh(x) = 57$.

10 Given that $f: x \rightarrow x + 2$, $g: x \rightarrow 2x^2 + 3$ and $h(x) \rightarrow \dfrac{1}{x}$, $x \neq 0$, find these functions.

a ff	**b** gg	**c** hh	**d** hgf
e hfg	**f** fgh	**g** fhg	**h** hgh

11 The composite function, fg, can be formed only if the range of g is a subset of (i.e. is contained within) the domain of f.

In the following examples this is not the case. Suggest the largest possible restricted domain of g which allows fg to be formed.

a $f(x) = \sqrt{x}$, $x \geqslant 0$
$g(x) = x + 2$, $x \in \mathbb{R}$

b $f(x) = 2x + 1$, $x \leqslant 20$
$g(x) = 5x$, $x \in \mathbb{R}$

c $f(x) = \dfrac{1}{x + 2}$, $x \geqslant 0$
$g(x) = 4 - x^2$, $x \in \mathbb{R}$

d $f(x) = \sqrt{x - 3}$, $x \geqslant 3$
$g(x) = \dfrac{2}{x}$, $x \in \mathbb{R}$, $x \neq 0$

1.3   Inverse functions

Consider again the example $f(x) = 2x + 1$ with domain $\{1, 2, 3, 4, 5\}$, and range $\{3, 5, 7, 9, 11\}$ (page 1).

The **inverse function** of f maps from the range of f back to the domain. The arrows would go in the opposite direction.

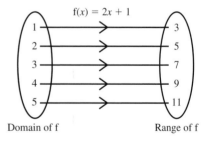

Since f has the effect of 'double and add one', the inverse function would be 'subtract one and halve'.

The inverse function is written f^{-1}.

In this case,

$$f^{-1}(x) = \frac{x - 1}{2}$$

or $f^{-1}: x \rightarrow \dfrac{x - 1}{2}$

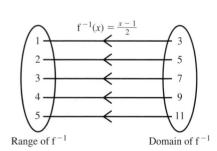

Note   The notation for inverse functions is unfortunate as it can be confused with reciprocals. Take care!

These are examples using $^{-1}$ for a reciprocal

$$x^{-1} = \frac{1}{x} \quad (f(x))^{-1} = \frac{1}{f(x)} \quad (\sin x)^{-1} = \frac{1}{\sin x}$$

whereas $f^{-1}(x)$ is the inverse of $f(x)$ and $\sin^{-1} x$ is inverse of $\sin x$.

For a unique inverse to exist the function must be one–one for the given domain.

Consider $f(x) = x^2$, $x \in \mathbb{R}$, a many–one function.

$$f(2) = 4$$
$$f(-2) = 4$$

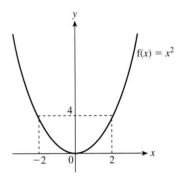

Two values of x, $+2$ and -2, have an image 4. Working backwards $f^{-1}(4)$ does not have a unique answer and hence $f(x) = x^2$ does not have an inverse function. However, if the domain of f is restricted to exclude negative numbers the inverse function $f^{-1}(x) = \sqrt{x}$ can then be defined.

Remember $\sqrt{}$ means take the $+$ve square root.

Note   One–one functions have inverses.

For many–one functions, an inverse can be defined by restricting the domain so that, for that part of the domain, the function is one–one.

Graphs of functions and their inverse functions

Consider $f(x) = 2x + 1$ and its inverse function $f^{-1}(x) = \dfrac{x-1}{2}$

$$f(3) = 7$$
$$f^{-1}(7) = 3$$

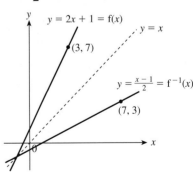

A point on the graph of $y = f(x)$ will be $(3, 7)$ and a point on the graph of $y = f^{-1}(x)$ will be $(7, 3)$.

In general if (a, b) lies on $y = f(x)$ then (b, a) lies on $y = f^{-1}(x)$.

For a function and its inverse, the roles of x and y are interchanged, so the two graphs are reflections of each other in the line $y = x$.

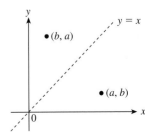

> **The graphs of a function and its inverse are reflections of each other in $y = x$, provided the scales on the axes are the same.**

To see what the graph of an inverse function looks like, turn over a page that has the graph of the function on it and rotate the page through 90° (clockwise). Hold the page to the light; the graph of the inverse function is seen.

Finding an inverse, i.e. working backwards, requires the same process as rearranging a formula or solving an equation.

For simple examples no formal method is needed.

Function $f(x)$	Process	Inverse process	Inverse function $f^{-1}(x)$	
kx $2x$	Multiply by k Double	Divide by k Halve	$\dfrac{x}{k}$ $\dfrac{x}{2}$	
$x + c$ $x + 1$	Add c Add 1	Subtract c Subtract 1	$x - c$ $x - 1$	
$ax + b$ $3x - 4$	Multiply by a and add b Multiply by 3 and subtract 4	Subtract b and divide by a Add four and divide by 3	$\dfrac{x - b}{a}$ $\dfrac{x + 4}{3}$	
$x^2, x \geqslant 0$	Square	Find square root	$\sqrt{x} \quad x \geqslant 0$	

For more complicated examples, the following method of finding an inverse is suggested. A simple example is used to illustrate the method.

> To find the inverse function, $f^{-1}(x)$ of $f(x)$
> - Put the function equal to y.
> - Rearrange to give x in terms of y.
> - Rewrite as $f^{-1}(x)$ replacing y by x.

Example 7   The function f is defined by $f(x) = 3x - 4$, $x \in \mathbb{R}$. Find the inverse function, f^{-1}, $x \in \mathbb{R}$.

Let $3x - 4 = y$

$\quad\quad\quad 3x = y + 4$

| Let the image of x be y. That is, the result of applying the function to x is y. |

$\therefore \quad\quad x = \dfrac{y + 4}{3}$

| To find how to go backwards, rearrange to give x in terms of y. |

So $f^{-1}(x) = \dfrac{x + 4}{3}$, $x \in \mathbb{R}$

| So given the image, y, $\frac{y+4}{3}$ must be worked out to get back to x. That is, add 4 to the number and divide by 3. *Note*: y in $\frac{y+4}{3}$ has been replaced by x. |

Example 8   Given $f(x) = \dfrac{1}{2 - x}$ for $x \in \mathbb{R}$, $x \neq 2$, find the inverse function, $f^{-1}(x)$.

Let $\dfrac{1}{2 - x} = y$

| Put the function equal to y. Rearrange to give x in terms of y. |

so $\quad 1 = y(2 - x)$

$\quad\quad\quad = 2y - xy$

| Multiply both sides by $2 - x$. |

$\quad\quad xy = 2y - 1$

| Isolate the term with x on one side and the rest of the terms on the other. |

$\quad\quad\quad x = \dfrac{2y - 1}{y}$

| Divide by y. |

$\therefore \quad f^{-1}(x) = \dfrac{2x - 1}{x}$

| Rewrite as $f^{-1}(x)$ replacing y by x. |

From the graph, for $f(x)$, the domain is $x \in \mathbb{R}$, $x \neq 2$, and the range is $y \in \mathbb{R}$, $y \neq 0$.

For the inverse function, $f^{-1}(x)$, the domain and range are reversed.

For $f^{-1}(x)$, the domain is $x \in \mathbb{R}$, $x \neq 0$, and the range is $y \in \mathbb{R}$, $y \neq 2$.

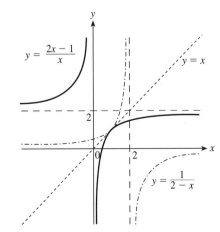

Practice in rearranging formulae, which is helpful for finding inverses, can be found on the CD-ROM accompanying Book 1 (A3.3).

> For a function and its inverse, the roles of x and y are interchanged, so the range of f is the domain of f^{-1}, and the range of f^{-1} is the domain of f.

Example 9 Given $f(x) = \dfrac{13 - 6x}{7}$, $x \in \mathbb{R}$

    a find the inverse function, $f^{-1}(x)$, $x \in \mathbb{R}$

    b verify the result using $x = 2$

    c show that $ff^{-1}(x) = f^{-1}f(x) = x$.

Solution **a** Let $\dfrac{13 - 6x}{7} = y$ Put the function equal to y.

                $13 - 6x = 7y$ Rearrange to give x in terms of y.

                    $6x = 13 - 7y$

    So $x = \dfrac{13 - 7y}{6}$ Rewrite as $f^{-1}(x)$ replacing y by x.

    and $f^{-1}(x) = \dfrac{13 - 7x}{6}$, $x \in \mathbb{R}$

  b $f(2) = \dfrac{13 - 6 \times 2}{7} = \dfrac{1}{7}$ Applying the function to 2 gives $\frac{1}{7}$; applying the inverse function to $\frac{1}{7}$ gives 2.

    $f^{-1}\left(\tfrac{1}{7}\right) = \dfrac{13 - 7 \times \frac{1}{7}}{6} = \dfrac{13 - 1}{6} = 2$ $f(2) = \frac{1}{7}$ and $f^{-1}\left(\frac{1}{7}\right) = 2$. The inverse function 'undoes' the function.

  c $f^{-1}(x) = \dfrac{13 - 7x}{6}$ From part **a**.

    $ff^{-1}(x) = \dfrac{13 - \dfrac{6 \times (13 - 7x)}{6}}{7}$ Substitute $\dfrac{13 - 7x}{6}$ in f.

          $= \dfrac{13 - (13 - 7x)}{7}$ *Note*: $13 - (13 - 7x) = 13 - 13 + 7x$

          $= x$

    $f(x) = \dfrac{13 - 6x}{7}$

    $f^{-1}f(x) = \dfrac{13 - \dfrac{7 \times (13 - 6x)}{7}}{6}$ Substitute $\dfrac{13 - 6x}{7}$ in f.

          $= \dfrac{13 - (13 - 6x)}{6}$

          $= x$

    So $ff^{-1}(x) = f^{-1}f(x) = x$.

Self-inverse functions

When $f(x) = f^{-1}(x)$ the function is said to be **self-inverse**.

Example 10 Show that $f(x) = \dfrac{x}{x-1}$, $x \in \mathbb{R}$, $x \neq 1$ is self-inverse.

Let $\dfrac{x}{x-1} = y$

> Put the function equal to y, and rearrange to give x in terms of y, by multiplying both sides by $x - 1$.

$x = y(x-1)$

$\quad = xy - y$

$y = xy - x$

> Take x outside the bracket.

$\quad = x(y-1)$

So $x = \dfrac{y}{y-1}$

> Rewrite as $f^{-1}(x)$ replacing y by x.

And $f^{-1}(x) = \dfrac{x}{x-1}$

Since $f(x) = f^{-1}(x)$ the function is self-inverse.

> Since a function, f, and its inverse, f^{-1}, undo each other $ff^{-1}(x) = f^{-1}f(x) = x$.
> If $f(x) = f^{-1}(x)$, then $f(x)$ is self-inverse.
> If $ff(x) = x$, then $f(x)$ is self-inverse.

Exercise 1C   *For this exercise, the domain is $\mathbb{R}$ unless stated otherwise.*

1 Find the inverse functions, $f^{-1}(x)$, of these functions, and state which of them have the property $f(x) = f^{-1}(x)$ (i.e. are self-inverse).

   a $f(x) = 3x$ **b** $f(x) = x + 4$ **c** $f(x) = x - 5$

   d $f(x) = \dfrac{x}{6}$ **e** $f(x) = \dfrac{1}{x}$, $x \neq 0$ **f** $f(x) = x^2$, $x \geqslant 0$

   g $f(x) = 3x - 4$ **h** $f(x) = 3 - 2x$ **i** $f(x) = 4 - x$

   j $f(x) = \dfrac{x+6}{5}$ **k** $f(x) = \dfrac{5}{x}$, $x \neq 0$ **l** $f(x) = \dfrac{1}{1-x}$, $x \neq 1$

2 Find the inverse of each of these functions, stating which functions are self-inverse.

   a $f : x \to \dfrac{2x+5}{3}$ **b** $f : x \to \dfrac{2x}{3} + 5$ **c** $f : x \to 2\left(\dfrac{x}{3} + 5\right)$

   d $f : x \to 2x + \frac{5}{3}$ **e** $f : x \to \frac{2}{3}(x+5)$ **f** $f : x \to \sqrt{x+4}$, $x \geqslant -4$

   g $f : x \to 7(x-4)$ **h** $f : x \to 4 - 7x$ **i** $f : x \to 3x^2 + 2$, $x \geqslant 0$

   j $f : x \to \dfrac{4}{7x}$, $x \neq 0$ **k** $f : x \to \dfrac{1}{x+1}$, $x \neq -1$ **l** $f : x \to \dfrac{1}{1 - \frac{1}{x}}$, $x \neq 0, 1$

2 m $f: x \to \dfrac{1}{x} + 1, \ x \neq 0$ **n** $f: x \to \dfrac{3}{5x}, \ x \neq 0$ **o** $f: x \to \dfrac{a}{bx}, \ x \neq 0$

p $f: x \to \frac{1}{2}\sqrt{x}, \ x > 0$ **q** $f: x \to \sqrt{2 - x}, \ x \leqslant 2$ **r** $f: x \to x^3 - 1$

s $f: x \to (3x + 1)^2 - 4, \ x \geqslant -\frac{1}{3}$ **t** $f: x \to \sqrt{2x - 1} + 5, \ x \geqslant \frac{1}{2}$

3 For each of these functions, by completing the square, or otherwise, find the range and the inverse function $f^{-1}(x)$. State the domain and range of $f^{-1}(x)$ and sketch, on the same axes, $y = f(x)$, $y = f^{-1}(x)$ and $y = x$.

a $f(x) = x^2 + 6x, \ x > -3$ **b** $f(x) = x^2 - 4x + 7, \ x > 2$

c $f(x) = 2 - 2x - x^2, \ x \geqslant -1$

4 a For $f(x) = 3x + 1$:
   i Find $f^{-1}(x)$ and state its domain.
   ii Sketch the graphs $y = f(x)$, $y = f^{-1}(x)$ and $y = x$ on the same axes, showing where the lines intersect.
   iii Show that $ff^{-1}(x) = f^{-1}f(x) = x$.
   iv Solve $f(x) = f^{-1}(x)$ and hence state the coordinates of the point of intersection in part **ii.**

b Repeat part **a** for $f(x) = 2 - 4x$.

5 $f(x) = 3x + 2$ and $g(x) = \dfrac{1}{x}$ with $x \neq 0$.

a Find $f^{-1}(x)$, $g^{-1}(x)$ and $gf(x)$

b Show that $(gf)^{-1}(x) = f^{-1}g^{-1}(x) = \dfrac{1}{3}\left(\dfrac{1}{x} - 2 \right)$.

6 Given that $f(x) = 2x - 5$ and $g(x) = 1 - x$ show that $(fg)^{-1}(x) = g^{-1}f^{-1}(x)$.

7 These functions are many–one and, therefore, do not have inverses. However, by restricting the domain, an inverse function can be found. Find a largest possible domain such that the functions are one–one, the corresponding ranges, and the inverse functions for the domains found. Assume $x \in \mathbb{R}$ and $x > -6$.

a $f(x) = x^2$

b $f(x) = (x + 2)^4$

c $f(x) = 4 - (x + 1)^2$

8 a Show that the function $g(x) = \dfrac{x}{x - 1}, \ x \neq 1$ is self-inverse.

b Sketch $y = \dfrac{x}{x - 1}$ and $y = x$ on the same axes and comment on the graphs.

9 **a** Copy the graph and, on the same axes, sketch $y = f^{-1}(x)$ and $y = f(x-2)$.

  b State the points at which the curves of
  $y = f^{-1}(x)$ and $y = f(x-2)$ cut the axes.

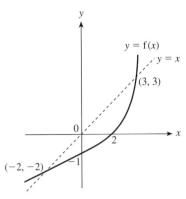

10 **a** Given that $f(x) = \dfrac{4}{x-1}$ with $x > 1$, find the inverse function $f^{-1}(x)$ and state its domain.

  b Sketch the graphs of $y = f(x)$, $y = f^{-1}(x)$ and $y = x$ on the same axes.

  c Solve $f(x) = f^{-1}(x)$, giving the answer correct to 3 significant figures.

11 For a function to have an inverse, it must be a one–one function.
Find, giving reasons, which of these have inverses.

  a $f: x \to 4 - x^2 \quad x \in \mathbb{R}$ **b** $f: x \to 4 - x^2 \quad x \in \mathbb{R} \quad x \leqslant 0$

  c $f: x \to \dfrac{1}{x} \qquad x \in \mathbb{R} \quad x \neq 0$ **d** $f: x \to \dfrac{1}{x-1} \qquad x \in \mathbb{R} \quad x \neq 1$

12 A function is defined as $f(x) = \dfrac{10-x}{x+2}$ with $x \neq -2$. Find

  a $f(6)$ **b** $f^{-1}(2)$ **c** k such that $f(k) = k$

13 **a** Given that $g(x) = \dfrac{2x+1}{x+2}$ with $x \neq -2$, find $g^{-1}(x)$ stating the domain of g^{-1}.

  b Solve $g(x) = g^{-1}(x)$.

14 **a** Given that $h(x) = -\dfrac{1}{x+1}$ with $x \neq -1$, find $h^{-1}(x)$ stating the domain of h^{-1}.

  b Show that $h(x) = h^{-1}(x)$ has no real solution.

15 Given $f(x) = \dfrac{2x+1}{a-x} \quad x \neq a$

  a Find $f^{-1}(x)$ and state its domain.

  b Find the value of a such that $f^{-1}(3) = 4$.

  c Find the value of a, $a \neq 0$, such that $f(x+a) = f(x)$ and explain the answer.

1.4 Even and odd functions

An **even function** is one where $f(x) = f(-x)$ for all values of x.

The graph is symmetrical about the y-axis.

Examples of even functions are $f(x) = x^2$, $f(x) = x^4$ and $f(x) = \cos x$.

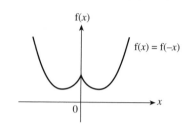

An **odd function** is one where $f(x) = -f(-x)$ for all values of x.

The graph has $180°$ rotational symmetry about the origin, $(0, 0)$.

Examples of odd functions are $f(x) = x$, $f(x) = x^3$ and $f(x) = \sin x$.

Most functions are neither even nor odd.

For example, $f(x) = x^2 + \dfrac{1}{x}$.

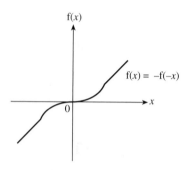

1.5 The modulus function

The notation $|x|$ means the magnitude of x, ignoring the sign. So, for example, $|6| = 6$ and $|-7| = 7$.

➤
$$|x| = x \quad \text{if } x \geqslant 0$$
$$|x| = -x \quad \text{if } x < 0$$

For $|x|$ read 'mod x'.

Note $|a| = |b| \Leftrightarrow a^2 = b^2$.

The modulus function can be used to express a range.

➤
$$|x| < a \Leftrightarrow -a < x < a \qquad |x| > a \Leftrightarrow x < -a \text{ or } x > a$$

$$|x| < 1 \Leftrightarrow -1 < x < 1 \qquad |x| \geqslant 1 \Leftrightarrow x \leqslant -1 \text{ or } x \geqslant 1$$

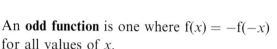

Remember 'Empty' circles indicate that x cannot be equal to -1 or $+1$. 'Full' circles indicate x can be equal to -1 or $+1$.

- $|x - a| = x - a$ for $x \geqslant a$ and $|x - a| = -(x - a) = a - x$ for $x < a$
- $|x - b| \leqslant a \Leftrightarrow -a \leqslant x - b \leqslant a$
- $|x - b| \leqslant a \Leftrightarrow -a + b \leqslant x \leqslant a + b$

Sketching functions involving modulus signs

To sketch $y = |f(x)|$

■ Sketch the curve $y = f(x)$, using a dashed line for points below the x-axis.

■ Reflect any part of the curve below the x-axis in the x-axis.

To sketch $y = f(|x|)$

■ Sketch $y = f(x)$ for x greater than or equal to 0.

■ Reflect the parts of the curve to the right of the y-axis in the y-axis.

> The solution is the solid line; the dashed line just acts as a guide.

Note Be careful not to confuse $y = |f(x)|$ and $y = f(|x|)$.

Examples 11 and 12 involve examples of $y = |f(x)|$.
Example 13 involves both $y = |f(x)|$ and $y = f(|x|)$.

Example 11 Sketch $y = |x|$.

The right-hand branch of $y = |x|$ has equation $y = x$.
The left-hand branch is the negative of x, i.e. $y = -x$.

To draw $y = |x|$, $y = x$ is drawn, with points below the x-axis (where y is $-$ve) shown as a dashed line. The dashed line is then reflected in the x-axis.

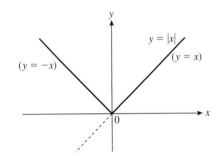

Example 12 Sketch $y = |3x + 1|$.

The graph of $y = |3x + 1|$ has the right-hand branch with equation $y = 3x + 1$.
The left-hand branch is the negative of $3x + 1$, i.e. $y = -3x - 1$.

To draw $y = |3x + 1|$, $y = 3x + 1$ is drawn, with points below the x-axis (where y is $-$ve) shown as a dashed line. The dashed line is then reflected in the x-axis.

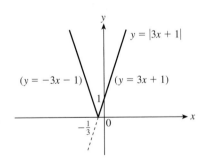

Example 13 $f(x) = (x - 1)^2 - 4$. Sketch the graphs of

a $y = f(x)$ **b** $y = |f(x)|$ **c** $y = f(|x|)$

Solution   **a**   When $x = 0$, $y = -3$.

> Curve cuts y-axis when $x = 0$.

So the curve cuts the y-axis at $y = -3$.

When $y = 0$ $(x - 1)^2 - 4 = 0$

> Curve cuts x-axis when $y = 0$.

$$(x - 1)^2 = 4$$
$$x - 1 = \pm 2$$
$$\therefore \qquad\qquad x = 1 \pm 2$$

So the curve cuts the x-axis at $x = 3$ and $x = -1$.

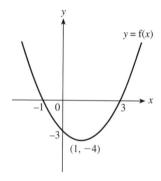

> $y = x^2$ has been translated 1 unit right and 4 down.

> Vertex $(0, 0)$ of $y = x^2$ has moved to $(1, -4)$.

b

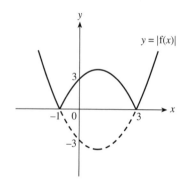

> For $f(x)$ +ve, i.e. for the curve above the x-axis $|f(x)| = f(x)$.

> For $f(x)$ −ve, i.e. for the curve below the x-axis $|f(x)| = -f(x)$.

> The portion of the curve below the x-axis, shown with a dotted line, must be reflected in the x-axis.

> Try drawing $y = |(x - 1)^2 - 4|$ on a graphical calculator. *Note:* $|x| = \text{ABS}(x)$.

c

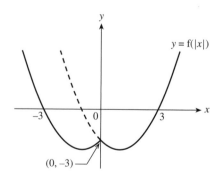

> For x +ve i.e. to the right of the y-axis $f(|x|) = f(x)$.

> For x −ve, i.e. to the left of the y-axis $f(|x|) = f(-x)$.

> So the curve to the right of the y-axis is reflected in the y-axis.

> Try drawing $y = (|x| - 1)^2 - 4$ on a graphical calculator.

Exercise 1D *For all sketches in this exercise, use the method above and then use a graphical calculator to check the solution. Note that in computer languages and on graphical calculators $|x|$ is usually written as ABS(x).*

1 Solve these inequalities and illustrate the solution on a number line.

    a $|x| < 2$ **b** $|x| \geqslant 3$ **c** $|x - 1| < 5$

2 Sketch these curves.

    a $y = |2x|$ **b** $y = |x - 4|$ **c** $y = |2x + 1|$ **d** $y = |2 - x|$

    e $y = |x - 2|$ **f** $y = |x| - 2$ **g** $y = 2 - |x|$

3 **a** Sketch $f(x) = |5 - 2x|$ for the domain $0 \leqslant x \leqslant 7$.

    b State the range of f.

    c Explain why $f(x)$ is not a one–one function.

    d Solve $f(x) = 4$.

4 Given that $f(x) = x^2 - 5x$ sketch, on separate diagrams, the graphs of

    a $y = |f(x)|$ **b** $y = f(|x|)$

5 Sketch, on the same axes, the graphs of $y = x^2 - x - 6$ and $y = |x|^2 - |x| - 6$.

6 Given that $f(x) = \cos x$ sketch, on separate diagrams, the graphs of

    a $y = |f(x)|$ **b** $y = f(|x|)$

7 Repeat Question 6 for $y = \sin x$ and $y = \tan x$.

Equations and inequalities with modulus signs

Equations and inequalities with modulus signs can be solved graphically or algebraically. If both sides are positive, the technique of squaring both sides can be used.

Example 14 Solve $|x - 1| = 3$ and $|x - 1| < 3$.

For the equality $|x - 1| = 3$

$$(x - 1)^2 = 3^2$$

$$x^2 - 2x + 1 = 9$$

$$x^2 - 2x - 8 = 0$$

$$(x + 2)(x - 4) = 0$$

$$\Rightarrow \qquad x = -2 \text{ or } x = 4$$

So $|x - 1| = 3$ when $x = -2$ and $x = 4$.

> $|x - 1| = 3$ could also be solved by finding the points of intersection of $y = |x - 1|$ and $y = 3$. However, since both sides are +ve, they can be squared.

For the inequality $|x - 1| < 3$

$$|x - 1| < 3$$

$$\Rightarrow -3 < x - 1 < 3$$

$$-2 < x < 4$$

> Use $|x| < a \Leftrightarrow -a < x < a$.

> The equality is solved when $x = -2$ and $x = 4$.

Example 15   *This example illustrates both methods of solution. In part **a** squaring cannot be used so the equation and the inequality are solved graphically. Part **b** could be solved using either method.*

Solve **a** $|2x - 3| = x + 3$ and $|2x - 3| \leqslant x + 3$

b $|2x - 3| = 5$ and $|2x - 3| \geqslant 5$

Solution   **a**   *For the equality $|2x - 3| = x + 3$*

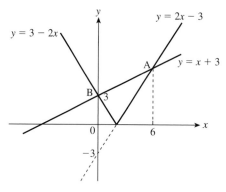

> Since $x + 3$ is not always +ve both sides *cannot* be squared.

> Sketch $y = |2x - 3|$ and $y = x + 3$ and find their points of intersection.

At A, $2x - 3 = x + 3$

$\therefore$ $x = 6$

At B, $3 - 2x = x + 3$

            $3x = 0$

$\therefore$ $x = 0$

So $|2x - 3| = x + 3$ when $x = 0$ and $x = 6$.

For the inequality $|2x - 3| \leqslant x + 3$

From the graph, $|2x - 3| \leqslant x + 3 \Rightarrow 0 \leqslant x \leqslant 6$

> The inequality holds when the graph of $y = |2x - 3|$ is below the graph of $y = x + 3$ or where the graphs intersect.

b   *For the equality $|2x - 3| = 5$*

$$(2x - 3)^2 = 5^2$$

$$4x^2 - 12x + 9 = 25$$

$$4x^2 - 12x - 16 = 0$$

$$x^2 - 3x - 4 = 0$$

$$(x - 4)(x + 1) = 0$$

$\Rightarrow$ $x = -1$ or $x = 4$

So $|2x - 3| = 5$ when $x = -1$ and $x = 4$.

> $|2x - 3| = 5$ could be solved by finding the points of intersection of $y = |2x - 3|$ and $y = 5$. However, since both sides are +ve, they can be squared.

For the inequality $|2x - 3| \geqslant 5$

$$|2x - 3| \geqslant 5$$

$\Rightarrow$ $2x - 3 \geqslant 5$ or $2x - 3 \leqslant -5$

            $2x \geqslant 8$ $2x \leqslant -2$

$\therefore$ $x \geqslant 4$ $x \leqslant -1$

So $x \leqslant -1$ or $x \geqslant 4$.

> Use $|x| \geqslant a \Leftrightarrow x \geqslant a$ or $x \leqslant -a$

Example 16 *This example shows both methods.*

Solve $|2x - 3| = |x|$ and $|2x - 3| > |x|$.

For the equality $|2x - 3| = |x|$

Graphically

At A, $2x - 3 = x$
$$x = 3$$

At B, $3 - 2x = x$
$$3x = 3$$
$$x = 1$$

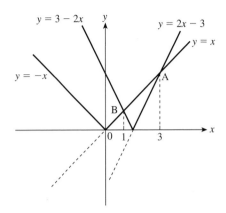

Algebraically
$$|2x - 3| = |x|$$
$$(2x - 3)^2 = x^2$$
$$4x^2 - 12x + 9 = x^2$$
$$3x^2 - 12x + 9 = 0$$
$$x^2 - 4x + 3 = 0$$
$$(x - 1)(x - 3) = 0$$
$$\Rightarrow \qquad x = 1 \quad \text{or} \quad x = 3$$

Solutions to $|2x - 3| = |x|$ are $x = 1$ and $x = 3$.

For the inequality $|2x - 3| > |x|$

From the graph, $|2x - 3| > |x| \Rightarrow x < 1$ or $x > 3$.

> The inequality holds where the graph of $y = |2x - 3|$ is above the graph of $y = |x|$.

Exercise 1E *In this exercise, both questions use the same expressions. Your answers to Question 1 should help you to answer Question 2.*

1 Solve these equations.

   a $|x + 1| = 4$ **b** $|2x - 1| = 5$

   c $|3 - x| = 0$ **d** $|1 - 2x| = 3$

   e $|6x - 5| = 7$ **f** $|x + 3| = |1 - x|$

   g $|x + 3| = |3x - 1|$ **h** $|2x + 1| = |1 - 4x|$

   i $|x + 3| = 2|x|$ **j** $3|2x - 1| = |x|$

2 Solve these inequalities.

   a $|x + 1| < 4$ **b** $|2x - 1| \geqslant 5$

   c $|3 - x| > 0$ **d** $|1 - 2x| \leqslant 3$

   e $|6x - 5| < 7$ **f** $|x + 3| \leqslant |1 - x|$

   g $|x + 3| \geqslant |3x - 1|$ **h** $|2x + 1| \leqslant |1 - 4x|$

   i $|x + 3| < 2|x|$ **j** $3|2x - 1| < |x|$

1.6 Transformation of graphs

Earlier in the course you learnt about the transformation of graphs by using translations, reflections and stretches. In this section we will consider combinations of such transformations.

For examples of how to sketch graphs using other techniques see the CD-ROM (E1.6).

Remember   All these transformations apply to $y = f(x)$. Assume $a > 0$.

- $y = f(x) + a$ is a translation $\begin{pmatrix} 0 \\ a \end{pmatrix}$.

  Adding a to the function moves the graph a units up.

- $y = f(x) - a$ is a translation $\begin{pmatrix} 0 \\ -a \end{pmatrix}$.

  Subtracting a from the function moves the graph a units down.

- $y = f(x - a)$ is a translation $\begin{pmatrix} a \\ 0 \end{pmatrix}$.

  Subtracting a from x (i.e. replacing x by $x - a$) moves the graph a units to the right.

- $y = f(x + a)$ is a translation $\begin{pmatrix} -a \\ 0 \end{pmatrix}$.

  Adding a to x (i.e. replacing x by $x + a$) moves the graph a units to the left.

- $y = f(x - a) + b$ is a translation $\begin{pmatrix} a \\ b \end{pmatrix}$.

- $y = -f(x)$ is a reflection in the x-axis.
- $y = f(-x)$ is a reflection in the y-axis.
- $y = af(x)$ is a stretch parallel to the y-axis with scale factor a.
- $y = f(ax)$ is a stretch parallel to the x-axis with scale factor $\dfrac{1}{a}$.

Example 17   Sketch, on the same axes, $y = \sqrt{x}$ and $y = \sqrt{x + 2}$, stating the values of x for which the graphs can be drawn.

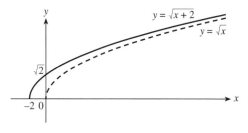

Note: $\sqrt{a}$ means the positive square root of a.

Replacing x by $x + 2$ shifts the curve **left** by **2** units.

$y = \sqrt{x}$ can be drawn for $x \geqslant 0$

$y = \sqrt{x + 2}$ can be drawn for $x \geqslant -2$

The graph cannot be drawn for values of x which make the number under the root sign negative.

Example 18   Sketch $y = f(x)$ given that $f(x) = \dfrac{1}{x-3} + 4$ for $x \neq 3$.

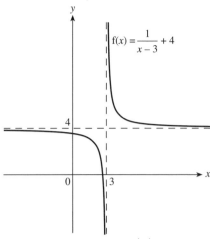

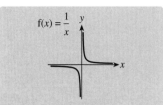

The graph of $\frac{1}{x}$ is translated 3 to the right and 4 up.
First draw the asymptotes, $x = 3$ and $y = 4$.
Then add the translated curve.

$f(x)$ is a translation $\begin{pmatrix} 3 \\ 4 \end{pmatrix}$ of $\dfrac{1}{x}$.

Example 19   Sketch the graph of $y = \dfrac{1}{x^2}$, $x \neq 0$. Describe the transformations required to obtain

the graph of $y = \dfrac{2}{(x+3)^2} + 2$ and sketch this graph on the same diagram.

Solution   The graph of $y = \dfrac{2}{(x+3)^2} + 2$ is obtained from the graph of $y = \dfrac{1}{x^2}$ by a stretch of

scale factor 2 parallel to the y-axis followed by a translation of $\begin{pmatrix} -3 \\ 2 \end{pmatrix}$.

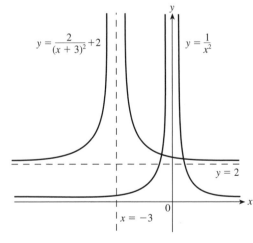

First draw the asymptotes, $x = -3$ and $y = 2$.
Then add the stretched and translated curve.

For Example 19, the stretch must be carried out *before* the translation.

The translation *followed* by the stretch would give $y = \dfrac{2}{(x+3)^2} + 4$, with a

horizontal asymptote at $y = 4$ rather than $y = 2$.

Example 20 Sketch $y = f(x)$ given that $f(x) = 2 + 2\sin 3x$.

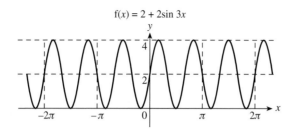

f(x) = 2 + 2sin 3x

$f(x)$ is a stretch parallel to the x-axis (scale factor $\frac{1}{3}$) of $\sin x$, followed by a stretch parallel to the y-axis (scale factor 2), followed by a translation $\begin{pmatrix} 0 \\ 2 \end{pmatrix}$.

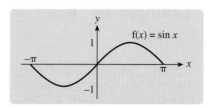

'Stretch' parallel to x-axis, scale factor $\frac{1}{3}$, for $f(x) = \sin 3x$.

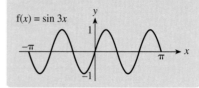

Stretch, parallel to the y-axis, scale factor 2, for $f(x) = 2\sin 3x$.

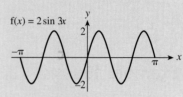

Translation 2 units up gives $f(x) = 2 + 2\sin 3x$.

For Example 20, the stretches must be carried out *before* the translation. The translation *followed* by the stretches would give $y = 2(\sin 3x + 2)$.

Example 21 For given functions, sketch the graphs, state the range, and state the period.

   a $y = 4\cos 3x$ $0° \leqslant x \leqslant 360°$

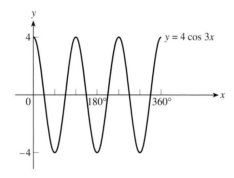

Range: $-4 \leqslant y \leqslant 4$

Period: $120°$

The curve $y = \cos x$ undergoes two stretches with one parallel to each of the axes. (The stretches can be done in either order.)

Stretching $y = \cos x$ parallel to the y-axis, scale factor 4, gives $y = 4\cos x$. Hence the range will be $-4 \leqslant y \leqslant 4$.

Stretching $y = 4\cos x$ parallel to the x-axis, scale factor $\frac{1}{3}$, gives $y = 4\cos 3x$. The graph is compressed towards the y-axis. Three periods of $y = 4\cos 3x$ fit into one of $y = 4\cos x$. The period of $y = \cos x$ is 360°, so the period of $y = 4\cos 3x$ is $\frac{360°}{3} = 120°$.

b $y = 1 - \tan\dfrac{x}{3}$ $-3\pi \leqslant x \leqslant 3\pi$ $x \neq \pm\dfrac{3\pi}{2}$

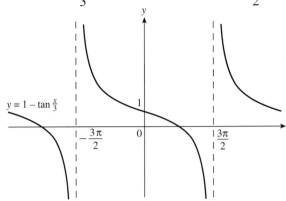

$y = 1 - \tan\frac{x}{3}$

> The curve $y = \tan x$ undergoes three transformations (the order is important).
> - Stretch parallel to x-axis scale factor 3 gives $y = \tan\dfrac{x}{3}$.
> - Reflection in x-axis gives $y = -\tan\dfrac{x}{3}$.
> - Move 1 unit up, i.e. translation $\begin{pmatrix}0\\1\end{pmatrix}$ gives $y = 1 - \tan\dfrac{x}{3}$.

> The range is any real value of y.

Range: $y \in \mathbb{R}$

Period: 3π

> The graph is stretched away from the y-axis.
> One period of $y = \tan x$ fits into a third of the period of $y = \tan\dfrac{x}{3}$.
> Since the period of $y = \tan x$ is π, the period of $y = 1 - \tan\dfrac{x}{3}$ is 3π.

c $y = \frac{1}{2}\sin(x + 30°)$ $-180° \leqslant x \leqslant 180°$

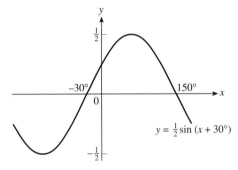

$y = \frac{1}{2}\sin(x + 30°)$

> The curve $y = \sin x$ undergoes two transformations (in either order).
> - Move 30° to the left, i.e. translation $\begin{pmatrix}-30°\\0\end{pmatrix}$, gives $y = \sin(x + 30°)$.
> - 'Stretch' parallel to y-axis scale factor $\frac{1}{2}$, gives $y = \frac{1}{2}\sin(x + 30°)$.

Range: $-\frac{1}{2} \leqslant y \leqslant \frac{1}{2}$

Period: $360°$

> The period is the same as for $y = \sin x$, i.e. 360°.

Note $\sin kx$ and $\cos kx$ have period $\dfrac{360°}{k}\left(\dfrac{2\pi}{k} \text{ radians}\right)$.

$\tan kx$ has period $\dfrac{180°}{k}\left(\dfrac{\pi}{k} \text{ radians}\right)$.

Exercise 1F

1 Sketch $y = |x|$ and, on the same axes

   a $y = |x| + 3$ **b** $y = |x - 3|$ **c** $y = -|x|$ **d** $y = 2|x|$

2 Given $f(x) = x^2$ sketch these graphs, each on a separate diagram, and find an expression, giving y in terms of x, for each function.

   a $y = f(x) - 1$ **b** $y = f(x + 1)$ **c** $y = f(2x)$ **d** $y = 2 + f(x + 3)$

3 Given $f(x) = \sqrt{x}$ for $x \geqslant 0$, sketch these graphs, each on a different diagram, stating the values of x for which each graph can be drawn, and find an expression, giving y in terms of x, for each of the functions.

   a $y = f(x) - 1$ **b** $y = f(x + 1)$ **c** $y = -f(x)$ **d** $y = 2 + f(x + 3)$

4 The function f is defined by $f(x) = |x| + 2$. Sketch, on the same axes, the graphs of

   a $y = f(x)$ **b** $y = f(x) - 1$ **c** $y = f(x + 1)$ **d** $y = -f(x)$

5 The function f is defined by $f(x) = |x| + 2$. Sketch, on the same axes, the graphs of

   a $y = f(x)$ **b** $y = f(2x)$ **c** $y = 2f(x)$

6 Sketch the graph of $y = f(x)$ where $f(x) = (x + 2)^3$.
On the same axes, show the graphs of $y = |f(x)|$ and $y = f(|x|)$.

7 Copy this sketch of $y = g(x)$ and, on the same axes, sketch
$y = -g(x)$, $y = g(-x)$ and $y = -g(-x)$.
Mark the images of points A and B under the
transformations, stating their coordinates.

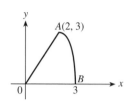

8 The graph of a function is reflected in the x-axis and then translated $\binom{3}{2}$. After
the transformations its equation is

$$y = 2 - \frac{1}{(x - 3)^2}$$

Find the original function.

9 a Copy the sketch of $y = f(x)$ and, on the
same axes, sketch $y = |f(x)|$ and $y = f(|x|)$.
  b Mark the points where the curves meet the
axes and state their coordinates.

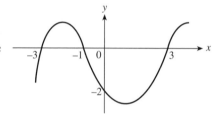

10 a Describe the geometrical transformations of the graph of $\cos\theta$ which would
give $3\cos(\theta - 30°) - 1$.
  b The graph of $y = \tan\theta$ undergoes these geometrical transformations. State
the equation of the resulting graph.
    i Stretch parallel to x-axis, scale factor 3, followed by a translation $\begin{pmatrix} 10° \\ -2 \end{pmatrix}$

    ii A translation $\begin{pmatrix} 60° \\ 4 \end{pmatrix}$, followed by a stretch parallel to the y-axis, scale
factor 2

11 For each of these functions, state the range, state whether the function is even,
odd or neither, and sketch the graph of $y = f(x)$.
  a $f(x) = 1 + \sin x$ **b** $f(x) = 2 + 3\cos x$ **c** $f(x) = 5\sin x + 10$
  d $f(x) = 1 - \cos x$ **e** $f(x) = -\sin 3x$ **f** $f(x) = 2\cos 2x$

12 For these equations sketch the graphs for $0 \leqslant x \leqslant 2\pi$, state the range, describe
the required transformation of $y = \tan x$ or $y = \cos x$, and state the period.
  a $y = \dfrac{1}{2}\cos\left(x - \dfrac{\pi}{4}\right) - 2$ **b** $y = 1 - \tan\dfrac{x}{2}$
  c $y = \cos\left(\dfrac{\pi}{3} - 2x\right)$ **d** $y = \tan(-3x)$

Exercise 1G (Review)

For this exercise, the domain is $\mathbb{R}$ unless otherwise stated.

1 The functions f and g are defined by $f(x) = x^3 + 2$ and $g(x) = x - 3$. Find
   a $gf(2)$ b $gf(x)$ c $fg(x)$ d $(fg)^{-1}(-6)$

2 The function f is defined by $f(x) = 3x^2 + 2$, $x \geqslant 0$. Find
   a $f(5)$ b $f'(5)$ c $f^{-1}(5)$ d $(f(5))^{-1}$

3 $f(x) = \dfrac{x}{5}$ and $g(x) = 7 - x$.
   a Find $f^{-1}(x)$, $g^{-1}(x)$, $fg(x)$ and $(fg)^{-1}(x)$.
   b Solve $fg(x) = (fg)^{-1}x$.

4 a Given $f(x) = 6 - 5x$ find $f^{-1}(x)$.
   b Solve $f(x) = f^{-1}(x)$.
   c Sketch, on the same axes, $y = f(x)$, $y = f^{-1}(x)$ and $y = x$, marking the points where the lines intersect and where they cut the axes.

5 Given that $f(x) = 2x$, $g(x) = x^2$ and $h(x) = x - 1$ match these expressions to the composite functions fgh, fhg, gfh, ghf, hfg and hgf.
   a $4x^2 - 1$ b $(2x - 1)^2$ c $4(x - 1)^2$
   d $2(x - 1)^2$ e $2(x^2 - 1)$ f $2x^2 - 1$

6 $f : x \rightarrow 10x + 2$, $g : x \rightarrow \dfrac{1}{x + 4}$, $x \neq -4$ and $h : x \rightarrow x^2 + 1$.
   a State, with reasons, whether the above functions are one–one or many–one.
   b Express fgh, in its simplest form, using the notation $fgh : x \rightarrow \ldots$.
   c State if $\mathbb{R}$ is a suitable domain for fgh.
   d Find the range of fgh and sketch the graph of $y = fgh(x)$.
   e Solve $fgh(x) = 3$.

7 a Show that $g(x) = \dfrac{2x - 1}{x - 3}$ can be expressed in the form $g(x) = \dfrac{a}{x - 3} + b$, where a and $b \in \mathbb{R}$ and $x \neq 3$.
   b The graph of $y = g(x)$ can be obtained by translating the graph of $y = \dfrac{a}{x}$.
      State the translation and hence sketch $y = g(x)$, showing the intercepts with the axes.
   c Solve $g(x) = x + 3$.

8 Sketch these curves.
   a $y = |x + 4|$ b $y = |2x - 1|$ c $y = |5 - x|$
   d $y = |3x|$ e $y = -3|2x|$ f $y = 4 + |x + 1|$

9 Given that $f(x) = x^2 + 3x$ sketch, on separate diagrams, the graphs of
   a $y = |f(x)|$ b $y = f(|x|)$

10 Sketch, on the same axes, the graphs of $y = \cos\left(x - \dfrac{\pi}{4}\right)$ and $y = \left|\cos\left(x - \dfrac{\pi}{4}\right)\right|$.

11 Solve these equations.

   a $|x + 3| = 2|x - 1|$ b $|2x - 4| = x$

   c $|3x + 2| = 2 - x$ d $|1 - 3x| - |2x + 1| = 0$

12 Use your answers to Question 11 to help you solve these inequalities.

   a $|x + 3| > 2|x - 1|$ b $|2x - 4| < x$

   c $|3x + 2| \geqslant 2 - x$ d $|1 - 3x| - |2x + 1| \leqslant 0$

13 Given that $f(x) = (x - 4)^2$ sketch, on separate axes, the graphs of

   a $y = f(x - 2)$ b $y = 2f(x) - 3$ c $y = f(|x|) + 5$

14 Given that $f(x) = x^2 - 9$ sketch, on separate axes, the graphs of

   a $y = f(x - 3) + 9$ b $y = 9 - f(x)$ c $y = |f(x)| + 4$

15 a Given that $f(x) = |x| + 1$ and $g(x) = x^2 + 4$, state the range of f and the range of g.

   b Find the functions, ff, gg and fg, stating their ranges.

There is a 'Test yourself' exercise (Ty1) and an 'Extension exercise' (Ext1) on the CD-ROM.

Key points

Functions

A **function** is a relationship between two sets: the **domain** (the set of 'inputs') and the **range** (the set of 'outputs').

A function is defined by
■ a rule connecting the domain and range sets and
■ the domain set.

The essential feature of a function is that, for each member of the domain, there is one and only one member of the range.

A function $y = f(x)$ is:
■ **one–one** if $a = b \Leftrightarrow f(a) = f(b)$
   i.e. for each value of x there is a unique value of y and vice versa
■ **many–one** if for some values of a and b, $(a \neq b)$, $f(a) = f(b)$
   i.e. more than one value of x maps to the same value of y.

The composite function, fg, can be formed only if the range of g is a subset of (i.e. is contained within) the domain of f.

The **composite function** fg means apply g first then f. $fg(x) = f(g(x))$.
The output from g becomes the input for f.

The **inverse function** f^{-1} 'undoes' f. So $ff^{-1}(x) = f^{-1}f(x) = x$.
■ If $f(x) = f^{-1}(x)$ the function $f(x)$ is **self-inverse**.
■ The range of f is the domain of f^{-1} and vice versa.

- The graphs of $y = f(x)$ and $y = f^{-1}(x)$ are reflections of each other in $y = x$, since for a function and its inverse function the roles of x and y are interchanged.
- To obtain the inverse function, rearrange $y = f(x)$ to give x in terms of y, then replace y by x.
- Only one–one functions have inverses. For many–one functions, an inverse only exists over a domain restricted so that the function is one–one for that domain.

Modulus function

$|x| = x$ if $x \geqslant 0$

$|x| = -x$ if $x < 0$

$|x| < a \Leftrightarrow -a < x < a$

$|x| > a \Leftrightarrow x < -a$ or $x > a$

$|x - a| = x - a$ for $x \geqslant a$ and $|x - a| = -(x - a) = a - x$ for $x < a$

$|x - b| \leqslant a \Leftrightarrow -a \leqslant x - b \leqslant a$

$|x - b| \leqslant a \Leftrightarrow -a + b \leqslant x \leqslant a + b$

Transformation of graphs

All these transformations apply to $y = f(x)$. Assume $a > 0$.

- $y = f(x) + a$ is a translation $\begin{pmatrix} 0 \\ a \end{pmatrix}$.

  Adding a to the function moves the graph a units up.

- $y = f(x) - a$ is a translation $\begin{pmatrix} 0 \\ -a \end{pmatrix}$.

  Subtracting a from the function moves the graph a units down.

- $y = f(x - a)$ is a translation $\begin{pmatrix} a \\ 0 \end{pmatrix}$.

  Subtracting a from x (i.e. replacing x by $x - a$) moves the graph a units to the right.

- $y = f(x + a)$ is a translation $\begin{pmatrix} -a \\ 0 \end{pmatrix}$.

  Adding a to x (i.e. replacing x by $x + a$) moves the graph a units to the left.

- $y = f(x - a) + b$ is a translation $\begin{pmatrix} a \\ b \end{pmatrix}$.

- $y = -f(x)$ is a reflection in the x-axis.
- $y = f(-x)$ is a reflection in the y-axis.
- $y = af(x)$ is a stretch parallel to the y-axis with scale factor a.
- $y = f(ax)$ is a stretch parallel to the x-axis with scale factor $\dfrac{1}{a}$.

For $y = f(|x|)$ sketch $y = f(x)$ for $x > 0$ and reflect the parts of the curve to the right of the y-axis in the y-axis.

For $y = |f(x)|$ sketch $y = f(x)$ with a dotted line where $y < 0$.
Reflect the dotted line in the x-axis.

2 Trigonometry

This chapter extends the work you have done on trigonometry in Chapters 16 and 17 of Book 1.

Trigonometric functions are central to our understanding of vibration and oscillatory motion, be it the sound produced when a guitar string is plucked, the way an earthquake wave travels through the earth or the variation in the length of the day over the year.

A graphical calculator, or computer graph sketching software, is a very valuable tool to help you with this topic. However, you must be able to sketch the necessary graphs without recourse to such technological aids.

After working through this chapter you should

■ *be familiar with secant, cosecant and cotangent, their graphs, symmetries and periodicity*

■ *be familiar with, and able to use,* $\tan \theta = \cot(90° - \theta)$, $\tan^2 \theta + 1 = \sec^2 \theta$ *and* $1 + \cot^2 \theta = \csc^2 \theta$

■ *be familiar with the notation for inverse trigonometric functions and their graphs*

■ *be able to solve trigonometric equations.*

The CD-ROM (E2.1) shows how to find the general solution of trigonometric equations.

2.1 The reciprocal functions: $y = \sec \theta$, $y = \csc \theta$ and $y = \cot \theta$

The reciprocal trigonometric functions are secant, cosecant and cotangent.

$$\sec \theta = \frac{1}{\cos \theta} \qquad \csc \theta = \frac{1}{\sin \theta} \qquad \cot \theta = \frac{1}{\tan \theta}$$

Their graphs can be sketched by looking at the graphs of $\cos \theta$, $\sin \theta$ and $\tan \theta$ and marking the reciprocal at each point, noting these facts:

■ The reciprocal of 1 is 1, and of -1 is -1.

■ The reciprocal of a +ve number is +ve, and of a −ve number is −ve.

■ A zero on a graph corresponds to a vertical asymptote on the reciprocal graph, because the graph shoots off to $+\infty$ or $-\infty$.

■ A vertical asymptote on a graph, where the graph shoots off to $+\infty$ or $-\infty$, corresponds to a zero on the reciprocal graph.

On the graph of $y = \sec \theta$:

■ Where $y = \cos \theta$ is zero, i.e. when $\theta = (2n + 1)90°$ for $n \in \mathbb{Z}$, $y = \dfrac{1}{\cos \theta} = \sec \theta$ has a vertical asymptote.

■ Where $y = \cos \theta$ has a local *maximum*, $y = \sec \theta$ has a local *minimum*, and *vice versa*.

■ The domain of $f(\theta) = \cos \theta$ is $\theta \in \mathbb{R}$. The domain of $f(\theta) = \sec \theta$ is $\theta \in \mathbb{R}$, $\theta \neq (2n + 1)90°$, $n \in \mathbb{Z}$.

■ The range of $f(\theta) = \cos \theta$ is $-1 \leqslant f(\theta) \leqslant 1$.

■ The range of $f(\theta) = \sec \theta$ is $f(\theta) \leqslant -1$ and $f(\theta) \geqslant 1$.

■ $y = \cos \theta$ has period $360°$, so $y = \sec \theta$ also has period $360°$.

■ $f(\theta) = \cos \theta$ and $f(\theta) = \sec \theta$ are both even functions.

Similar observations can be made on $y = \operatorname{cosec} \theta$ and $y = \cot \theta$.

$y = \sec \boldsymbol{\theta}$

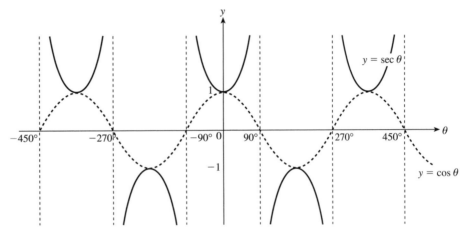

$y = \operatorname{cosec} \boldsymbol{\theta}$

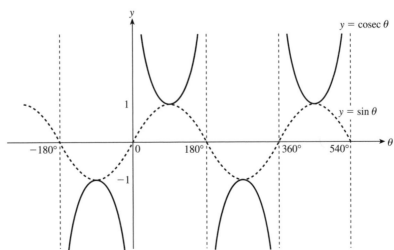

$y = \cot \theta$

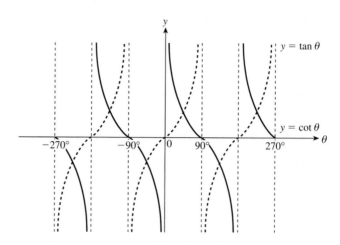

This table summarises the properties of the six ratios.

Comparing the six trigonometrical functions ($n \in \mathbb{Z}$)

Function	Period	Even/odd	Zeros occur at $\theta =$	Asymptotes occur at $\theta =$	Domain	Range
$y = \cos \theta$	$360°$	Even	$(2n + 1)90°$	–	$\theta \in \mathbb{R}$	$-1 \leqslant y \leqslant 1$
$y = \sec \theta$	$360°$	Even	–	$(2n + 1)90°$	$\theta \in \mathbb{R}$ $\theta \neq (2n + 1)90°$	$y \leqslant -1, y \geqslant 1$
$y = \sin \theta$	$360°$	Odd	$180n°$	–	$\theta \in \mathbb{R}$	$-1 \leqslant y \leqslant 1$
$y = \operatorname{cosec} \theta$	$360°$	Odd	–	$180n°$	$\theta \in \mathbb{R}$ $\theta \neq 180n°$	$y \leqslant -1, y \geqslant 1$
$y = \tan \theta$	$180°$	Odd	$180n°$	$(2n + 1)90°$	$\theta \in \mathbb{R}$ $\theta \neq (2n + 1)90°$	$y \in \mathbb{R}$
$y = \cot \theta$	$180°$	Odd	$(2n + 1)90°$	$180n°$	$\theta \in \mathbb{R}$ $\theta \neq 180n°$	$y \in \mathbb{R}$

Identities involving the reciprocal functions

Remember $\dfrac{\sin \theta}{\cos \theta} = \tan \theta$ and $\sin^2 \theta + \cos^2 \theta = 1$

- $\dfrac{\sin \theta}{\cos \theta} = \tan \theta \Leftrightarrow \dfrac{\cos \theta}{\sin \theta} = \cot \theta$

- From the diagram, $\tan \theta = \dfrac{b}{a} = \cot (90° - \theta)$

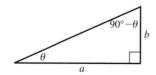

- Dividing the identity $\sin^2 \theta + \cos^2 \theta = 1$ by $\cos^2 \theta$ gives
  $$\tan^2 \theta + 1 = \sec^2 \theta$$

- Dividing the identity $\sin^2 \theta + \cos^2 \theta = 1$ by $\sin^2 \theta$ gives
  $$1 + \cot^2 \theta = \operatorname{cosec}^2 \theta$$

> $$\frac{\cos\theta}{\sin\theta} = \cot\theta \qquad\qquad \tan\theta = \cot(90° - \theta)$$
>
> $$\tan^2\theta + 1 = \sec^2\theta \qquad\qquad 1 + \cot^2\theta = \csc^2\theta$$

Remember $\sin\theta = \cos(90° - \theta)$ and $\cos\theta = \sin(90° - \theta)$

From the properties of $\sin\theta$, $\cos\theta$ and $\tan\theta$, results about their reciprocals can be deduced. For example, $\sin(180° - \theta) = \sin\theta$. So (for $\sin\theta \neq 0$)

$$\frac{1}{\sin(180° - \theta)} = \frac{1}{\sin\theta}$$

$$\csc(180° - \theta) = \csc\theta$$

Calculators do not evaluate the reciprocal trigonometric functions directly.

To find, for example, $\cot 71°$, find $\dfrac{1}{\tan 71°}$.

Example 1   If $x = a\csc\theta$ and $y = a\cot\theta$, simplify $\sqrt{x^2 - y^2}$.

$$\sqrt{x^2 - y^2} = \sqrt{a^2\csc^2\theta - a^2\cot^2\theta}$$

> Substitute for x and y.

$$= \sqrt{a^2(\csc^2\theta - \cot^2\theta)}$$

But $\csc^2\theta - \cot^2\theta = 1$

> Use the identity: $1 + \cot^2\theta = \csc^2\theta$

$$\therefore\quad \sqrt{x^2 - y^2} = \sqrt{a^2}$$

$$= a$$

Example 2   If $\sin\theta = \frac{12}{13}$ and θ is obtuse, find the value of $\cot\theta$.

To find $\cot\theta$, both the sign and the magnitude are needed.

θ is obtuse $\therefore \cot\theta$ is $-$ve.

S	A
T	C

To find the magnitude, two methods are possible:

Method 1: Using the identity $\cot^2\theta = \csc^2\theta - 1$

$$\cot^2\theta = \csc^2\theta - 1$$

> $\sin\theta = \frac{12}{13}$ so $\csc\theta = \frac{13}{12}$

$$= \left(\tfrac{13}{12}\right)^2 - 1$$

$$= \tfrac{169}{144} - 1$$

$$= \tfrac{25}{144}$$

$$\therefore\quad \cot\theta = \pm\sqrt{\tfrac{25}{144}} = \pm\tfrac{5}{12}$$

But $\cot\theta$ is $-$ve, so $\cot\theta = -\tfrac{5}{12}$.

> *Note*: It is not necessary to find θ and doing so would lose accuracy.

Method 2: Using Pythagoras' theorem

$$x^2 = 13^2 - 12^2$$
$$= 169 - 144$$
$$= 25$$
$$\therefore \quad x = 5$$

But $\cot\theta$ is $-$ve, so $\cot\theta = -\frac{5}{12}$.

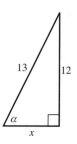

Draw a right-angled triangle with $\sin\alpha = \frac{12}{13}$. (Ignore the sign, only the size is of interest.) α is the acute equivalent angle.

Example 3 Eliminate θ from $x = a\sec\theta$ and $y = a\cot\theta$.

$$x = a\sec\theta \Rightarrow \sec\theta = \frac{x}{a}$$

$$y = a\cot\theta \Rightarrow \cot\theta = \frac{y}{a} \Rightarrow \tan\theta = \frac{a}{y}$$

Substituting in $\tan^2\theta + 1 = \sec^2\theta$ gives

$$\left(\frac{a}{y}\right)^2 + 1 = \left(\frac{x}{a}\right)^2$$

$$\frac{x^2}{a^2} - \frac{a^2}{y^2} = 1$$

The identity $\tan^2\theta + 1 = \sec^2\theta$ will enable θ to be eliminated.

Example 4 Solve $\sec(2\theta - 45°) = 2$ for $0° < \theta < 180°$.

$$\sec(2\theta - 45°) = 2 \Rightarrow \cos(2\theta - 45°) = \tfrac{1}{2}$$

$$\therefore \quad 2\theta - 45° = \ldots -60°, 60°, 300°, 420°, \ldots$$

So $\quad\quad 2\theta = \ldots -15°, 105°, 345°, 465°, \ldots$

and $\quad \therefore \quad \theta = \ldots -7.5°, 52.5°, 172.5°, 232.5°, \ldots$

In the range $0° < \theta < 180°$, $\theta = 52.5°$ or $\theta = 172.5°$

$\sec x = \dfrac{1}{\cos x}$

List solutions for $\cos x = \tfrac{1}{2}$.

Add 45° and then divide by 2.

Give solutions in the range stated.

Example 5 *Methods for proving identities, as in this example, are summarised on page 204.*

Prove the identity

$$\frac{\tan\theta + \cot\theta}{\sec\theta + \operatorname{cosec}\theta} = \frac{1}{\sin\theta + \cos\theta}$$

$$\frac{\tan\theta + \cot\theta}{\sec\theta + \operatorname{cosec}\theta} = \frac{\frac{\sin\theta}{\cos\theta} + \frac{\cos\theta}{\sin\theta}}{\frac{1}{\cos\theta} + \frac{1}{\sin\theta}}$$

$$= \frac{\sin^2\theta + \cos^2\theta}{\sin\theta + \cos\theta}$$

$$= \frac{1}{\sin\theta + \cos\theta}$$

Rewrite the LHS in terms of $\sin\theta$ and $\cos\theta$.

Multiply numerator and denominator by $\cos\theta\sin\theta$.

Use the identity: $\sin^2\theta + \cos^2\theta = 1$

Example 6 (Extension)

Using radians for any angle required, state the domain, range and period of $f(x) = 2\sec 3x$.

Remember π radians $= 180°$

$$f(x) = 2\sec 3x$$

Use $\sec\theta = \dfrac{1}{\cos\theta}$

$$= \frac{2}{\cos 3x}$$

When the numerator is zero, the function is undefined.

$f(x)$ is not defined when $\cos 3x = 0$.

So the domain of f is all real values of x, excluding those which make $\cos 3x$ zero.

$\cos 3x = 0$ when
$$3x = (2n+1)\frac{\pi}{2}$$
$$x = (2n+1)\frac{\pi}{6}$$

$\cos 3x$ has period $\dfrac{2\pi}{3}$, so $f(x)$ has period $\dfrac{2\pi}{3}$.

Period of $\cos kx$ is $\dfrac{2\pi}{k}$

$\cos 3x$ has a local maximum of 1

$-1 \leqslant \cos\theta \leqslant 1$

$\therefore$ $\sec 3x$ has a local minimum of 1.

So $f(x) = 2\sec 3x$ has a local minimum of 2.

Hence $f(x) \geqslant 2$.

Similarly $f(x) \leqslant -2$.

$\therefore$ domain is $x \in \mathbb{R}$, $x \neq (2n+1)\dfrac{\pi}{6}$, range is $f(x) \leqslant -2$, $f(x) \geqslant 2$ and period is $\dfrac{2\pi}{3}$.

Example 7

a Describe the graph of $y = 2 + \frac{1}{2}\sec\theta$ in terms of transformations of the graph of $y = \sec\theta$.

b Find in radians, the two smallest positive roots of $2 + \frac{1}{2}\sec\theta = \tan^2\theta$.

Solution

a $y = \frac{1}{2}\sec\theta$ is a stretch of $y = \sec\theta$ parallel to the y-axis, scale factor $\frac{1}{2}$.

Note: The scale factor is numerically less than one, so the curve is compressed towards the x-axis.

So $y = 2 + \frac{1}{2}\sec\theta$ is a stretch parallel to the y-axis scale factor $\frac{1}{2}$, followed by a translation $\binom{0}{2}$.

b $2 + \frac{1}{2}\sec\theta = \tan^2\theta$

Using $1 + \tan^2\theta = \sec^2\theta$

To express the equation in terms of one trigonometrical ratio, use the identity which links $\sec\theta$ and $\tan\theta$.

$$2 + \frac{1}{2}\sec\theta = \sec^2\theta - 1$$

So $\sec^2\theta - \frac{1}{2}\sec\theta - 3 = 0$

$$2\sec^2\theta - \sec\theta - 6 = 0$$

Recognise a quadratic in $\sec\theta$.

$$(\sec\theta - 2)(2\sec\theta + 3) = 0$$

If preferred, substitute y for $\sec\theta$ giving $2y^2 - y - 6 = 0$.

$\Rightarrow$ $\sec\theta = 2$ or $\sec\theta = -\frac{3}{2}$

$\therefore$ $\cos\theta = \frac{1}{2}$ or $\cos\theta = -\frac{2}{3}$

Use $\sec\theta = \dfrac{1}{\cos\theta}$

When $\cos\theta = \frac{1}{2}$, $\theta = \frac{\pi}{3}$.

When $\cos\theta = -\frac{2}{3}$, $\theta \approx 2.3$.

So, the two smallest positive roots are

$\frac{\pi}{3}$ and 2.3 (correct to 1 d.p.).

Select the two smallest positive roots.
Check that the calculator is in radian mode.

Note: $\frac{\pi}{3}$ is exact, 2.3 is correct to 1 d.p.

Remember In Book 1 (see Section 16.5, page 254) you saw how to find the exact values of trigonometric ratios of certain special angles.

θ in degrees	$\sin\theta$	$\cos\theta$	$\tan\theta$	θ in radians
$0°$	0	1	0	0
$30°$	$\frac{1}{2}$	$\frac{\sqrt{3}}{2}$	$\frac{1}{\sqrt{3}}$	$\frac{\pi}{6}$
$45°$	$\frac{1}{\sqrt{2}}$	$\frac{1}{\sqrt{2}}$	1	$\frac{\pi}{4}$
$60°$	$\frac{\sqrt{3}}{2}$	$\frac{1}{2}$	$\sqrt{3}$	$\frac{\pi}{3}$
$90°$	1	0	Undefined	$\frac{\pi}{2}$

In this chapter, exact values are given when the above angles are involved. It is however quite acceptable to give the decimal equivalents in your answer.

Exercise 2A

1 If $x = a\operatorname{cosec}\theta$, $y = b\cot\theta$ and $z = c\sec\theta$ simplify

   a $\sqrt{x^2 - a^2}$ **b** $b^2 + y^2$ **c** $z^2 - c^2$

   d $\dfrac{x}{x^2 - a^2}$ **e** $\dfrac{y}{b^2 + y^2}$ **f** $\dfrac{\sqrt{z^2 - c^2}}{z}$

2 If $\cos\theta = -\frac{8}{17}$ and θ is obtuse, find the values of

   a $\cot\theta$ **b** $\operatorname{cosec}\theta$

3 If $\tan\theta = \frac{7}{24}$ and θ is reflex, find the values of

   a $\operatorname{cosec}\theta$ **b** $\sec\theta$

4 Eliminate θ from these pairs of equations.

   a $x = a\cot\theta$, $y = b\operatorname{cosec}\theta$ **b** $x = a\sec\theta$, $y = b\tan\theta$

   c $x = a\tan\theta$, $y = b\cos\theta$ **d** $x = a\cot\theta$, $y = b\sin\theta$

5 Solve for $-180° \leqslant \theta \leqslant 180°$

   a $\operatorname{cosec}\theta = 2$ **b** $\sec 2\theta = 3$ **c** $\cot\theta = 1$

   d $\sec(\theta - 30°) = 2$ **e** $\operatorname{cosec}\frac{1}{2}\theta = \sqrt{2}$ **f** $\operatorname{cosec} 2\theta = -1$

   g $\cot(2\theta + 10°) = \sqrt{3}$ **h** $2\cot(30° - \theta) + 1 = 0$ **i** $\sec(\theta - 150°) = 4$

   j $\operatorname{cosec} 3\theta = -5$ **k** $\sec(17° - 2\theta) = -2$ **l** $\cot(\theta + 40°) = -3$

6 Solve for $-\pi < \theta < \pi$

    a $\sec\theta = 2\cos\theta$ **b** $\cot\theta = 5\cos\theta$ **c** $\tan\theta = 4\cot\theta + 3$

    d $5\sin\theta + 6\operatorname{cosec}\theta = 17$ **e** $4\sin\theta + \operatorname{cosec}\theta = 4$ **f** $10\cos\theta + 1 = 2\sec\theta$

7 Solve for $0 \leqslant \theta \leqslant 2\pi$

    a $\operatorname{cosec}^2 2\theta = 4$ **b** $\cot^2 \frac{1}{2}\theta = 0.9$ **c** $\sec^2\theta = 3\tan\theta - 1$

    d $\operatorname{cosec}^2\theta = 3 + \cot\theta$ **e** $3\tan^2\theta + 5 = 7\sec\theta$ **f** $2\cot^2\theta + 8 = 7\operatorname{cosec}\theta$

8 Prove these identities.

    a $\sec^2\theta - \operatorname{cosec}^2\theta = \tan^2\theta - \cot^2\theta$ **b** $\sec\theta + \tan\theta = \dfrac{1}{\sec\theta - \tan\theta}$

    c $\sec\theta + \operatorname{cosec}\theta\cot\theta = \sec\theta\operatorname{cosec}^2\theta$ **d** $\sin^2\theta(1 + \sec^2\theta) = \sec^2\theta - \cos^2\theta$

    e $\dfrac{1 - \cos\theta}{\sin\theta} = \dfrac{1}{\operatorname{cosec}\theta + \cot\theta}$

9 **a** Sketch, on the same axes, $y = \operatorname{cosec} x$ and $y = 1 + \operatorname{cosec} x$, for $0 \leqslant x \leqslant 2\pi$.

    b State the domains, range and period of the functions f and g where
$f(x) = \operatorname{cosec} x$ and $g(x) = 1 + \operatorname{cosec} x$.

10 Sketch

    a $y = \sec 2x$ **b** $y = 3\sec 2x$ **c** $y = \operatorname{cosec} \frac{1}{2}x$ **d** $y = 2\operatorname{cosec} \frac{1}{2}x$

11 **a** State the geometrical transformations of the graph of $y = \sec x$ needed to give the graph of $y = 2\sec\left(x - \dfrac{\pi}{4}\right)$.

    b Sketch, on the same axes, the graphs of $y = \sec x$ and $y = 2\sec\left(x - \dfrac{\pi}{4}\right)$, for $-\pi < x < \pi$.

12 State the geometrical transformations of the graph of $y = \operatorname{cosec} x$ required to give the graph of $y = 1 - \operatorname{cosec} x$.

13 Expressing any angles required in radians, state the domains, ranges and periods of the function f for each of the following.

    a $f(x) = \cos 4x$ **b** $f(x) = 2 + \sec x$ **c** $f(x) = 3\operatorname{cosec}\left(x + \dfrac{\pi}{2}\right)$

    d $f(x) = 4\sec\dfrac{x}{2}$ **e** $f(x) = \cot 2x$ **f** $f(x) = \operatorname{cosec} x + \cot x$

2.2   Inverse trigonometrical functions

$\sin\theta = x \Rightarrow \theta$ is the angle whose sine is x,
i.e. $\theta = \sin^{-1} x$.

Similarly $\cos\theta = x \Rightarrow \theta = \cos^{-1} x$

           $\tan\theta = x \Rightarrow \theta = \tan^{-1} x$

> Do not confuse
> $\sin^{-1} x$ with $\dfrac{1}{\sin x} = (\sin x)^{-1}$.

> $\sin^{-1}$, $\cos^{-1}$ and $\tan^{-1}$ can also be written as arcsin, arccos and arctan, but most calculators use the notation $\sin^{-1}$, $\cos^{-1}$, $\tan^{-1}$.

$y = \sin x$ is a many–one relationship. For any value of x, there is only one value of y; for any value of $y\,(-1 \leqslant y \leqslant 1)$ there are many values of x, e.g.

$$x = \frac{\pi}{6} \Rightarrow y = \sin\frac{\pi}{6} = \frac{1}{2}$$

but $\quad y = \frac{1}{2} \Rightarrow x = \sin^{-1}\frac{1}{2} = \frac{\pi}{6},\ \frac{5\pi}{6},\ \frac{13\pi}{6},\,\ldots$

Because $y = \sin x$ is a many–one function, the inverse relationship, $y = \sin^{-1} x$, is a one–many relationship. The word 'relationship' is used because $y = \sin^{-1} x$ is *not a* function.

Remember Only one–one and many–one relationships are functions.

The graph of $y = \sin^{-1} x$ is the reflection of $y = \sin x$ in the line $y = x$ (providing the scales on the axes are the same).

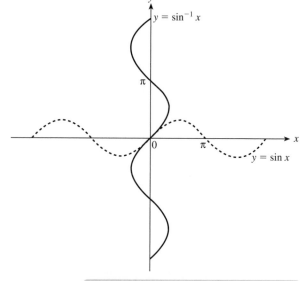

Remember With inverses, the roles of x and y are interchanged so the graph of $y = \sin^{-1} x$ is a sine curve on the y-axis.

To see what the graph of $y = \sin^{-1} x$ looks like, turn over a page with the graph of $y = \sin x$ on it and rotate the page through 90° clockwise. Hold the page to the light; the graph of $y = \sin^{-1} x$ is seen.

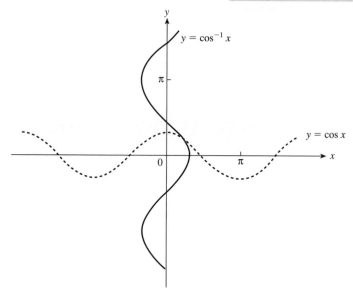

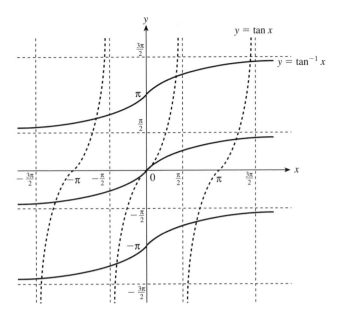

For the function f where $f(x) = \sin x$ to have an inverse, its domain has to be restricted so that it is one–one.

By restricting the domain to $-\dfrac{\pi}{2} \leqslant x \leqslant \dfrac{\pi}{2}$,

the function f where $f(x) = \sin x$ is one–one with range $-1 \leqslant y \leqslant 1$.

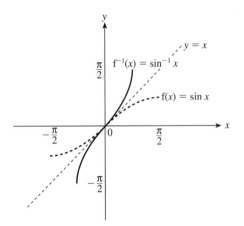

The inverse function f^{-1} where $f^{-1}(x) = \sin^{-1} x$ does exist; its domain is $-1 \leqslant x \leqslant 1$,

its range is $-\dfrac{\pi}{2} \leqslant y \leqslant \dfrac{\pi}{2}$.

For any value of x where $-1 \leqslant x \leqslant 1$, $\sin^{-1} x$ has a unique value in the range

$$-\frac{\pi}{2} \leqslant y \leqslant \frac{\pi}{2}$$

This is called its **principal value** (PV).

> Evaluating an inverse trigonometrical function on a calculator (using arcsin or $\sin^{-1}$, etc.) gives one value, the PV.

The range of principal values for sin⁻¹, cos⁻¹ and tan⁻¹

Function	Domain	Range for PVs		Sketch
		Radians	Degrees	
$y = \sin^{-1} x$	$-1 \leqslant x \leqslant 1$	$-\dfrac{\pi}{2} \leqslant y \leqslant \dfrac{\pi}{2}$	$-90° \leqslant y \leqslant 90°$	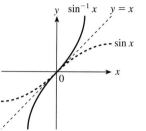
$y = \cos^{-1} x$	$-1 \leqslant x \leqslant 1$	$0 \leqslant y \leqslant \pi$	$0 \leqslant y \leqslant 180°$	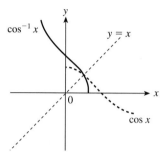
$y = \tan^{-1} x$	$x \in \mathbb{R}$	$-\dfrac{\pi}{2} < y < \dfrac{\pi}{2}$	$-90° < y < 90°$	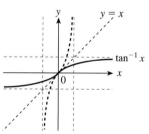

In each case, the PV range is chosen to be in the region of the origin and to cover all possible values that the function can take.

For the rest of this chapter, it is assumed that all inverse trigonometric functions take their PVs.

Example 8

At this level, trigonometrical work is usually in radians. So always assume an answer is required in radians, unless stated otherwise.

Find   **a** $\sin^{-1}\frac{1}{2}$ **b** $\tan^{-1}(-1)$ **c** $\cos^{-1}(0.7)$

Solution   **a** $\sin^{-1}\frac{1}{2} = \dfrac{\pi}{6}$

> Using the special triangles, find the angle θ in the PV range, $-\dfrac{\pi}{2} \leqslant \theta \leqslant \dfrac{\pi}{2}$, whose sine is $\frac{1}{2}$.

b $\tan^{-1}(-1) = -\dfrac{\pi}{4}$

> Again, use the special triangles.

c $\cos^{-1}(0.7) = 0.795$ (3 s.f.)

> Use the calculator function $\cos^{-1}$ or arccos.

Example 9   Find the exact value of

    a $\tan(\tan^{-1} 0.3)$ **b** $\cos\left(\sin^{-1}\frac{2}{3}\right)$ **c** $\sin(\tan^{-1} x)$

Solution   **a** $\tan(\tan^{-1} 0.3) = 0.3$

> $\tan$ and $\tan^{-1}$ are inverses of each other. One 'undoes' the other.

    b $\cos\left(\sin^{-1}\frac{2}{3}\right)$

        Let $\sin^{-1}\frac{2}{3} = A$ so $\sin A = \frac{2}{3}$.

> The angle whose sine is $\frac{2}{3}$ in the PV range is an acute angle.

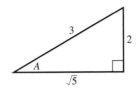

> Use Pythagoras' theorem to find the third side of the triangle.

> $$\sin A = \frac{2}{3} \Rightarrow \cos A = \frac{\sqrt{5}}{3}$$

Method 1: Using Pythagoras' theorem

$$\therefore \quad \cos\left(\sin^{-1}\frac{2}{3}\right) = \frac{\sqrt{5}}{3}$$

Method 2: Using the identity $\sin^2 A + \cos^2 A = 1$

$$\cos^2 A = 1 - \sin^2 A$$

> Rearrange the identity.

$$\therefore \quad \cos A = \sqrt{1 - \left(\frac{2}{3}\right)^2}$$

> $\sin^{-1}\frac{2}{3}$ is acute $\Rightarrow \cos A$ is $+$ve, so the $+$ve root is required.

$$= \sqrt{1 - \frac{4}{9}}$$

$$= \sqrt{\frac{5}{9}}$$

$$= \frac{1}{3}\sqrt{5}$$

    c $\sin(\tan^{-1} x)$

      Let $\tan^{-1} x = y$

      Then $\tan y = x$

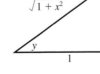

> Use Pythagoras' theorem to find the third side of the triangle.

$$\therefore \quad \sin y = \frac{x}{\sqrt{1 + x^2}}$$

So $\sin(\tan^{-1} x) = \dfrac{x}{\sqrt{1 + x^2}}$

> As PVs are used
> $x > 0 \Rightarrow \tan^{-1} x > 0$
> $\Rightarrow \sin(\tan^{-1} x) > 0$
> $x < 0 \Rightarrow \tan^{-1} x < 0$
> $\Rightarrow \sin(\tan^{-1} x) < 0$
> There is no ambiguity with the sign.

1 Write down, *without* using a calculator, the principal values (PVs), in radians, of

    a $\sin^{-1}\left(\dfrac{\sqrt{3}}{2}\right)$ **b** $\cos^{-1}\left(\dfrac{1}{\sqrt{2}}\right)$ **c** $\tan^{-1} 1$ **d** $\sin^{-1}\left(-\dfrac{1}{2}\right)$

    e $\cos^{-1}\left(-\dfrac{\sqrt{3}}{2}\right)$ **f** $\tan^{-1}(-1)$ **g** $\sin^{-1}(-1)$ **h** $\cos^{-1}(-1)$

    i $\tan^{-1} 0$ **j** $\cos^{-1} 0$

2 On the same axes, sketch $y = \sec x$ and $y = |\sec x|$.

3 **a**   Sketch, on the same axes, $y = \tan x$ and $y = \cot x$.

    b   Sketch, on the same axes, $y = |\tan x|$ and $y = |\cot x|$.

    c   Hence show that the solutions of the equation $|\tan x| = |\cot x|$ are of the

         form $x = (2n+1)\dfrac{\pi}{4}$, where $n \in \mathbb{Z}$.

4 Find, *without* using a calculator, the value of

    a $\sin(\sin^{-1} 0.7)$ **b** $\sin^{-1}\left(\sin\dfrac{\pi}{7}\right)$ **c** $\sin\left(\cos^{-1}\dfrac{1}{2}\right)$

    d $\sec\left(\sin^{-1}\dfrac{\sqrt{3}}{2}\right)$ **e** $\tan\left(\cos^{-1} -\dfrac{\sqrt{3}}{2}\right)$ **f** $\cos^{-1}\left(\sin\dfrac{\pi}{6}\right)$

5 Show that $\sin^{-1} x + \cos^{-1} x = \dfrac{\pi}{2}$.

6 Simplify $\cos(\sin^{-1} x)$.

---

1 If $\sin\theta = \frac{3}{5}$ and θ is obtuse, find the values of

    a $\cot\theta$ **b** $\sec\theta$

2 Solve for $0° \leqslant \theta < 360°$

    a $\operatorname{cosec} 2\theta = 4$ **b** $\cot\theta = -1$

    c $\cot(\theta - 20°) = \sqrt{3}$ **d** $\sqrt{3}\sec(2\theta - 10°) = 2$

    e $\sec(\frac{1}{2}\theta + 45°) = -1$ **f** $\operatorname{cosec} 2\theta = 2$

3 Solve for $-\pi < \theta < \pi$

    a $\operatorname{cosec}\theta = 4\sin\theta$ **b** $\cot\theta = \tan\theta$

    c $\sec^2\theta = 2$ **d** $2\sec^2\theta = 3\tan\theta + 1$

    e $2\cot^2 - 1 = 5\operatorname{cosec}\theta$ **f** $6\tan^2\theta = 4 - 11\sec\theta$

4 Describe a sequence of geometrical transformations that maps the graph of $y = \cot x$ onto the graph of $y = 4 + \cot 3x$.

5 Sketch on the same axes the graphs of $y = \operatorname{cosec} x$ and $y = \operatorname{cosec} |x|$.

There is a 'Test yourself' exercise (Ty2) and an 'Extension exercise' (Ext2) on the CD-ROM.

➤ Key points

Reciprocal functions

$$\sec\theta = \frac{1}{\cos\theta} \qquad \csc\theta = \frac{1}{\sin\theta} \qquad \cot\theta = \frac{1}{\tan\theta} = \frac{\cos\theta}{\sin\theta}$$

$$\tan\theta = \cot(90° - \theta) \qquad \tan^2\theta + 1 = \sec^2\theta \qquad 1 + \cot^2\theta = \csc^2\theta$$

Inverse trigonometrical functions

$\sin^{-1} x = y \Rightarrow x = \sin y$.

$\sin^{-1} x$ can be written as arcsin x.

	Domain	Principal value range
$y = \sin^{-1} x$	$-1 \leqslant x \leqslant 1$	$-\dfrac{\pi}{2} \leqslant y \leqslant \dfrac{\pi}{2}$
$y = \cos^{-1} x$	$-1 \leqslant x \leqslant 1$	$0 \leqslant y \leqslant \pi$
$y = \tan^{-1} x$	$x \in \mathbb{R}$	$-\dfrac{\pi}{2} < y < \dfrac{\pi}{2}$

The graphs of the inverse trigonometrical functions can be sketched by reflecting the graphs of $y = \sin x$ etc... over a restricted domain in the line $y = x$.

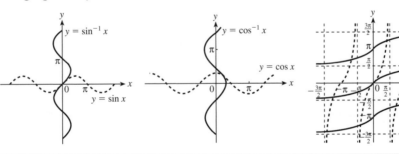

The functions e^x and ln x

Earlier in your mathematical studies you met the number π and learnt something of its importance, not just in relation to the area and circumference of a circle but also in its link with trigonometric functions. In this chapter you will meet an equally important number, e, and learn of its importance as a base for logarithms.

After working through this chapter you should be

■ *familiar with the exponential function e^x, its inverse function, ln x and their graphs*

■ *able to solve equations involving the exponential function or natural logarithms.*

3.1 The function e^x and its inverse, ln x

Earlier in the course exponential functions, for example 2^x, 3^x, $\left(\frac{5}{2}\right)^x$, where the variable is in the index (or exponent), and their inverses, log functions, were introduced.

Remember $f(x) = 2^x \Leftrightarrow f^{-1}(x) = \log_2 x$

These diagrams show the graphs of exponential functions $f(x) = a^x$ for $a = 2$, 3 and 4, and superimposed on each graph with a dotted line is its gradient function, $f'(x)$.

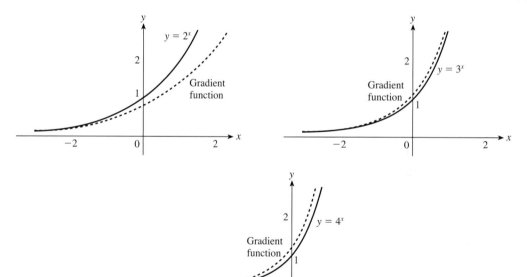

Each curve for $f(x)$ passes through (0, 1), since $f(0) = a^0 = 1$. For each function, the gradient function is also exponential of the form $f'(x) = ma^x$, where m is to be determined.

■ For $f(x) = 2^x$, $f'(x)$ is below $f(x)$, implying that $m < 1$.

■ For $f(x) = 3^x$ and $f(x) = 4^x$, $f'(x)$ is above $f(x)$, implying that $m > 1$.

> $f'(0) = m \times a^0 = m$, so m is the value of the gradient of a^x at (0, 1).

To investigate the value of m, consider $f(x) = a^x$, where a is a positive constant.

Consider δx, a small change in x so that

$$f(x + \delta x) - f(x) = a^{x + \delta x} - a^x$$

$$= a^x(a^{\delta x} - 1)$$

> When δx is a small change in x,
> $$f'(x) = \lim_{\delta x \to 0}\left(\frac{f(x + \delta x) - f(x)}{\delta x}\right)$$

$$f'(x) = \lim_{\delta x \to 0}\left(\frac{a^x(a^{\delta x} - 1)}{\delta x}\right)$$

$$= a^x \lim_{\delta x \to 0}\left(\frac{a^{\delta x} - 1}{\delta x}\right)$$

Writing $m = \lim_{\delta x \to 0}\left(\dfrac{a^{\delta x} - 1}{\delta x}\right)$ gives $f'(x) = a^x \times m$.

Consider $a = 2$, i.e. $f(x) = 2^x$.

Values correct to 7 decimal places of $\dfrac{2^{\delta x} - 1}{\delta x}$ as $\delta x \to 0$ are shown in this table:

δx	0.05	0.005	0.0005	0.000 05
$\dfrac{2^{\delta x} - 1}{\delta x}$	0.705 298 5	0.694 349 7	0.693 267 3	0.693 159 2

$m \approx 0.693$, implying that $\dfrac{d}{dx}(2^x) \approx 0.693(2^x)$.

Similarly, it can be shown that $\dfrac{d}{dx}(3^x) \approx 1.10(3^x)$ and $\dfrac{d}{dx}(4^x) \approx 1.39(4^x)$.

The function e^x^

The gradient at $(0, 1)$ on $y = 2^x$ is approximately 0.693 (<1). On $y = 3^x$ it is approximately 1.10 (>1). Intuitively, there is a number between 2 and 3 whose gradient at $(0, 1)$ is exactly 1. This number, represented by the letter e, is one of the most important numbers in mathematics. The number e is irrational, and $e = 2.718\,282$ (to 6 d.p.) For the special function, e^x, **the exponential function**, $\dfrac{d}{dx}(e^x) = e^x$.

The function ln x

Earlier in the course exponential functions and their inverses, logarithmic functions, were introduced.

Recall that $x = a^y \Leftrightarrow \log_a x = y$. Hence, $x = e^y \Leftrightarrow \log_e x = y$.

The logarithm to base e of x can be written as $\ln x$ rather than $\log_e x$; such logarithms are called natural or Napierian logarithms (after John Napier (1550–1617), a Scottish mathematician).

Graphs of e^x and $\ln x$

As x increases, the graph of e^x eventually becomes steeper than any polynomial graph. Conversely, the graph of $\ln x$, the reflection of e^x in the line $y = x$, tends slowly to infinity, more slowly than any polynomial graph.

For e^x the domain is $x \in \mathbb{R}$ and the range is $y \in \mathbb{R}$, $y > 0$.

For the inverse function, $\ln x$, the domain is $x \in \mathbb{R}$, $x > 0$ and the range is $y \in \mathbb{R}$.

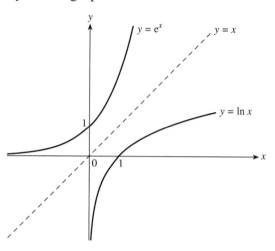

Note that the graphs of e^x and $\ln x$ are reflections of each other in $y = x$.

The x-axis is an asymptote to $y = e^x$.
The y-axis is an asymptote to $y = \ln x$.

The graphs of e^x and $\ln x$ can be transformed using the methods of Chapter 1.

Example 1   Sketch $y = f(x)$ where $f(x) = 1 - e^x$. State the domain and range of the function and any asymptotes of the curve.

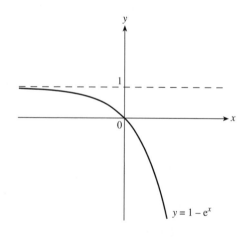

> $y = -e^x$ is a reflection of $y = e^x$ in the x-axis. $y = 1 - e^x$ is a translation of
> $$\begin{pmatrix} 0 \\ 1 \end{pmatrix} \text{ of } y = -e^x.$$
> (1 is added to the function.)
>
> The x-axis is an asymptote to $y = e^x$ and $y = -e^x$ so $y = 1$ will be an asymptote to the translated curve.

Domain: $x \in \mathbb{R}$.   Range: $y \in \mathbb{R}$, $y < 1$.   Asymptote: $y = 1$.

Note   To sketch, for example, $y = e^{3x+2}$ write $y = e^2 e^{3x}$.
$y = e^{3x}$ is a stretch of $y = e^x$ scale factor $\frac{1}{3}$ parallel to the x-axis.
$y = e^2 e^{3x}$ is a stretch of $y = e^{3x}$ scale factor e^2 parallel to the y-axis.

Example 2   Given that $f(x) = \ln(x + 2)$, find $f^{-1}(x)$ and sketch the graphs of $y = f(x)$ and $y = f^{-1}(x)$ on the same axes, stating the equations of any asymptotes. State the range of f and the domain and range of f^{-1}.

Let $\ln(x+2) = y$

Use $x = e^y \Leftrightarrow \ln x = y$

so $\qquad x + 2 = e^y$

$\qquad\qquad x = e^y - 2$

$\therefore \qquad f^{-1}(x) = e^x - 2$

Rewrite as $f^{-1}(x)$ replacing y by x.

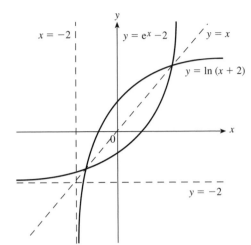

$y = \ln(x+2)$ is the graph of $y = \ln x$ translated by $\begin{pmatrix} -2 \\ 0 \end{pmatrix}$. The y-axis is an asymptote to $y = \ln x$ and so $x = -2$ is an asymptote to the translated curve.

The graph of the inverse function, $y = e^x - 2$, is the reflection of the original graph in the line $y = x$. It follows that $y = -2$ is an asymptote to this graph.

The range of $f(x)$ is $y \in \mathbb{R}$. $x = -2$ is an asymptote to the graph of $y = f(x)$.
The domain of $f^{-1}(x)$ is $x \in \mathbb{R}$ and the range is $y \in \mathbb{R}$, $y > -2$.
$y = -2$ is an asymptote to the graph of $y = f^{-1}(x)$.

Example 3   Describe the transformations required to sketch the graph of $y = 4 + 3\ln(x+5)$ starting from the graph of $y = \ln x$. Sketch the graph of $y = 4 + 3\ln(x+5)$.

The graph of $y = 4 + 3\ln(x+5)$ is obtained from the graph of $y = \ln x$ by a stretch of scale factor 3 parallel to the y-axis followed by a translation of $\begin{pmatrix} -5 \\ 4 \end{pmatrix}$.

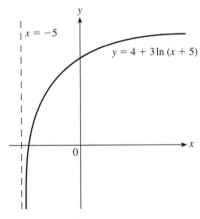

First draw the asymptote, $x = -5$. Then add the stretched and translated curve.

For Example 3, the stretch must be carried out *before* the translation.
The translation *followed* by the stretch would give $y = 12 + 3\ln(x+5)$.

Exercise 3A   **1**   Use your calculator to find, correct to 3 significant figures

    a e^3 **b** $\ln 4$ **c** $e^{-2.7}$ **d** $\ln \frac{1}{4}$

2 Simplify

a $e^{\ln 4}$ **b** $e^{\ln x}$ **c** $\ln e^3$ **d** $\ln e^x$

e $\ln(e^{3-x})$ **f** $e^{\ln(2x+1)}$ **g** $e^{2\ln 2}$ **h** $e^{-3\ln 4}$

3 Sketch these functions, stating the range of f in each case.

a $f(x) = 1 + e^x, \quad x \in \mathbb{R}$ **b** $f(x) = 2 + \ln x, \quad x \in \mathbb{R}, \, x > 0$

c $f(x) = e^{x-1}, \quad x \in \mathbb{R}$ **d** $f(x) = \ln(4 + x), \quad x \in \mathbb{R}, \, x > -4$

4 The function f is defined by $f(x) = ke^x$, $x \in \mathbb{R}$ and k is a positive constant.

  a State the range of f.

  b Find $f\left(\ln \frac{1}{k}\right)$.

  c Find $f^{-1}(x)$ and state the domain of f^{-1}.

  d Sketch on the same axes $y = f(x)$ and $y = f^{-1}(x)$.

  e Given that the graphs of $y = f(x)$ and $y = f^{-1}(x)$ cut the axes at A and B, find the length of AB.

5 For each of these functions, the domain is $x \in \mathbb{R}$.

Find $f^{-1}(x)$, and sketch the graphs of $f(x)$ and $f^{-1}(x)$ on the same axes, showing any intersection with the axes.

State the range of f and the domain and range of f^{-1}.

  a $f(x) = \ln x + 5, \, x > 0$ **b** $f(x) = e^x + 4$ **c** $f(x) = \frac{1}{2}e^x$

6 For each of these functions, find $f^{-1}(x)$, and sketch the graphs of $f(x)$ and $f^{-1}(x)$ on the same axes, showing any intersection with the axes. State the range of f and the domain and range of f^{-1}.

  a $f(x) = \ln(x - 2), \, x > 2$ **b** $f(x) = e^{x+1}$ **c** $f(x) = 3\ln(1 - x), \, x < 1$

7 Sketch, on separate diagrams, the graphs of

  a $y = |\ln x|$ **b** $y = e^{|x|}$ **c** $y = |e^x - 1|$

8 Sketch, on the same axes, the graphs of $y = e^x$ and $y = 2e^{x-1} - 1$.

9 Sketch, on separate diagrams, the graphs of

  a $y = e^{2x-1} + 1$ **b** $y = e^{3x+2} - 4$ **c** $y = e^{-x+1} + 2$

3.2 Solving equations involving exponential functions and natural logarithms

The techniques used earlier in the course to solve equations involving exponential or logarithmic functions can also be used when the base is e.

■ To solve an equation of the form $e^{ax+b} = p$ take natural logarithms of each side.

■ To solve an equation of the form $\ln(ax + b) = q$ rewrite the equation as $ax + b = e^q$.

Example 4   Solve $e^{2x-1} = 75$.

$$e^{2x-1} = 75$$

Taking natural logarithms of both sides

$$\ln(e^{2x-1}) = \ln 75$$

$$(2x-1)\ln e = \ln 75 \qquad \boxed{\text{Use } \log a^n = n\log a}$$
$$\qquad\qquad\qquad\qquad\qquad \boxed{\ln e = 1}$$

$$2x - 1 = \ln 75$$

$$2x = \ln 75 + 1$$

$$x = \frac{\ln 75 + 1}{2}$$

$$x = 2.66 \text{ (3 s.f.)}$$

Example 5   Solve $3e^{2x} - 5e^x = 2$, giving your answer exactly.

$$3e^{2x} - 5e^x = 2$$

Let $y = e^x$.

Then $3y^2 - 5y = 2 \qquad \boxed{\text{If } y = e^x \text{ then } y^2 = (e^x)^2 = e^{2x}}$

$$3y^2 - 5y - 2 = 0$$

$$(3y + 1)(y - 2) = 0$$

$$\Rightarrow \quad y = -\tfrac{1}{3} \text{ or } y = 2$$

Substituting for y

$$e^x = -\tfrac{1}{3} \text{ or } e^x = 2$$

There is no solution for $e^x < 0$ but $e^x = 2$ gives $x = \ln 2$.

Example 6   Solve $\ln(2x - 5) = 3.2$

$$\ln(2x - 5) = 3.2 \qquad \boxed{\ln x = y \Rightarrow x = e^y}$$

$$2x - 5 = e^{3.2}$$

$$2x = e^{3.2} + 5$$

$$x = \frac{e^{3.2} + 5}{2}$$

$$x = 14.8 \text{ (3 s.f.)}$$

Exercise 3B

1 Solve for x.

a $e^x = 6.2$ **b** $e^x = 16.4$ **c** $2e^x - 1 = 83$

d $4e^{3x} = 170$ **e** $e^{-x} = 2$ **f** $e^{-x} = 7$

g $e^{-4x} = 8$ **h** $e^{-5x} + 4 = 5$

2 Solve for x.

a $e^{2x+3} = 42$ b $e^{3x-2} = 100$ c $e^{5x+2} = 85$

d $e^{5-x} = 10$ e $e^{14-3x} = 170$ f $4e^{3x+2} = 78$

3 Solve for x.

a $\ln x = 2$ b $\ln(x+3) = 1.6$ c $\ln(3x-5) = 2.9$

d $\ln(4x+3) = 4$ e $\ln(4-x) = 2$ f $\ln(6-3x) = -3$

g $\ln(x+4) = \ln(2x-3)$ h $\ln(2x+5) = \ln(14-x)$

4 Solve for x.

a $e^{3x+2} = e^{2x-1}$ b $e^{4x+1} = e^{1-x}$

c $e^{2x} - 3e^x - 4 = 0$ d $2e^{2x} + 3e^x - 5 = 0$

e $e^{2x} - 9 = 0$ f $4e^{2x} + 3e^x = 1$

g $e^{2x} - 5e^x = -6$ h $12e^{2x} + 6 = 17e^x$

Exercise 3C (Review)

1 Sketch these functions, stating the range in each case.

a $f(x) = 4 + e^x$ b $f(x) = 3 + \ln x$

c $f(x) = e^{3x} - 5$ d $f(x) = 2 + \ln(x-3)$

2 Describe the geometrical transformations required to sketch the graph of $y = e^{6x-2} - 4$ starting from the graph of $y = e^x$. Sketch the graph of $y = e^{6x-2} - 4$.

3 Describe the geometrical transformations required to sketch the graph of $y = 5 + 2\ln(x-3)$ starting from the graph of $y = \ln x$. Sketch the graph of $y = 5 + 2\ln(x-3)$.

4 Solve for x.

a $e^{5x+1} = 125$ b $e^{4x-2} = e^{6x-9}$

c $e^{4-3x} = 0.2$ d $e^{0.1x-3} = 510$

5 Solve for x.

a $\ln(3x-20) = \ln(2x-10)$ b $\ln(3x+2) = 4$

c $2\ln(5x-1) = 9$ d $\frac{1}{4}\ln(4-3x) = 1.2$

There is an 'Extension exercise' (Ext3) on the CD-ROM.

➤ Key points

The functions e^x and ln x

■ The function f where $f(x) = e^x$ is called **the exponential function**.

■ If $f(x) = e^x$ then the inverse function of f is the natural logarithm function f^{-1} where $f^{-1}(x) = \log_e x$ or $\ln x$.

■ The graphs of e^x and $\ln x$ are reflections of each other in $y = x$.

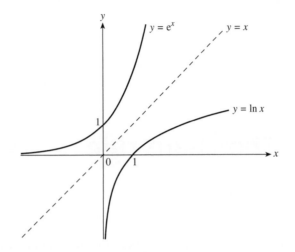

■ To solve an equation of the form $e^{ax+b} = p$ take natural logarithms of each side.

■ To solve an equation of the form $\ln(ax + b) = q$ rewrite the equation as $ax + b = e^q$.

Differentiation

Earlier in the course you learnt about the language and notation of differentiation including how to find the derivative of x^n and of expressions involving sums and differences of powers of x. In this chapter you will see how to differentiate a number of other common functions and how to deal with products, quotients and composite functions.

After working through this chapter you should be able to

- *use the chain rule to differentiate composite functions*
- *find the derivatives of products and quotients*
- *find the derivatives of exponential, logarithmic and trigonometric functions.*

The CD-ROM (E4.1) shows how differentiation can be used to obtain approximate values for the change in one variable when another, connected, variable is changed.

4.1 The chain rule

The **chain rule** is used to differentiate composite functions such as

$$(3x+1)^8, \qquad e^{3+2x}, \qquad \ln(2x^2-4), \qquad \sqrt{4x+3}$$

It is also known as the **function of a function rule** or the **composite function rule** and is one of the most useful results in calculus.

Consider how to differentiate powers of the function $3x+1$:

$$(3x+1)^2 \quad (3x+1)^3 \quad (3x+1)^4 \quad (3x+1)^8$$

For each of these, it is possible to multiply out the brackets, differentiate term by term and then factorise the result:

If $\quad f(x) = (3x+1)^2 = 9x^2 + 6x + 1$

then $\quad f'(x) = 18x + 6$

$$= 6(3x+1)$$

so $\quad f(x) = (3x+1)^2 \Rightarrow f'(x) = 6(3x+1)$

If $\quad f(x) = (3x+1)^3$

$$= 27x^3 + 27x^2 + 9x + 1$$

then $\quad f'(x) = 81x^2 + 54x + 9$

$$= 9(9x^2 + 6x + 1)$$

$$= 9(3x+1)^2$$

so $\quad f(x) = (3x+1)^3 \Rightarrow f'(x) = 9(3x+1)^2$

In a similar way, it can also be shown that

$$f(x) = (3x+1)^4 \Rightarrow f'(x) = 12(3x+1)^3$$

$$f(x) = (3x+1)^5 \Rightarrow f'(x) = 15(3x+1)^4$$

$$f(x) = (3x+1)^8 \Rightarrow f'(x) = 24(3x+1)^7$$

It is very laborious to multiply out all the brackets and then factorise the derivative. In fact, there is no need to do this, since the pattern emerging in the above derivatives can be obtained directly using the chain rule.

The chain rule can also be used to differentiate functions that cannot be expanded into a finite number of terms, for example

$$\frac{1}{\sqrt{4x+5}} \qquad e^{3x+1} \qquad \ln(5x+2)$$

Suppose y is a function of t and t is a function of x.

If δy, δt and δx are corresponding increments in the variables y, t and x, then

$$\frac{\delta y}{\delta x} = \frac{\delta y}{\delta t} \times \frac{\delta t}{\delta x} \qquad ①$$

When δy, δt and δx tend to zero,

$$\frac{\delta y}{\delta x} \to \frac{dy}{dx}$$

$$\frac{\delta y}{\delta t} \to \frac{dy}{dt}$$

and $\qquad \dfrac{\delta t}{\delta x} \to \dfrac{dt}{dx}$

so equation ① becomes

$$\frac{dy}{dx} = \frac{dy}{dt} \times \frac{dt}{dx}$$

> **If y is a function of t and t is a function of x, then**
>
> $$\frac{dy}{dx} = \frac{dy}{dt} \times \frac{dt}{dx}$$

Informally in words: to find dy/dx, express y as a function of t, differentiate y with respect to t, then multiply by dt/dx.

Example 1 Find $\dfrac{dy}{dx}$ when **a** $y = 6(4x-2)^3$ **b** $y = \sqrt{4x^2 - 1}$

Solution **a** $y = 6(4x-2)^3$

Let $t = 4x - 2$ then $y = 6t^3$.

Letting $t = 4x - 2$ enables y to be written as a function of t and both t and y can now be differentiated.

So $\qquad \dfrac{dt}{dx} = 4$

Differentiate t with respect to x and y with respect to t.

and $\qquad \dfrac{dy}{dt} = 18t^2$

Re-write t in terms of x.

$\qquad\qquad = 18(4x-2)^2$

By the chain rule:

$$\frac{dy}{dx} = \frac{dy}{dt} \times \frac{dt}{dx}$$

$$= 18(4x - 2)^2 \times 4$$

$$= 72(4x - 2)^2$$

b $\quad y = \sqrt{4x^2 - 1} = (4x^2 - 1)^{\frac{1}{2}}$

> Write y in index form first, then write the contents of the bracket as t so y is a function of t.

Let $t = 4x^2 - 1$ then $y = t^{\frac{1}{2}}$.

So $\quad \dfrac{dt}{dx} = 8x$

and $\quad \dfrac{dy}{dt} = \frac{1}{2} t^{-\frac{1}{2}}$

$$= \tfrac{1}{2}(4x^2 - 1)^{-\frac{1}{2}}$$

$$\frac{dy}{dx} = \frac{dy}{dt} \times \frac{dt}{dx}$$

> Apply the chain rule.

$$= \tfrac{1}{2}(4x^2 - 1)^{-\frac{1}{2}} \times 8x$$

> It is good style to write the answer in the format given in the question.

$$= \frac{4x}{\sqrt{4x^2 - 1}}$$

Example 2 Find $\dfrac{dy}{dt}$ when $y = (t^2 + 3t - 6)^5$.

> t is given as one of the variables, so another letter must be used for the substitution.

Let $u = t^2 + 3t - 6$ then $y = u^5$.

> Define u as a function of t by letting the contents of the bracket be u. y is now a function of u.

So $\quad \dfrac{du}{dt} = 2t + 3$

> Differentiate u with respect to t and y with respect to u.

and $\quad \dfrac{dy}{du} = 5u^4 = 5(t^2 + 3t - 6)^4$

$$\frac{dy}{dt} = \frac{dy}{du} \times \frac{du}{dt}$$

> Notice the pattern: $(2t + 3)$ is the derivative of $(t^2 + 3t - 6)$.

$$= 5(t^2 + 3t - 6)^4 \times (2t + 3)$$

> Though not essential, the expression is often rewritten with the simpler bracket first.

$$= 5(2t + 3)(t^2 + 3t - 6)^4$$

Example 3 Find $\dfrac{dy}{dx}$ when $y = (3x^2 + 1)^8$.

Let $t = 3x^2 + 1$ then $y = t^8$.

So $\dfrac{dt}{dx} = 6x$

and $\dfrac{dy}{dt} = 8t^7 = 8(3x^2 + 1)^7$

$\dfrac{dy}{dx} = 8(3x^2 + 1)^7 \times 6x$

$= 48x(3x^2 + 1)^7$

> $6x$ is the derivative of $3x^2 + 1$.

Note With practice, it is possible to find the derivative without showing all the working. This fits in with an alternative way of writing the chain rule.

➤
$$\boxed{\dfrac{d}{dx}\, \mathbf{g}(\mathbf{f}(x)) = \mathbf{g}'(\mathbf{f}(x)) \times \mathbf{f}'(x)}$$

> This is sometimes written
> $$\dfrac{d}{dx}\, f(g(x)) = f'(g(x)) \times g'(x)$$

Differentiation of powers of f(x)

In Example 3

$$\dfrac{d}{dx}(\square)^8 = 8(\square)^7 \times \text{derivative of } (\square)$$

i.e. $\dfrac{d}{dx}(f(x))^8 = 8(f(x))^7 \times f'(x)$

A general rule can be obtained for differentiating powers of $f(x)$, i.e. expressions of the type $(f(x))^n$:

➤
$$\boxed{\begin{array}{l}\dfrac{d}{dx}\,(\mathbf{f}(x))^n = n(\mathbf{f}(x))^{n-1} \times \mathbf{f}'(x) \\[2mm] \textbf{If the expression is multiplied by a} \\ \textbf{constant } \mathbf{k}, \textbf{ then} \\[2mm] \dfrac{d}{dx}\, k(\mathbf{f}(x))^n = kn(\mathbf{f}(x))^{n-1} \times \mathbf{f}'(x)\end{array}}$$

> This looks very complicated but it is in fact easy to apply and the working can usually be done mentally.
> Informally in words: To differentiate a function to the power n, write down $n \times$ (the function to the power $(n-1)$) $\times$ (derivative of the function).

Example 4   Find $\dfrac{dy}{dx}$ when

    a $y = \dfrac{1}{(x^4 + 5)^2}$ **b** $y = 6(x^2 - 2)^3$

Solution   **a** $y = \dfrac{1}{(x^4 + 5)^2}$

> Write y in index form first, so $y = (\square)^{-2}$
> $\dfrac{dy}{dx} = -2(\square)^{-3} \times$ derivative of $(\square)$.

$$= (x^4 + 5)^{-2}$$

$$\dfrac{dy}{dx} = -2(x^4 + 5)^{-3} \times 4x^3$$

> Tidy up the expression.

$$= -8x^3(x^4 + 5)^{-3}$$

> Write in the format given in the question.

$$= -\dfrac{8x^3}{(x^4 + 5)^3}$$

    b $y = 6(x^2 - 2)^3$

> $y = 6(\square)^3$ so $\dfrac{dy}{dx} = 6 \times 3(\square)^2 \times$ derivative of $(\square)$

$$\dfrac{dy}{dx} = 18(x^2 - 2)^2 \times 2x$$

$$= 36x(x^2 - 2)^2$$

A specific application is when $f(x)$ is a **linear function**, i.e. $f(x) = ax + b$. For example

$$\dfrac{d}{dx}(3x + 1)^8 = 8(3x + 1)^7 \times 3$$

$$= 24(3x + 1)^7$$

$$\dfrac{d}{dx}3(1 - 5x)^4 = 3 \times 4(1 - 5x)^3 \times (-5)$$

$$= -60(1 - 5x)^3$$

$$\dfrac{d}{du}\left(\dfrac{1}{2(6u - 5)^2}\right) = \dfrac{d}{du}\tfrac{1}{2}(6u - 5)^{-2}$$

$$= \tfrac{1}{2} \times (-2)(6u - 5)^{-3} \times 6$$

$$= -\dfrac{6}{(6u - 5)^3}$$

➤ $\boxed{\dfrac{d}{dx}(ax + b)^n = na(ax + b)^{n-1}}$

Note   This is a particularly useful format to remember when integrating (Section 5.4).

Example 5   The tangent to the curve $y = \sqrt{2x - 1}$ at P(5, 3) crosses the y-axis at A and the x-axis at B. Find the area of triangle OAB.

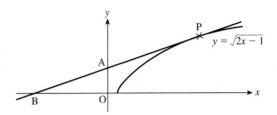

$$y = \sqrt{2x - 1}$$

Writing y in index form, $y = (\Box)^{\frac{1}{2}}$

$$= (2x - 1)^{\frac{1}{2}}$$

$$\frac{dy}{dx} = \tfrac{1}{2}(2x - 1)^{-\frac{1}{2}} \times 2$$

To find the gradient of the tangent, find $\frac{dy}{dx} = \frac{1}{2}(\Box)^{-\frac{1}{2}} \times$ derivative of $(\Box)$.

$$= \frac{1}{\sqrt{2x - 1}}$$

When $x = 5$

$$\frac{dy}{dx} = \frac{1}{\sqrt{10 - 1}} = \tfrac{1}{3}$$

This gives the gradient of the tangent at $x = 5$.

Equation of tangent at (5, 3):

Use $y - y_1 = m(x - x_1)$

$$y - 3 = \tfrac{1}{3}(x - 5)$$

$$3(y - 3) = x - 5$$

$$3y - 9 = x - 5$$

$$3y = x + 4$$

Write the equation of the line in a convenient format.

At A, $x = 0$, so $3y = 4$

Find the coordinates of A and B.

$$y = \tfrac{4}{3}$$

At B, $y = 0$, so $0 = x + 4$

$$x = -4$$

$\therefore$ A is $\left(0, \tfrac{4}{3}\right)$ and B is $(-4, 0)$.

Draw a sketch showing O, A and B.

Area OAB $= \tfrac{1}{2} \times 4 \times \tfrac{4}{3} = 2\tfrac{2}{3}$

The area of triangle OAB is $2\tfrac{2}{3}$ square units.

The use of $\dfrac{dy}{dx} = \dfrac{1}{\left(\dfrac{dx}{dy}\right)}$

In some circumstances x may be given as a function of y rather than y as a function of x. However, it is still possible to find the gradient $\dfrac{dy}{dx}$ by making use of a special case of the chain rule.

Suppose that x is given as a function of y and δx and δy are corresponding increments in the variables x and y, then

$$\frac{\delta y}{\delta x} \times \frac{\delta x}{\delta y} = 1 \qquad \textcircled{1}$$

When δx and δy tend to zero, $\dfrac{\delta y}{\delta x} \to \dfrac{dy}{dx}$ and $\dfrac{\delta x}{\delta y} \to \dfrac{dx}{dy}$

so equation $\textcircled{1}$ becomes

$$\frac{dy}{dx} \times \frac{dx}{dy} = 1 \text{ or, alternatively,}$$

➤ $$\boxed{\frac{dy}{dx} = \frac{1}{\left(\dfrac{dx}{dy}\right)}}$$

Example 6 Find $\dfrac{dy}{dx}$ when   **a** $x = y^2$ **b** $x = y^3 + 4\sqrt{y}$

In each case express your answer as a function of y.

Solution   **a** $\qquad\qquad x = y^2$

So $\quad \dfrac{dx}{dy} = 2y$

but $\quad \dfrac{dy}{dx} = \dfrac{1}{\left(\dfrac{dx}{dy}\right)}$

$\therefore \quad \dfrac{dy}{dx} = \dfrac{1}{2y}$

> Although the derivative can be written as $\dfrac{1}{2\sqrt{x}}$ it is more usual to leave x as a function of y as this is the form of the original expression.

b
$$x = y^3 + 4\sqrt{y}$$

Write y in index form.

$$= y^3 + 4y^{\frac{1}{2}}$$

So $\dfrac{\mathrm{d}x}{\mathrm{d}y} = 3y^2 + 2y^{-\frac{1}{2}}$

but $\dfrac{\mathrm{d}y}{\mathrm{d}x} = \dfrac{1}{\left(\dfrac{\mathrm{d}x}{\mathrm{d}y}\right)}$

$\therefore \quad \dfrac{\mathrm{d}y}{\mathrm{d}x} = \dfrac{1}{3y^2 + 2y^{-\frac{1}{2}}}$

Multiply both the numerator and denominator by $y^{\frac{1}{2}}$ to simplify the expression.

$$= \dfrac{y^{\frac{1}{2}}}{3y^{\frac{5}{2}} + 2}$$

Exercise 4A

1 Find $\dfrac{\mathrm{d}y}{\mathrm{d}x}$ when

a $y = (2x + 5)^3$ **b** $y = 2(3x - 5)^4$ **c** $y = (4 - 3x)^4$

d $y = \sqrt{6 - 4x}$ **e** $y = \dfrac{2}{\sqrt{3x + 1}}$ **f** $y = 2(4x + 5)^{\frac{1}{3}}$

g $y = \dfrac{1}{(1 - x)^{\frac{3}{2}}}$ **h** $y = \dfrac{1}{(5x - 2)^3}$ **i** $y = \dfrac{3}{5(2 - 5x)^3}$

2 Differentiate with respect to x

a $(3x^2 + 2)^6$ **b** $\frac{1}{2}(3x^2 + 5x)^4$ **c** $\sqrt[4]{1 - 5x^2}$

d $\dfrac{1}{6x^3 - 2}$ **e** $\dfrac{1}{\sqrt{4x^2 - 9}}$ **f** $\dfrac{2}{6x^2 + 3x - 1}$

3 If $A = y + \dfrac{1}{y + 3}$, find $\dfrac{\mathrm{d}A}{\mathrm{d}y}$, simplifying the answer.

4 Find $\dfrac{\mathrm{d}y}{\mathrm{d}x}$ in terms of y when

a $x = y^3 + 7$ **b** $x = 2y^4 - 5$ **c** $x = 4y^5 - 8\sqrt{y}$

d $x = \dfrac{1}{4y^2}$ **e** $x = (y + 1)(2y - 3)$ **f** $x = \dfrac{1}{y} - \dfrac{1}{y^2}$

5 Find the equation of the tangent to the curve $y = (5x + 3)^4$ at the point $(-1, 16)$, giving your answer in the form $ax + by + c = 0$.

6 A curve has equation $y = (x + 3)^3 - 4x - 12$.
Find the coordinates of points on the curve with gradient 8.

7 The curve $y = \sqrt{2x + 4}$ crosses the y-axis at A$(0, a)$.

    a Sketch the curve, stating the value of a.

    b Find the value of $\dfrac{dy}{dx}$ at A.

    c State the gradient of the normal at A.

    d Find an equation of the normal at A.

8 For each of these curves, find the coordinates of any stationary points and state whether the stationary point is a maximum, a minimum or a point of inflexion.

    a $y = (2x - 3)^6$ **b** $y = x + \dfrac{1}{x}$ **c** $y = (x + 3)^4 - 4x$ **d** $y = \left(\tfrac{1}{2}x - 1\right)^3$

9 The displacement, s, of a body from the origin at time t is given by $s = (2t - 3)^5$.

    a The velocity, v at time t is given by $\dfrac{ds}{dt}$. Find the velocity when $t = 2$.

    b The acceleration at time t is given by $\dfrac{dv}{dt}$. Find the acceleration when $t = 1$.

10 Find an equation of the tangent to the curve

$$y = \frac{2}{x^2 - 3}$$

at the point $\left(3, \tfrac{1}{3}\right)$.

11 The curve $y = (2x + 1)^3$ crosses the x-axis at A and the y-axis at P.

    a Find the coordinates of A and P.

    b Find an equation of the normal to the curve at P.

    c The normal at P cuts the x-axis at B. Find the area of triangle APB.

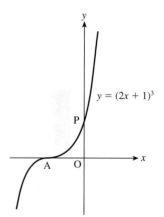

12 If $f(x) = \tfrac{1}{3}(x + 1)^{-3}$, find $f'(x)$ when $x = 1$.

13 Given that

$$y = \sqrt{3x - 1} + \frac{4}{\sqrt{3x - 1}}$$

show that

$$\frac{dy}{dx} = \frac{9x - 15}{2(3x - 1)^{\frac{3}{2}}}$$

14 Find the coordinates of the stationary points on the curve $y = (x^2 - 4)^4$.

15 a Show that
$$\frac{2x+1}{x^2+x-2} = \frac{1}{x-1} + \frac{1}{x+2}$$

b Show that
$$\frac{\mathrm{d}}{\mathrm{d}x}\left(\frac{2x+1}{x^2+x-2}\right) + \frac{1}{(x-1)^2} + \frac{1}{(x+2)^2} = 0$$

16 Given that $y = (2x - 1)^4$, show that
$$\left(\frac{\mathrm{d}^2 y}{\mathrm{d}x^2}\right)^2 = 2304y$$

17 The diagram shows the graph of
$y = (2x - 3)^3 - 12x$

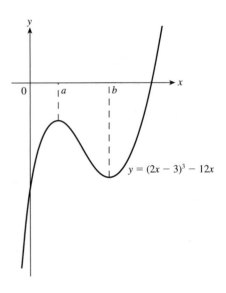

a The curve has a maximum point when $x = a$ and a minimum point when $x = b$. Show that $a = \frac{1}{2}\left(3 - \sqrt{2}\right)$ and find the exact value of b.

b Show that an equation of the tangent at $(0.5, -14)$ is $y = 12x - 20$.

c There is another point on the curve where the tangent is parallel to the tangent at $(0.5, -14)$. Find the coordinates of this point.

4.2 Differentiation of the exponential function and natural logarithms

In Chapter 3 you saw that for an exponential function $f(x) = a^x$ its gradient function was of the form $f'(x) = ma^x$, where m was a constant. For the special function $f(x) = e^x$, *the exponential function*, $m = 1$ and $f'(x) = e^x$.

➤ $$\frac{\mathrm{d}}{\mathrm{d}x}(e^x) = e^x$$

Differentiation of functions of the type $e^{f(x)}$

The chain rule will now be used to differentiate e^{kx}.

Let $y = e^{kx}$ and $t = kx$, then $y = e^t$.

> y is expressed as a function of t and t as a function of x.

$$y = e^t \qquad t = kx$$

> Both functions can be differentiated.

$$\frac{dy}{dt} = e^t \qquad \frac{dt}{dx} = k$$

> Differentiate y with respect to t and t with respect to x.

By the chain rule,

$$\frac{dy}{dx} = \frac{dy}{dt} \times \frac{dt}{dx}$$

$$= e^t \times k$$

$$= ke^{kx}$$

> Substitute for t.

$$\boxed{\frac{d}{dx}(e^{kx}) = ke^{kx}}$$

Example 7 Find $\dfrac{dy}{dx}$, given y.

a $y = e^{6x} + e^{-\frac{x}{3}}$

> Apply $\dfrac{d}{dx}(e^{kx}) = ke^{kx}$ with $k = 6$ and $k = -\frac{1}{3}$.

$$\frac{dy}{dx} = 6e^{6x} - \frac{1}{3}e^{-\frac{x}{3}}$$

b $y = 8e^{-4x}$

> Apply $\dfrac{d}{dx}(e^{kx}) = ke^{kx}$ with $k = -4$.

$$\frac{dy}{dx} = 8e^{-4x} \times (-4)$$

$$= -32e^{-4x}$$

The chain rule can also be used to find the derivatives of more complicated expressions, for example

$$\frac{d}{dx}\left(e^{3x+2}\right) \qquad \frac{d}{dx}\left(4e^{x^2+5x-2}\right)$$

Example 8 Find $\dfrac{dy}{dx}$ when **a** $y = e^{3x+2}$ **b** $y = 4e^{x^2+5x-2}$

Solution **a** $y = e^{3x+2}$

Let $t = 3x + 2$ then $y = e^t$

> Letting t be the exponent allows y to be written as a function of t.

So $\dfrac{dt}{dx} = 3$

> Differentiate t with respect to x and y with respect to t.

and $\dfrac{dy}{dt} = e^t = e^{3x+2}$

Using the chain rule:

$$\frac{dy}{dx} = \frac{dy}{dt} \times \frac{dt}{dx}$$

$$= e^{3x+2} \times 3$$

$$= 3e^{3x+2}$$

Note: $\frac{dy}{dx} = $ [derivative of $(3x+2)$] $\times e^{3x+2}$

b $\quad y = 4e^{x^2+5x-2}$

Let $t = x^2 + 5x - 2$ then $y = 4e^t$

So $\quad \dfrac{dt}{dx} = 2x + 5$

and $\quad \dfrac{dy}{dt} = 4e^t = 4e^{x^2+5x-2}$

$$\frac{dy}{dx} = \frac{dy}{dt} \times \frac{dt}{dx}$$

$$= 4e^{x^2+5x-2} \times (2x + 5)$$

$$= 4(2x + 5)e^{x^2+5x-2}$$

$\dfrac{dy}{dx} = 4 \times$ [derivative of $(x^2 + 5x - 2)$] $\times e^{x^2+5x-2}$

$$\frac{d}{dx}\left(e^{f(x)}\right) = f'(x)e^{f(x)}$$

If the expression is multiplied by a constant k, then

$$\frac{d}{dx}\left(ke^{f(x)}\right) = kf'(x)e^{f(x)}$$

In words: to differentiate an exponential function, multiply the exponential function by the derivative of the exponent.

Remember $\quad \dfrac{d}{dx}\left(e^{\square}\right) = $ derivative of $(\square) \times \left(e^{\square}\right)$

Example 9 $\quad$ **a** $\quad \dfrac{d}{dx}\left(e^{3x^2}\right) = 6xe^{3x^2}$

$f(x) = 3x^2$ so $f'(x) = 6x$

$\quad$ **b** $\quad \dfrac{d}{dt}\left(e^{\frac{1}{t}}\right) = -\dfrac{1}{t^2}e^{\frac{1}{t}}$

$f(t) = \dfrac{1}{t}$ so $f'(t) = -\dfrac{1}{t^2}$

$\quad$ **c** $\quad \dfrac{d}{dy}\left(\dfrac{1}{e^{y^3}}\right) = \dfrac{d}{dy}\left(e^{-y^3}\right) = -3y^2e^{-y^3} = -\dfrac{3y^2}{e^{y^3}}$

Write in index form first.

A specific example is when the exponent is a linear function, i.e. $f(x) = ax + b$.
For example

$$\frac{d}{dx}\left(e^{5x-1}\right) = 5e^{5x-1}$$

$$\frac{d}{dx}\left(4e^{3x+5}\right) = 12e^{3x+5}$$

$$\frac{d}{dx}\left(2e^{5-3x}\right) = -6e^{5-3x}$$

$$\frac{d}{dx}\left(e^{ax+b}\right) = ae^{ax+b}$$

Example 10   Determine the nature of any stationary points on the curve $f(x) = e^{3x-1} - 3x$.

$f(x) = e^{3x-1} - 3x$ Differentiate to find $f'(x)$.

$f'(x) = 3e^{3x-1} - 3$

$f'(x) = 0$ when $3e^{3x-1} - 3 = 0$ At a stationary point, $f'(x) = 0$.

$e^{3x-1} = 1$ Either take logs to base e and use $\ln 1 = 0$ or use $e^0 = 1$.

$3x - 1 = 0$

$x = \frac{1}{3}$

When $x = \frac{1}{3}$, $y = e^0 - 1 = 0$.

There is a stationary point at $\left(\frac{1}{3}, 0\right)$. Check the nature of the stationary point.

$f''(x) = 9e^{3x-1}$ so $f''\left(\frac{1}{3}\right) = 9 > 0$. If $f''(x) > 0$, the stationary point is a minimum point.

$\therefore$ $\left(\frac{1}{3}, 0\right)$ is a minimum point.

Exercise 4B

1 Differentiate with respect to x

  a e^x **b** e^{4x} **c** e^{-x} **d** e^{-2x} **e** $-3e^{-5x}$ **f** $e^{3x/2}$

2 Differentiate with respect to x

  a e^{4x-5} **b** $3e^{6-2x}$ **c** $\dfrac{e^{3x}}{3}$ **d** $e(e^{3x})$ **e** $\dfrac{1}{2e^{6x+1}}$ **f** xe^3

3 By simplifying each expression first, find the derivative of each of these.

  a $e^{4x-3} \times e^{3x+2}$ **b** $\dfrac{e^{2x+1}}{e^x}$

4 $f(x) = e^{ax+b}$

  a Find $f''(x)$.

  b Write down an expression for $f^n(x)$, the nth derivative of $f(x)$.

5 The diagram shows a sketch of the curve $y = e^{2x-3}$.

  a A has coordinates $(a, 1)$. Find the value of a.

  b Find an equation of the tangent at A.

  c B has coordinates $(0, b)$. Find the value of b.

  d Find an equation of the normal at B.

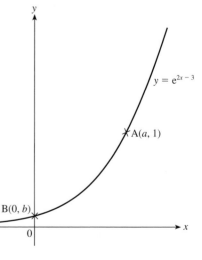

6 Determine the nature of any stationary points on the curve $y = x - e^{x-1}$.

7 Find the range of values for which the function f where $f(x) = 2x + e^{1-2x}$ is increasing.

8 Find an equation of the tangent to the curve

$$y = \frac{1}{e^{4x+2}}$$

at $x = -0.5$.

9 **a** Find an equation of the normal to the curve $y = e^{3x-2}$ at the point (1, e).

  b The normal cuts the y-axis at A and the x-axis at B.
   Find the area of triangle OAB where O is the origin.

10 **a** Sketch the curve $y = 2e^{3x+2}$.

  b M lies on the curve and has x coordinate $-\frac{1}{3}$. Find the gradient at M and hence find an equation of the tangent to the curve at M.

  c The tangent at M meets the x-axis at A and the y-axis at B.
   Show that M is the midpoint of AB.

11 Show that the curve $y = e^{3x} - 2x$ has a minimum point at
$\left(\frac{1}{3}\ln\frac{2}{3}, \frac{2}{3}\left(1 - \ln\frac{2}{3}\right)\right)$.

12 Given that $y = e^{2x} + e^{-x}$, find

  a the intercept of the curve on the y-axis

  b any stationary points of the curve and determine their nature.

13 The function f where $f(x) = e^{ax+b}$ is such that $f(0) = \dfrac{1}{e}$ and $f'(x) = 5f(x)$. Find the values of a and b.

14 Find

  a $\dfrac{d}{dx}\left(4e^{x^2-2}\right)$ **b** $\dfrac{d}{dt}\left(e^{2t^3+3t}\right)$ **c** $\dfrac{d}{du}\left(\dfrac{1}{2e^{-u^2}}\right)$ **d** $\dfrac{d}{dx}\left(e^{\sqrt{x}}\right)$

15 Find the coordinates of the stationary point on the curve $y = e^{x^2} + e^{-x^2}$.

16 A particle moves in a straight line so that t seconds after leaving a fixed point O, its displacement, s metres, is given by $s = 6 - 6e^{-2t}$.

  a Find the displacement after 10 seconds.

  b The velocity, v is given by $\dfrac{ds}{dt}$. Calculate the initial velocity.

  c The acceleration is given by $\dfrac{dv}{dt}$. Calculate the initial acceleration.

Differentiation of ln x

If $y = \ln x$, $e^y = x$.

Expressing $y = \ln x$ in index form.

Differentiating with respect to y,

$$e^y = \frac{dx}{dy}$$

But
$$\frac{dy}{dx} = \frac{1}{\left(\dfrac{dx}{dy}\right)}$$

$\therefore$
$$\frac{dy}{dx} = \frac{1}{e^y} = \frac{1}{x}$$

So
$$\frac{d}{dx}(\ln x) = \frac{1}{x}$$

$$\boxed{\frac{d}{dx}(\ln x) = \frac{1}{x}}$$

Differentiation of ln kx

If $\quad y = \ln kx$ where k is constant

then $\quad y = \ln k + \ln x$

Using $\log ab = \log a + \log b$

Differentiating with respect to x,

$$\frac{dy}{dx} = \frac{1}{x}$$

$\ln k$ is constant and so its derivative is zero.

So

$$\boxed{\frac{d}{dx}(\ln kx) = \frac{1}{x}}$$

Differentiation of ln (f(x))

The chain rule can be used to differentiate logarithmic functions, to find, for example

$$\frac{d}{dx}\ln(3x + 2) \qquad \frac{d}{dx}\ln(x^2 - 5) \qquad \frac{d}{dx}(4\ln(7 - 2x))$$

Example 11   Differentiate $y = \ln(x^2 - 5)$ with respect to x.

$$y = \ln(x^2 - 5)$$

Let $t = x^2 - 5$,   then $y = \ln t$.

| Letting the contents of the bracket be t allows y to be written as a function of t.

So $\dfrac{dt}{dx} = 2x$

and $\dfrac{dy}{dt} = \dfrac{1}{t} = \dfrac{1}{x^2 - 5}$

$\dfrac{dy}{dx} = \dfrac{dy}{dt} \times \dfrac{dt}{dx}$

| Use the chain rule.

$\qquad = \dfrac{1}{x^2 - 5} \times 2x$

$\qquad = \dfrac{2x}{x^2 - 5}$

| Notice the format here:
$\dfrac{d}{dx}\ln(x^2 - 5) = \dfrac{\text{derivative of } (x^2 - 5)}{(x^2 - 5)}$

Using the chain rule, a general result can be obtained:

➤
$$\frac{d}{dx}(\ln(f(x))) = \frac{f'(x)}{f(x)}$$

If the expression is multiplied by a constant k, then

$$\frac{d}{dx}(k\ln(f(x))) = k\frac{f'(x)}{f(x)}$$

| In words: to differentiate a natural log function, differentiate the function, then divide by the function.

Remember $\dfrac{d}{dx}\ln(\square) = \dfrac{\text{derivative of } (\square)}{(\square)}$

Example 12   **a** $\dfrac{d}{dx}(4\ln(7 - 2x^3)) = 4 \times \dfrac{-6x^2}{7 - 2x^3}$

| $\dfrac{d}{dx}4\ln(\square) = 4 \times \dfrac{\text{derivative of } (\square)}{(\square)}$

$\qquad\qquad\qquad\qquad\quad = -\dfrac{24x^2}{7 - 2x^3}$

b $\dfrac{d}{dx}(\ln(3x^5 + 2)) = \dfrac{15x^4}{3x^5 + 2}$

| $f(x) = 3x^5 + 2,\; f'(x) = 15x^4$

When $f(x)$ is a linear function, i.e. $f(x) = ax + b$, then

➤
$$\frac{d}{dx}(\ln(ax + b)) = \frac{a}{ax + b}$$

Example 13   The curve $y = \ln(2x - 4)$ crosses the x-axis at A. The normal at A crosses the y-axis at B.

Find the coordinates of A and B.

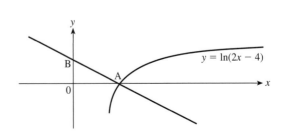

Solution At A, $y = 0$, so

$$0 = \ln(2x - 4)$$

$$2x - 4 = 1$$

$$\therefore \quad x = 2.5$$

$\ln 1 = 0$

So, A has coordinates $(2.5, 0)$.

$$y = \ln(2x - 4)$$

$$\frac{dy}{dx} = \frac{2}{2x - 4}$$

Simplify: $\dfrac{2}{2x - 4} = \dfrac{2}{2(x - 2)}$ and then cancel.

$$= \frac{1}{x - 2}$$

When $x = 2.5$

$$\frac{dy}{dx} = \frac{1}{2.5 - 2} = 2$$

The product of the gradients of perpendicular lines is -1.

So the gradient of the normal at A is $-\frac{1}{2}$.

Equation of normal at A:

Use $y - y_1 = m(x - x_1)$

$$y - 0 = -\tfrac{1}{2}(x - 2.5)$$

$$y = -\tfrac{1}{2}x + 1.25$$

At B, $x = 0$, so $y = 1.25$.

So, B has coordinates $(0, 1.25)$.

Simplifying a logarithmic function before differentiating

If a logarithmic function can be simplified, then doing so *before* differentiating will make the working much easier to carry out.

Example 14 Differentiate with respect to x

    a $\ln(6x^2)$ **b** $\ln\sqrt{5x + 6}$

Remember $\ln(ab) = \ln a + \ln b$ and $\ln(a^n) = n \ln a$.

Solution **a** Let $y = \ln(6x^2)$

Take care: $\ln(6x^2) \neq 2\ln(6x)$

$$= \ln 6 + \ln x^2$$

$$= \ln 6 + 2\ln x$$

$\ln 6$ is a constant, so its derivative is zero.

$$\frac{dy}{dx} = \frac{2}{x}$$

Differentiating without simplifying gives
$\dfrac{dy}{dx} = \dfrac{12x}{6x^2} = \dfrac{2}{x}$

b Let $y = \ln \sqrt{5x + 6}$

$$= \ln (5x + 6)^{\frac{1}{2}}$$

$$= \tfrac{1}{2} \ln (5x + 6)$$

$$\frac{dy}{dx} = \frac{1}{2} \times \frac{5}{5x + 6}$$

$$= \frac{5}{2(5x + 6)}$$

> Simplify the log expression first.

> Differentiating directly gives
> $$\frac{dy}{dx} = \frac{\tfrac{1}{2}(5x + 6)^{-\frac{1}{2}} \times 5}{(5x + 6)^{\frac{1}{2}}}$$
> which is more complicated to simplify.

Exercise 4C

1 Differentiate

a $\ln x$ **b** $\ln 6x$ **c** $\ln x^3$ **d** $\ln 3x^2$ **e** $-7 \ln x$ **f** $2 \ln \dfrac{x}{4}$

2 Differentiate with respect to x

a $\ln (2x + 3)$ **b** $\ln \left(\dfrac{x - 1}{2} \right)$ **c** $4 \ln (5 - 2x)$ **d** $\dfrac{\ln (2 - 3x)}{2}$ **e** $3 \ln \left(\dfrac{4x + 3}{2} \right)$

3 Simplify each expression, then differentiate with respect to x.

a $\ln (5x^2)$ **b** $\ln \left(\dfrac{4}{x} \right)$ **c** $\ln \sqrt[4]{x}$

d $\ln (2x + 8)^2$ **e** $\ln \left(\dfrac{3x + 1}{2x - 5} \right)$ **f** $4 \ln \sqrt{5x + 6}$

g $\ln (4(2x + 1)^2)$ **h** $\ln \left(\dfrac{3}{(3x - 1)^2} \right)$ **i** $\ln (e^{4x})$

4 Find an equation of the tangent to the curve $y = \ln (4x - 2)$ at the point $(1, \ln 2)$.

5 Find an equation of the normal to the curve $y = 3 - \ln (5x + 1)$ at the point where $x = 0$.

6 The curve $y = \ln (8x + 2)$ cuts the x-axis at A and the y-axis at B.

a Sketch the curve $y = \ln (8x + 2)$.

b Find the coordinates of A and B.

c The tangents at A and B intersect at T. Find the coordinates of T.

7 a Find the gradient of $y = 2 \ln x - x^2$ at the point where $x = 2$.

b Find the coordinates of the stationary point on $y = 2 \ln x - x^2$.

8 If $\dfrac{d}{dx}(\ln \sqrt{ax + b}) = \dfrac{3}{ax + 1}$, find the values of a and b.

9 $f(x) = \ln (2x - 1) - x$. Show that the maximum value of $f(x)$ is $\ln 2 - 1.5$.

10 Find the coordinates of the point of intersection of $y = \ln (3x - 1)$ and $y = \ln (x + 2)$. Find also the gradient of each curve at the point of intersection.

11 Find $f'(x)$.

a $f(x) = 2 \ln (x^2 - 3)$ **b** $f(x) = \ln (x^3 - 3x)$ **c** $f(x) = 4 \ln \left(5 - \dfrac{2}{x} \right)$

12 Find the derivative of each of these, simplifying your answer where possible.

    a $\ln(6x^2 + 3x)$ **b** $\ln(6x^2 \times 3x)$

    c $\ln((3x + 2)(5x - 1))$ **d** $\ln(3x^2 + 5)^2$

    e $\ln(x^2 - 4) - \ln(x + 2)$ **f** $\ln(2x^3 + 3) + \ln(2x + 3)$

13 Show that the minimum value of $x^2 - \ln x$ is $\frac{1}{2}(1 + \ln 2)$.

14 Find the coordinates of the stationary points on the curve $y = x^2 + \ln(2x + 3)$.

15 An equation of the normal to the curve $y = \ln(2x + 1)^3$ at $(a, 0)$ is $ky + x = 0$. Find the values of a and k.

4.3 Differentiation of products and quotients

Many mathematical expressions are the product of two functions, i.e. they are formed by multiplying two functions together, for example

$$y = xe^{-x} \quad y = 4x^2 \ln 5x \quad y = (2x + 3)^4 (5x - 1)^3$$

Other expressions are formed by dividing one function by another, for example

$$y = \frac{2x + 3}{5x - 1} \qquad y = \frac{2x}{e^{3x}} \qquad y = \frac{\ln x}{2x - 3}$$

These products and quotients can be differentiated using the product rule and quotient rule.

The product rule

The product rule is derived as follows:

Consider $y = f(x) \times g(x)$ and let $u = f(x)$ and $v = g(x)$.

Then $y = uv$ ①

Let x increase by a small amount δx, with corresponding increases in y, u and v of δy, δu and δv, so that

$$y + \delta y = (u + \delta u)(v + \delta v)$$

i.e. $y + \delta y = uv + u\delta v + v\delta u + \delta u \delta v$ ②

$$\delta y = u\delta v + v\delta u + \delta u \delta v$$

$$\frac{\delta y}{\delta x} = \frac{u\delta v}{\delta x} + \frac{v\delta u}{\delta x} + \frac{\delta u \delta v}{\delta x}$$

i.e. $\dfrac{\delta y}{\delta x} = u\dfrac{\delta v}{\delta x} + v\dfrac{\delta u}{\delta x} + \delta u\dfrac{\delta v}{\delta x}$

Let $\delta x \to 0$.

Then $\delta u \to 0$ and $\delta v \to 0$,

$$\frac{\delta y}{\delta x} \to \frac{dy}{dx} \qquad \frac{\delta v}{\delta x} \to \frac{dv}{dx} \qquad \text{and} \qquad \frac{\delta u}{\delta x} \to \frac{du}{dx}$$

> Subtract ① from ②.

> Divide each term by δx.

$$\therefore \quad \frac{dy}{dx} = u\frac{dv}{dx} + v\frac{du}{dx} + 0 \times \frac{dv}{dx} = u\frac{dv}{dx} + v\frac{du}{dx}$$

➤ | **If $y = uv$, where u and v are functions of x, then**

$$\frac{dy}{dx} = u\frac{dv}{dx} + v\frac{du}{dx}$$ |

> Informally in words:
> To differentiate a product of two functions, write down first function × derivative of second + second function × derivative of first.

The product rule can also be written without introducing u and v:

➤ | $$\frac{d}{dx}(f(x)\,g(x)) = f(x)\,g'(x) + g(x)\,f'(x)$$ |

Example 15   Find the first derivatives of   **a** $y = xe^{-x}$ **b** $y = 4x^2 \ln 5x$

Solution   **a** $y = x \times e^{-x}$

Let $u = x$, then

$$\frac{du}{dx} = 1$$

Let $v = e^{-x}$ then

$$\frac{dv}{dx} = -e^{-x}$$

$$\frac{dy}{dx} = u\frac{dv}{dx} + v\frac{du}{dx}$$

$$= x(-e^{-x}) + e^{-x} \times 1$$

$$= -xe^{-x} + e^{-x}$$

$$= e^{-x}(1 - x)$$

> Define u and v, both as functions of x.
> $$\overset{u}{\searrow} \quad \overset{v}{\searrow}$$
> $$y = x \times e^{-x}$$

> Simplify your answer by factorising if possible.

b $y = 4x^2 \ln 5x$

Let $u = 4x^2$, then

$$\frac{du}{dx} = 8x$$

Let $v = \ln 5x$ then

$$\frac{dv}{dx} = \frac{1}{x}$$

$$\frac{dy}{dx} = u\frac{dv}{dx} + v\frac{du}{dx}$$

$$= 4x^2 \times \frac{1}{x} + \ln 5x \times 8x$$

$$= 4x(1 + 2\ln 5x)$$

> Use $\dfrac{d}{dx}\ln kx = \dfrac{1}{x}$

Example 16   The diagram shows the graph $y = (2x + 3)^2 (5x - 1)^3$.

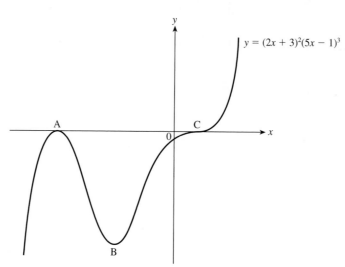

$y = (2x + 3)^2(5x - 1)^3$

a   Show that $\dfrac{dy}{dx} = (2x + 3)(5x - 1)^2(50x + 41)$.

b   Given that $\dfrac{dy}{dx} = 0$ at A, B and C, find the x coordinates of A, B and C.

c   Find the range of values of x for which $\dfrac{dy}{dx} < 0$.

Solution   **a**   Consider $y = (2x + 3)^2 (5x - 1)^3$

Let $u = (2x + 3)^2$, then

$$\frac{du}{dx} = 2(2x + 3) \times 2 = 4(2x + 3)$$

> Use the chain rule to find $\dfrac{du}{dx}$ and $\dfrac{dv}{dx}$.

Let $v = (5x - 1)^3$, then

$$\frac{dv}{dx} = 3(5x - 1)^2 \times 5 = 15(5x - 1)^2$$

$y = uv \Rightarrow \quad \dfrac{dy}{dx} = u\dfrac{dv}{dx} + v\dfrac{du}{dx}$

> Use the product rule.

$\dfrac{dy}{dx} = (2x + 3)^2 \times 15(5x - 1)^2 + (5x - 1)^3 \times 4(2x + 3)$

> Simplify carefully.

$= 15(2x + 3)^2(5x - 1)^2 + 4(5x - 1)^3(2x + 3)$

> Take out factors as early as possible.

$= (2x + 3)(5x - 1)^2(15(2x + 3) + 4(5x - 1))$

$= (2x + 3)(5x - 1)^2(30x + 45 + 20x - 4)$

$= (2x + 3)(5x - 1)^2(50x + 41)$

b $\dfrac{dy}{dx} = 0$ when

> The factorised form enables the equation to be solved easily.

$(2x + 3)(5x - 1)^2(50x + 41) = 0$

$\Rightarrow \quad 2x + 3 = 0 \quad$ or $\quad 5x - 1 = 0 \quad$ or $\quad 50x + 41 = 0$

$\qquad x = -1.5 \qquad\qquad x = 0.2 \qquad\qquad x = -0.82$

The x-coordinates of A, B and C are -1.5, -0.82 and 0.2 respectively.

c From the diagram, $\dfrac{dy}{dx} < 0$ when $-1.5 < x < -0.82$.

Look for the x values where the gradient is negative.

Note With practice, the product rule can be applied without showing all the working.

Example 17 Find   **a** $\dfrac{d}{dx}(2x+3)\ln 3x$ **b** $\dfrac{d}{dx}x^2 e^{3x-2}$

Solution   **a** $\dfrac{d}{dx}(2x+3)\ln 3x = (2x+3)\times\dfrac{1}{x} + \ln 3x \times 2$

Use $\dfrac{d}{dx}(\ln 3x) = \dfrac{1}{x}$

$$= \dfrac{2x+3}{x} + 2\ln 3x$$

b $\dfrac{d}{dx}x^2 e^{3x-2} = x^2 3e^{3x-2} + 2xe^{3x-2}$

Factorise.

$$= xe^{3x-2}(3x+2)$$

The quotient rule

The quotient rule is derived as follows:

Writing $u = f(x)$ and $v = g(x)$, then

$$y = \dfrac{u}{v} \qquad\qquad ①$$

and

$$y + \delta y = \dfrac{u + \delta u}{v + \delta v} \qquad\qquad ②$$

Substitute ① into ②.

$$\delta y = \dfrac{u + \delta u}{v + \delta v} - \dfrac{u}{v}$$

$$= \dfrac{v(u + \delta u) - u(v + \delta v)}{v(v + \delta v)}$$

$$= \dfrac{vu + v\delta u - uv - u\delta v}{v^2 + v\delta v}$$

Divide throughout by δx.

$$\dfrac{\delta y}{\delta x} = \dfrac{v\dfrac{\delta u}{\delta x} - u\dfrac{\delta v}{\delta x}}{v^2 + v\delta v}$$

Let $\delta x \to 0$, then $\delta u \to 0$, $\delta v \to 0$ and

$$\dfrac{\delta y}{\delta x} \to \dfrac{dy}{dx} \qquad \dfrac{\delta v}{\delta x} \to \dfrac{dv}{dx} \quad \text{and} \quad \dfrac{\delta u}{\delta x} \to \dfrac{du}{dx}$$

$$\therefore \quad \dfrac{dy}{dx} = \dfrac{v\dfrac{du}{dx} - u\dfrac{dv}{dx}}{v^2}$$

> If $y = \dfrac{u}{v}$, where u and v are functions of x, then
>
> $$\dfrac{dy}{dx} = \dfrac{v\dfrac{du}{dx} - u\dfrac{dv}{dx}}{v^2}$$

Informally in words:
To differentiate a quotient, write down (bottom function × derivative of top − top function × derivative of bottom) ÷ bottom².

The quotient rule can also be written without introducing u and v:

> $$\dfrac{d}{dx}\left(\dfrac{f(x)}{g(x)}\right) = \dfrac{g(x)f'(x) - f(x)g'(x)}{(g(x))^2}$$

Example 18 Differentiate $y = \dfrac{3x}{5x+1}$ with respect to x.

Consider $y = \dfrac{3x}{5x+1}$

Let $u = 3x$, then

$$\dfrac{du}{dx} = 3$$

Let $v = 5x + 1$, then

$$\dfrac{dv}{dx} = 5$$

$y = \dfrac{u}{v} \quad \Rightarrow \quad \dfrac{dy}{dx} = \dfrac{v\dfrac{du}{dx} - u\dfrac{dv}{dx}}{v^2}$

$$\dfrac{dy}{dx} = \dfrac{(5x+1)\times 3 - 3x \times 5}{(5x+1)^2}$$

$$= \dfrac{15x + 3 - 15x}{(5x+1)^2}$$

$$= \dfrac{3}{(5x+1)^2}$$

Express y in the form $y = \dfrac{u}{v}$ i.e. $y = \dfrac{\overset{u}{3x}}{\underset{v}{5x+1}}$

Simplify the expression.

It is also possible to differentiate a quotient by expressing it as a product and then using the product rule. For example

$$y = \dfrac{3x}{5x+1}$$

can be written as

$$y = 3x(5x+1)^{-1}$$

$$\frac{dy}{dx} = 3x \times (-5)(5x+1)^{-2} + (5x+1)^{-1} \times 3$$

$$= \frac{-15x}{(5x+1)^2} + \frac{3}{5x+1}$$

$$= \frac{-15x + 3(5x+1)}{(5x+1)^2}$$

$$= \frac{3}{(5x+1)^2}$$

The result is the same as that obtained using the quotient rule. It is a matter of personal preference which rule is used, but usually the algebra manipulation involved in simplifying the derivative is less complicated when the quotient rule is used.

Example 19   Using the quotient rule, differentiate $y = \dfrac{2x}{e^{3x}}$ with respect to x.

$$y = \frac{2x}{e^{3x}}$$

> Use the quotient rule with $u = 2x$ and $v = e^{3x}$.

$$\frac{dy}{dx} = \frac{e^{3x} \times 2 - 2x \times 3e^{3x}}{(e^{3x})^2}$$

> Take out factors.

$$= \frac{2e^{3x}(1 - 3x)}{(e^{3x})^2}$$

> Cancel e^{3x} on top and bottom.

$$= \frac{2(1 - 3x)}{e^{3x}}$$

> *The algebraic manipulation can be complicated when using the quotient rule and it is necessary to simplify carefully.*

Example 20   **a**   Find $\dfrac{dy}{dx}$ when $y = \dfrac{\ln x}{2x - 3}$.

b   The equation of the tangent to the curve $y = \dfrac{\ln x}{2x - 3}$ at $(2, \ln 2)$ can be written as $y = mx + c$. Find the exact values of m and c.

Solution $y = \dfrac{\ln x}{2x - 3}$

> Use the quotient rule with $u = \ln x$ and $v = 2x - 3$.

a $\dfrac{dy}{dx} = \dfrac{(2x-3) \times \dfrac{1}{x} - \ln x \times 2}{(2x - 3)^2}$ ①

> Simplify the numerator:
>
> $(2x - 3) \times \dfrac{1}{x} - 2\ln x = \dfrac{(2x-3) - 2x\ln x}{x}$

$$= \frac{2x - 3 - 2x\ln x}{x(2x - 3)^2}$$

> Use $\dfrac{a}{b} \div c = \dfrac{a}{b} \times \dfrac{1}{c} = \dfrac{a}{bc}$

b When $x = 2$, $\dfrac{\mathrm{d}y}{\mathrm{d}x} = \dfrac{1 - 4\ln 2}{2} = \dfrac{1}{2} - 2\ln 2$

Equation of tangent at $(2, \ln 2)$:

> Use $y - y_1 = m(x - x_1)$

$y - \ln 2 = \left(\tfrac{1}{2} - 2\ln 2\right)(x - 2)$

$y - \ln 2 = \left(\tfrac{1}{2} - 2\ln 2\right)x - \left(\tfrac{1}{2} - 2\ln 2\right) \times 2$

> Rearrange into the form $y = mx + c$.

$y = \left(\tfrac{1}{2} - 2\ln 2\right)x + 5\ln 2 - 1$

$\therefore \quad m = \tfrac{1}{2} - 2\ln 2$ and $c = 5\ln 2 - 1$.

Note If just part **b** is requested, there is no need to simplify the expression for $\dfrac{\mathrm{d}y}{\mathrm{d}x}$.

The gradient at $(2, \ln 2)$ can be found by substituting $x = 2$ directly into ①,

giving $\dfrac{\mathrm{d}y}{\mathrm{d}x} = \tfrac{1}{2} - 2\ln 2$ straight away.

Example 21

When differentiating an expression in quotient form, it is advisable to check whether it can be simplified first. This example compares the two methods.

Find

$$\frac{\mathrm{d}}{\mathrm{d}x}\left(\frac{3x^4 - 2x^3 + 4}{x}\right)$$

Method 1: Simplifying first

$$\frac{\mathrm{d}}{\mathrm{d}x}\left(\frac{3x^4 - 2x^3 + 4}{x}\right) = \frac{\mathrm{d}}{\mathrm{d}x}\left(3x^3 - 2x^2 + 4x^{-1}\right) = 9x^2 - 4x - 4x^{-2}$$

Method 2: Using the quotient rule

$$\frac{\mathrm{d}}{\mathrm{d}x}\left(\frac{3x^4 - 2x^3 + 4}{x}\right) = \frac{x(12x^3 - 6x^2) - (3x^4 - 2x^3 + 4) \times 1}{x^2}$$

$$= \frac{12x^4 - 6x^3 - 3x^4 + 2x^3 - 4}{x^2}$$

$$= \frac{9x^4 - 4x^3 - 4}{x^2}$$

$$= 9x^2 - 4x - 4x^{-2}$$

> *Note*: There is less algebraic manipulation when the expression is simplified first.

Exercise 4D

1 Using the product rule, differentiate these with respect to x, simplifying the answer where possible.

   a $x^2(1 + x)^3$ **b** $x(x^2 + 1)^4$ **c** $(x + 1)^2(x^2 - 1)$ **d** $x^3(5x + 1)^2$

   e xe^{5x} **f** $4x^3e^{2x}$ **g** $x\ln x$ **h** $x^2\ln(x + 3)$

2 Find the x-coordinates of the stationary points on the curve $y = (x^2 - 1)\sqrt{1 + x}$.

3 Find an equation of the tangent to the curve $y = (2x + 3)e^{1 - 5x}$ at the point where the curve crosses the y-axis.

4 The equation of the normal to the curve $y = (x + 2)\ln(2x - 3)$ at $x = 2$ is $ax + by = c$. Find a, b and c.

5 Determine the nature of any stationary points on these curves.

   a $y = xe^{-2x}$ **b** $y = x^2e^{-x}$

6 Find the minimum value of $f(x) = x^3 \ln x$.

7 Find $\dfrac{d}{dx}\left((x^2 - 3)(x + 1)^2\right)$, simplifying your result.

8 Using the quotient rule, differentiate with respect to x

  a $\dfrac{x}{x + 1}$ **b** $\dfrac{x}{x - 1}$ **c** $\dfrac{x}{x^2 + 1}$ **d** $\dfrac{4 + x}{4 - x}$

  e $\dfrac{1 + 5x}{5 - x}$ **f** $\dfrac{x}{(1 + 2x)^2}$ **g** $\dfrac{2x + 1}{(x + 2)^2}$ **h** $\dfrac{x}{(x + 3)^4}$

9 Differentiate with respect to x

  a $\dfrac{e^x}{x}$ **b** $\dfrac{e^{2x+1}}{4x - 3}$ **c** $\dfrac{3x}{e^{1-4x}}$ **d** $\dfrac{e^{5x+2}}{e^{4x-1}}$

  e $\dfrac{\ln x}{x}$ **f** $\dfrac{\ln x^2}{x}$ **g** $\dfrac{\ln x}{x^2}$ **h** $\dfrac{x^2}{\ln x}$

10 If $f(x) = \dfrac{(x - 3)^2}{(x + 2)^2}$, find $f'(-1)$.

11 Find the coordinates of any stationary points on the curve $y = \dfrac{x^2}{3x - 1}$.

12 A curve has equation $y = \dfrac{3x}{2x - 3}$.

  a Find the equation of the normal to the curve at the origin.

  b The normal at the origin cuts the curve again at P. Find the coordinates of P.

13 Find the second derivative of e^{x^3}.

14 $\dfrac{d}{dx}\left((x + 1)^3(2x - 1)^4\right) = (x + 1)^2(2x - 1)^3(ax + b)$.

  a Find the values of a and b.

  b Find the x-coordinates of the stationary points on the curve $y = (x + 1)^3(2x - 1)^4$.

  c Find the equation of the tangent at the point where the curve crosses the y-axis.

15 Show that if $y = \dfrac{1 + 2x}{1 - 4x}$

  then $\dfrac{d^2y}{dx^2} = \dfrac{k}{(1 - 4x)^3}$

  for k constant, and find the value of k.

16 Find $\dfrac{d}{dx}\left(\dfrac{x}{\sqrt{1 + x^2}}\right)$.

17 Show that the maximum value of $\dfrac{\ln x}{x^3}$ is $\dfrac{1}{3}e^{-1}$.

4.4 Differentiation of trigonometric functions

Differentiation of sin *x* and cos *x*

Consider the function $y = \sin x$, where x is in radians.

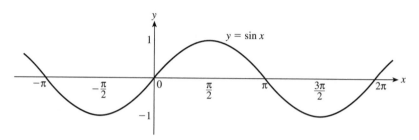

An intuitive idea of the gradient function can be gained by considering the value of the gradient at various points on the curve.

■ When $x = -\dfrac{\pi}{2}, \dfrac{\pi}{2}, \dfrac{3\pi}{2}, \ldots$, the gradient is zero.

■ For $-\dfrac{\pi}{2} < x < \dfrac{\pi}{2}$, the gradient is positive. It increases from zero at $x = -\dfrac{\pi}{2}$, reaches a maximum at $x = 0$, then decreases to zero at $x = \dfrac{\pi}{2}$.

■ For $\dfrac{\pi}{2} < x < \dfrac{3\pi}{2}$, the gradient is negative. It decreases from zero at $x = \dfrac{\pi}{2}$, reaches a minimum at $x = \pi$, then increases to zero at $x = \dfrac{3\pi}{2}$.

The graph of the gradient function is the cosine curve.

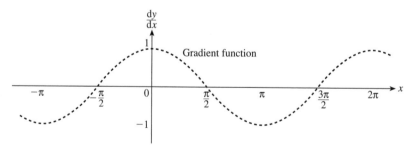

Consider now the function $y = \cos x$ and its gradient function.

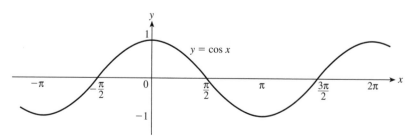

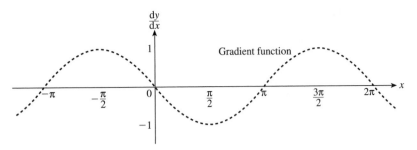

The gradient function for $y = \cos x$ is not the sine curve, but the negative of it.

➤
$$\frac{d}{dx}(\sin x) = \cos x$$

$$\frac{d}{dx}(\cos x) = -\sin x$$

Remember   These results hold only for x in radians.

⌀   To prove these two results, theory relating to small angles is required. This is covered by material on the CD-ROM (E12.1).

The derivatives of $\sin f(x)$ and $\cos f(x)$

The chain rule can be used to differentiate composite functions of $\sin x$ and $\cos x$.

Example 22   Differentiate with respect to x

    a $y = \sin 2x$

    b $y = 3\cos(5x^2 - 1)$

Solution **a** $y = \sin 2x$

    Let $t = 2x$, then $y = \sin t$

    So $\dfrac{dt}{dx} = 2$

    and $\dfrac{dy}{dt} = \cos t$

          $= \cos 2x$

    $\dfrac{dy}{dx} = \dfrac{dy}{dt} \times \dfrac{dt}{dx}$

          $= \cos 2x \times 2$

          $= 2\cos 2x$

b $y = 3\cos(5x^2 - 1)$

Let $t = 5x^2 - 1$, then $y = 3\cos t$

So $\dfrac{dt}{dx} = 10x$

and $\dfrac{dy}{dt} = -3\sin t$

$= -3\sin(5x^2 - 1)$

$\dfrac{dy}{dx} = \dfrac{dy}{dt} \times \dfrac{dt}{dx}$

$= -3\sin(5x^2 - 1) \times 10x$

$= -30x\sin(5x^2 - 1)$

With practice, the derivative can be written down straight away using the general results obtained by the chain rule:

➤
$$\dfrac{d}{dx}(\sin f(x)) = f'(x)\cos(f(x))$$

$$\dfrac{d}{dx}(\cos f(x)) = -f'(x)\sin(f(x))$$

Example 23 **a** $\dfrac{d}{dx}(4\sin 3x) = 12\cos 3x$ **b** $\dfrac{d}{dx}\left(\tfrac{1}{2}\cos(\pi - 4x)\right) = 2\sin(\pi - 4x)$

c $\dfrac{d}{dx}\left(\sin(3x^2 - 2)\right) = \cos(3x^2 - 2) \times \dfrac{d}{dx}(3x^2 - 2)$

$= \cos(3x^2 - 2) \times 6x$

$= 6x\cos(3x^2 - 2)$

d $\dfrac{d}{dx}\left(6\cos(\pi - \tfrac{1}{2}x)\right) = 6 \times \left(-\sin(\pi - \tfrac{1}{2}x)\right) \times \left(-\tfrac{1}{2}\right)$

$= 3\sin(\pi - \tfrac{1}{2}x)$

Example 24 Find $\dfrac{dy}{dx}$ for $x = 4\cos 5y$.

$x = 4\cos 5y$

so $\dfrac{dx}{dy} = -20\sin 5y$

but $\dfrac{dy}{dx} = \dfrac{1}{\left(\dfrac{dx}{dy}\right)}$

$\therefore$ $\dfrac{dy}{dx} = -\dfrac{1}{20\sin 5y}$

$= -\tfrac{1}{20}\operatorname{cosec} 5y$

> If $\dfrac{dy}{dx}$ is to be expressed in terms of x
> use $\sin 5y = \sqrt{1 - \cos^2 5y} = \sqrt{1 - \left(\tfrac{x}{4}\right)^2}$

Example 25   If $y = \sin n\theta$, show that $\dfrac{d^2y}{d\theta^2} + n^2y = 0$.

$$y = \sin n\theta \Rightarrow \quad \frac{dy}{d\theta} = n\cos n\theta$$

$$\frac{d^2y}{d\theta^2} = n \times (-n\sin n\theta)$$

> $\sin n\theta = y$, so substitute it here.

$$\frac{d^2y}{d\theta^2} = -n^2y$$

> The result can also be shown by substitution:
> $$\text{LHS} = \frac{d^2y}{d\theta^2} + n^2y$$
> $$= -n^2\sin n\theta + n^2\sin n\theta = 0 = \text{RHS}$$

So $\dfrac{d^2y}{d\theta^2} + n^2y = 0$

Example 26   The diagram shows the graph of $y = e^{\sin x}$.

A and B are two of the points where the gradient is zero.
Find the coordinates of A and B.

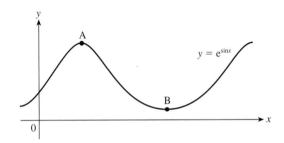

Solution $y = e^{\sin x}$

$$\frac{dy}{dx} = \cos x\, e^{\sin x}$$

> Use $\dfrac{d}{dx}\left(e^{f(x)}\right) = f'(x)\,e^{f(x)}$

$$\frac{dy}{dx} = 0 \text{ when } \cos x = 0.$$

> $e^{f(x)} > 0$ for all $f(x)$ so $e^{f(x)} \neq 0$.

$$\cos x = 0 \Rightarrow \quad x = \frac{\pi}{2}, \frac{3\pi}{2}, \frac{5\pi}{2}, \frac{7\pi}{2}, \ldots$$

> Solve the trigonometric equation, remembering to work in radians.

At A, $x = \frac{\pi}{2}, y = e^{\sin\frac{\pi}{2}} = e^1 = e$

At B, $x = \frac{3\pi}{2}, y = e^{\sin\frac{3\pi}{2}} = e^{-1} = \frac{1}{e}$

A is $\left(\dfrac{\pi}{2}, e\right)$ and B is $\left(\dfrac{3\pi}{2}, \dfrac{1}{e}\right)$.

> Leave these as exact answers.

Angle x in degrees

The angle must be in radians when differentiating trigonometric functions. So, if x is given in degrees, it must be changed to radians, where $x^\circ = \dfrac{\pi}{180}x$ radians.

Example 27 $\dfrac{d}{dx}(\sin x^\circ) = \dfrac{d}{dx}\left(\sin\left(\dfrac{\pi}{180}x\right)\right)$

$$= \frac{\pi}{180}\cos\left(\frac{\pi}{180}x\right)$$

$$= \frac{\pi}{180}\cos x^\circ$$

Example 28   Differentiate $y = 2\sin^3 x$ with respect to x.

$y = 2\sin^3 x = 2(\sin x)^3$

> Write $\sin^3 x$ as $(\sin x)^3$; it is easier to see the substitution when y is written in this form.

Let $t = \sin x$, then $y = 2t^3$

So $\quad \dfrac{dt}{dx} = \cos x$

and $\quad \dfrac{dy}{dt} = 6t^2$

$\qquad\quad = 6(\sin x)^2$

$\dfrac{dy}{dx} = \dfrac{dy}{dt} \times \dfrac{dt}{dx}$

$\qquad = 6(\sin x)^2 \times \cos x$

$\qquad = 6\sin^2 x \cos x$

> Without showing the chain rule working:
> $\dfrac{d}{dx}\left(2 \times \left(\Box^3\right)\right) = 2 \times 3\left(\Box^2\right) \times$ derivative of $\Box$

Exercise 4E

1 Differentiate with respect to x

   a $\cos 3x$ **b** $\sin 5x$ **c** $3\sin x$ **d** $2\cos(3x - 1)$

   e $-\frac{1}{2}\sin 4x$ **f** $\dfrac{\cos 4x}{2}$ **g** $\sin\left(\dfrac{\pi}{2} - x\right)$ **h** $3\cos 2x + 2\sin 6x$

2 Find the gradient of tangent to the curve $y = \cos 2x - \cos x$ at the point where $x = \dfrac{\pi}{4}$.

3 Find an equation of the tangent to the curve $y = 2\sin 6x$ at $\left(\dfrac{\pi}{6},\ 0\right)$.

4 Find the x-coordinates of the stationary points on the curve $y = x + \sin 2x$ for $0 \leqslant x \leqslant \pi$.

5 Find the first derivative of

   a $\sin(x^2)$ **b** $\cos\sqrt{x}$ **c** $\sin\left(\dfrac{1}{x}\right)$ **d** $\cos x°$ **e** $\sin(\pi x)°$

6 Differentiate with respect to x

   a $e^{\sin 3x}$ **b** $4e^{2\cos x}$ **c** $\ln(\sin 2x)$ **d** $\ln(\cos^2 x)$ **e** $\cos(e^x)$

7 Find $\dfrac{dy}{dx}$ in terms of y when

   a $x = 5\sin 3y$ **b** $x = \frac{1}{4}\cos(2y - 1)$

   c $x = -\frac{1}{2}\cos 4y$ **d** $x = 2\sin 3y + 3\cos 2y$

8 Find $\dfrac{d^2y}{dx^2}$ when $y = 3\sin x - 4\cos x$.

9 A particle is moving along a straight line so that, at time t seconds after leaving a fixed point O, its velocity $v\,\text{ms}^{-1}$ is given by $v = 10\sin\frac{1}{2}t$.

   a Find the time when the acceleration, given by $\dfrac{dv}{dt}$, is first zero.

   b Find the velocity at this instant.

10 A particle is moving along a straight line and its displacement, x metres, from a fixed point O on the line at time t seconds, is given by $x = 4\cos t + 3\sin t$.

  a Find the velocity of the particle when $t = \dfrac{\pi}{4}$.

  b Find the displacement when the velocity is first zero.

  c Find the acceleration when $t = \dfrac{\pi}{3}$.

11 This diagram shows a sketch of part of the curve $y = \cos^3 x$.

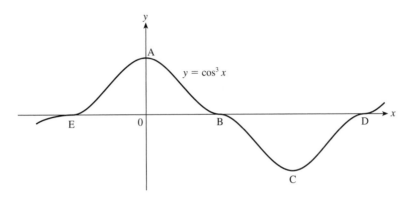

Find the coordinates of A, B, C, D and E.

Trigonometric differentiation involving the product and quotient rules

Example 29 Differentiate with respect to x

  a $y = x^2 \cos 3x$

  b $\dfrac{3\sin 5x}{x^3}$

Solution **a** $y = x^2 \cos 3x$

$$\frac{dy}{dx} = x^2 \times 3(-\sin 3x) + \cos 3x \times 2x$$

$$= x(2\cos 3x - 3x\sin 3x)$$

> Use the product rule:
> $$y = uv \Rightarrow \frac{dy}{dx} = u\frac{dv}{dx} + v\frac{du}{dx}$$
> with $u = x^2$ and $v = \cos 3x$.

> Simplify the answer by factorising, if possible.

  b $y = \dfrac{3\sin 5x}{x^3}$

$$\frac{dy}{dx} = \frac{x^3 \times 15\cos 5x - 3\sin 5x \times 3x^2}{(x^3)^2}$$

$$= \frac{3x^2(5x\cos 5x - 3\sin 5x)}{x^6}$$

$$= \frac{3(5x\cos 5x - 3\sin 5x)}{x^4}$$

> Use the quotient rule:
> $$y = \frac{u}{v} \Rightarrow \frac{dy}{dx} = \frac{v\dfrac{du}{dx} - u\dfrac{dv}{dx}}{v^2}$$
> with $u = 3\sin 5x$ and $v = x^3$.

> Simplify the expression carefully, taking out any factors and cancelling terms, if possible.

Example 30   If $x = e^{-t} \sin 2t$, show that

$$\frac{d^2 x}{dt^2} + 2\frac{dx}{dt} + 5x = 0$$

$$x = e^{-t} \sin 2t \qquad ① $$

> Use the product rule, with $u = e^{-t}$ and $v = \sin 2t$.

$$\frac{dx}{dt} = e^{-t} \times 2\cos 2t + \sin 2t \times (-e^{-t})$$

$$= 2e^{-t} \cos 2t - e^{-t} \sin 2t \quad ②$$

> From ① $e^{-t} \sin 2t = x$, so substitute it in ② to obtain ③.

$$= 2e^{-t} \cos 2t - x \qquad ③$$

> Differentiate again, using the product rule to differentiate $2e^{-t} \cos 2t$ with $u = 2e^{-t}$ and $v = \cos 2t$.

$$\frac{d^2 x}{dt^2} = 2e^{-t}(-2\sin 2t) + \cos 2t(-2e^{-t}) - \frac{dx}{dt}$$

> $Note:$ $\frac{d}{dt}(x) = \frac{dx}{dt}$

$$= -4e^{-t} \sin 2t - 2e^{-t} \cos 2t - \frac{dx}{dt}$$

> From ③ $2e^{-t}\cos 2t = \frac{dx}{dt} + x$

$$= -4x - \left(\frac{dx}{dt} + x\right) - \frac{dx}{dt}$$

$$= -5x - 2\frac{dx}{dt}$$

$$\therefore \quad \frac{d^2 x}{dt^2} + 2\frac{dx}{dt} + 5x = 0$$

> Alternatively, differentiating ② and then substituting for $\frac{d^2 x}{dt^2}, \frac{dx}{dt}$ and x in the given expression, verifies the result.

Differentiation of tan x, cot x, sec x and cosec x

Consider $y = \tan x = \dfrac{\sin x}{\cos x}$.

Using the quotient rule with $u = \sin x$ and $v = \cos x$ gives

$$\frac{dy}{dx} = \frac{\cos x \times \cos x - \sin x \times (-\sin x)}{\cos^2 x}$$

$$= \frac{\cos^2 x + \sin^2 x}{\cos^2 x}$$

> Use $\cos^2 x + \sin^2 x = 1$

$$= \frac{1}{\cos^2 x}$$

> Use $\sec x = \dfrac{1}{\cos x}$

$$= \sec^2 x$$

So $\dfrac{d}{dx}(\tan x) = \sec^2 x.$

Now consider $y = \sec x = \dfrac{1}{\cos x} = (\cos x)^{-1}$.

Using the chain rule:

$$\frac{dy}{dx} = (-1) \times (\cos x)^{-2} \times (-\sin x)$$

$$= \frac{\sin x}{\cos^2 x}$$

$$= \frac{1}{\cos x} \times \frac{\sin x}{\cos x}$$

$$= \sec x \tan x$$

So $\quad \dfrac{d}{dx}(\sec x) = \sec x \tan x$.

$$\boxed{\begin{aligned} &\frac{d}{dx}(\tan x) = \sec^2 x \\[2mm] &\frac{d}{dx}(\sec x) = \sec x \tan x \end{aligned}}$$

Similarly, using

$$\operatorname{cosec} x = \frac{1}{\sin x} = (\sin x)^{-1} \quad \text{and} \quad \cot x = \frac{\cos x}{\sin x}$$

the derivatives of $\operatorname{cosec} x$ and $\cot x$ can be obtained.

$$\boxed{\begin{aligned} &\frac{d}{dx}(\operatorname{cosec} x) = -\operatorname{cosec} x \cot x \\[2mm] &\frac{d}{dx}(\cot x) = -\operatorname{cosec}^2 x \end{aligned}}$$

Note: The derivatives of the trigonometric functions beginning with 'co' ($\cos x$, $\cot x$, $\operatorname{cosec} x$) have a negative sign, while the derivatives of the others ($\sin x$, $\tan x$, $\sec x$) do not have a negative sign.

Example 31 *The derivatives of composite functions can be found by the chain rule.*

a $\quad \dfrac{d}{dx}\left(\tan\left(\dfrac{\pi}{4} + 5x\right)\right) = 5\sec^2\left(\dfrac{\pi}{4} + 5x\right)$

b $\quad \dfrac{d}{dx}(\sec(2x + 3)) = 2\sec(2x + 3)\tan(2x + 3)$

c $\quad \dfrac{d}{dx}(\operatorname{cosec}(x^2)) = -2x\operatorname{cosec}(x^2)\cot(x^2)$

d $\quad \dfrac{d}{dx}(\cot(5x)) = -5\operatorname{cosec}^2(5x)$

Example 32
(Extension)

The graph shows, for $0 < x < \dfrac{\pi}{2}$, the curve

$y = f(x)$, where $f(x) = \tan x + 3 \cot x$.

Find the minimum value of $f(x)$ in this range.

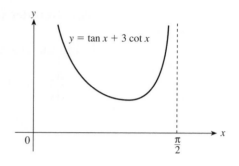

Solution

$$f(x) = \tan x + 3 \cot x$$

$$f'(x) = \sec^2 x - 3 \operatorname{cosec}^2 x$$

Find where $f'(x) = 0$.

$$f'(x) = 0 \Rightarrow \quad \sec^2 x - 3 \operatorname{cosec}^2 x = 0$$

$$\sec^2 x = 3 \operatorname{cosec}^2 x$$

$$\frac{1}{\cos^2 x} = \frac{3}{\sin^2 x}$$

$$\frac{\sin^2 x}{\cos^2 x} = 3$$

Use $\tan x = \dfrac{\sin x}{\cos x}$

$$\tan^2 x = 3$$

Solve the trigonometric equation.

$$\tan x = \pm\sqrt{3}$$

$$x = \pm\frac{\pi}{3}, \ \pm\frac{2\pi}{3}, \ \dots$$

The graph shows that $f\left(\dfrac{\pi}{3}\right)$ gives the minimum value of $f(x)$ in the specified range.

In the range $0 < x < \dfrac{\pi}{2}$, $f'(x) = 0$ when $x = \dfrac{\pi}{3}$.

$$f\left(\frac{\pi}{3}\right) = \tan\left(\frac{\pi}{3}\right) + 3 \cot\left(\frac{\pi}{3}\right)$$

Use $\cot\left(\dfrac{\pi}{3}\right) = \dfrac{1}{\tan\left(\frac{\pi}{3}\right)}$

$$= \sqrt{3} + \frac{3}{\sqrt{3}}$$

Note: $\dfrac{a}{\sqrt{a}} = \sqrt{a}$

$$= 2\sqrt{3}$$

Minimum value of $f(x)$ is $2\sqrt{3}$.

Exercise 4F

1 Differentiate with respect to x

a $x \cos x$ **b** $x \sin 2x$ **c** $x^2 \sin x$ **d** $\sin x \cos x$

e $\dfrac{\sin x}{x}$ **f** $\dfrac{\cos 2x}{x}$ **g** $\dfrac{x}{\sin x}$ **h** $\dfrac{x^2}{\cos x}$

i $e^x \sin 2x$ **j** $e^{-3x} \cos 2x$ **k** $\dfrac{e^x}{\sin x}$ **l** $\dfrac{\cos x}{e^x}$

2 Show that $y = x \sin x$ has a stationary point when $x = -\tan x$.

3 Find $\dfrac{d}{dx}(2 \cos x + 2x \sin x - x^2 \cos x)$.

4 Show that the gradient of the curve

$$y = \frac{e^x}{\cos x}$$

is zero when $x = \frac{3}{4}\pi$.

5 Find the value of x between 0 and π for which $e^{-x}\cos x$ is stationary.

6 Differentiate with respect to x

a $y = \ln(x^2 \cos x)$ **b** $y = \ln(\sin x \cos x)$ **c** $y = \ln\left(\dfrac{1 + \cos x}{1 - \cos x}\right)$

7 Differentiate with respect to x

a $\tan 5x$ **b** $\frac{1}{2}\tan\left(\frac{1}{2}x\right)$ **c** $\dfrac{2\tan 3x}{6}$ **d** $\dfrac{\sin 4x}{\cos 4x}$ **e** $\ln(\tan 4x)$

f $\ln(\tan^2 x)$ **g** $x \tan x$ **h** $\dfrac{\tan x}{x}$ **i** $\sqrt{\tan x}$ **j** $\tan\left(\dfrac{\pi}{4} + \pi x\right)$

8 **a** Find an equation of the tangent to the curve $y = \tan x$ when $x = \dfrac{\pi}{4}$.

   b Find an equation of the normal to the curve $y = \tan^2 x$ when $x = \dfrac{\pi}{4}$.

9 Find $\dfrac{dy}{dx}$ in terms of y when

a $x = \tan 2y$ **b** $x = -\frac{1}{6}\cot 3y$

c $x = \frac{1}{2}\operatorname{cosec} 2y$ **d** $x = \frac{1}{4}\sec(4y + 1)$

10 Find the gradient of the tangent to the curve $y = \ln(\tan x)$ when $x = \dfrac{\pi}{4}$.

11 **a** Differentiate $\cot x$ by writing it as $\dfrac{\cos x}{\sin x}$ and using the quotient rule.

   b Differentiate $\cot x$ by writing it as $(\tan x)^{-1}$ and using the chain rule.

12 Differentiate with respect to x

a $\operatorname{cosec} 5x$ **b** $2\cot 3x$ **c** $-4\sec 3x$

d $\sec^3 x$ **e** $\dfrac{\sec x}{\operatorname{cosec} x}$ **f** $\tan(e^x)$

13 Find an equation of the tangent to the curve $y = 2\sec x - \tan x$ at $x = 0$.

14 Show that if $y = \ln(\sec x + \tan x)$, then $\dfrac{dy}{dx} = \sec x$.

1 Show that

$$\frac{d}{dx}\left((4x-1)\sqrt{4x-1}\right) = 6\sqrt{4x-1}$$

2 Given that

$$y = 5x + 2 + \frac{1}{5x+2}$$

show that $\frac{d^2y}{dx^2}$ is of the form $\frac{c}{(5x+2)^n}$ and find the values of c and n.

3 Given that

$$\frac{d}{dx}\left((2x+3)^4 + (2x+3)^5\right) = 2(2x+3)^3(ax+b)$$

find the values of a and b.

4 A curve has equation

$$y = \frac{1}{x^2+4}$$

a Find an equation of the tangent when $x = 2$, giving the answer in the form $y = mx + c$.

b Find the gradient of the normal when $x = -1$.

5 $y = e^{2x} - e^x$

a Find any stationary points on the curve.

b Sketch the curve.

6 Find the range of values of x for which f, defined by $f(x) = x - 3\ln x$, $x > 0$, is an increasing function.

7 For the curve $y = e^{3x}$, find

a the gradient when $x = 2$

b the coordinates of the points where the tangent to the curve at $x = 2$ meets the axes.

8 Given that $y = e^x - x^3$ show that $\frac{dy}{dx} = 1$ and $\frac{d^2y}{dx^2} = 1$ when $x = 0$.

9 $y = 4x + \frac{1}{x}$

a Find the coordinates of any stationary points on the curve and state the nature of the point(s).

b Write down the equation of an asymptote to the curve.

10 The curve $y = \ln(x^2 - 3)$ crosses the x-axis at A and B.

a Find the coordinates of A and B.

b The normals at A and B meet at P. Find the coordinates of P.

11 Find the coordinates of the maximum and minimum points on the curve $y = x^2(5 - x)^3$, distinguishing between them.

12 Differentiate with respect to x

   a $\tan(x^2)$

   b $\tan\sqrt{x}$

13 Differentiate $y = \operatorname{cosec} x$ by writing it as $(\sin x)^{-1}$ and using the chain rule.

14 Differentiate with respect to x

   a $(2x + 3)^4(x - 1)^3$

   b $\dfrac{(2x + 3)^4}{(x - 1)^3}$

15 Differentiate with respect to x

   a $x^3 \sin x$

   b $x^3 e^x$

   c $x e^{-x^2}$

16 The curve

$$y = \frac{2x + 1}{2x - 1}$$

crosses the x-axis at A and the y-axis at B.
Find the point of intersection of the tangents to the curve at A and B.

17 For each of these curves, find an equation of the tangent to the curve at the point with x-coordinate specified.

   a $y = 2\cos 4x$ at $x = \dfrac{\pi}{8}$

   b $y = 3\tan x$ at $x = -\dfrac{\pi}{3}$

18 Find θ, where $0 \leqslant \theta \leqslant 2\pi$, such that the gradient of the curve $y = -\cos\frac{1}{4}\theta$ is 0.25.

19 If $f(x) = \cos x \tan x$, find $f'\left(\dfrac{\pi}{3}\right)$.

20 If $y = a\sin nx + b\cos nx$, show that

$$\frac{d^2 y}{dx^2} + n^2 y = 0$$

There is a 'Test yourself' exercise (Ty4) and an 'Extension exercise' (Ext4) on the CD-ROM.

Key points

Chain rule

The chain rule is used to differentiate composite functions.

If y is a function of t and t is a function of x, then $\dfrac{dy}{dx} = \dfrac{dy}{dt} \times \dfrac{dt}{dx}$

$$\frac{dy}{dx} = \frac{1}{\left(\dfrac{dx}{dy}\right)}$$

$$\frac{d}{dx}\, g(f(x)) = g'(f(x)) \times f'(x)$$

$$\frac{d}{dx}\, (f(x))^n = n(f(x))^{n-1} \times f'(x)$$

$$\frac{d}{dx}\, k(f(x))^n = kn(f(x))^{n-1} \times f'(x)$$

$$\frac{d}{dx}\, (ax+b)^n = na(ax+b)^{n-1}$$

Exponential functions

$$\frac{d}{dx}(e^x) = e^x$$

$$\frac{d}{dx}(e^{f(x)}) = f'(x)e^{f(x)}$$

$$\frac{d}{dx}(ke^{f(x)}) = kf'(x)e^{f(x)}$$

$$\frac{d}{dx}(e^{ax+b}) = ae^{ax+b}$$

Logarithmic functions

$$\frac{d}{dx}(\ln kx) = \frac{1}{x}$$

$$\frac{d}{dx}(\ln(f(x))) = \frac{f'(x)}{f(x)}$$

$$\frac{d}{dx}(k\ln(f(x))) = k\frac{f'(x)}{f(x)}$$

$$\frac{d}{dx}(\ln(ax+b)) = \frac{a}{ax+b}$$

Simplify a log expression before differentiating if possible.

$$\frac{d}{dx}\log_e x = \frac{d}{dx}\ln x = \frac{1}{x}$$

Product rule

If $y = uv$, where u and v are functions of x, then $\dfrac{dy}{dx} = u\dfrac{dv}{dx} + v\dfrac{du}{dx}$

$\dfrac{d}{dx}(f(x)g(x)) = f(x)g'(x) + g(x)f'(x)$

Quotient rule

If $y = \dfrac{u}{v}$, where u and v are functions of x, then $\dfrac{dy}{dx} = \dfrac{v\dfrac{du}{dx} - u\dfrac{dv}{dx}}{v^2}$

$\dfrac{d}{dx}\left(\dfrac{f(x)}{g(x)}\right) = \dfrac{g(x)f'(x) - f(x)g'(x)}{(g(x))^2}$

Trigonometric functions

$\dfrac{d}{dx}(\sin x) = \cos x$ $\qquad$ $\dfrac{d}{dx}(\sec x) = \sec x \tan x$

$\dfrac{d}{dx}(\cos x) = -\sin x$ $\qquad$ $\dfrac{d}{dx}(\operatorname{cosec} x) = -\operatorname{cosec} x \cot x$

$\dfrac{d}{dx}(\tan x) = \sec^2 x$ $\qquad$ $\dfrac{d}{dx}(\cot x) = -\operatorname{cosec}^2 x$

$\dfrac{d}{dx}(\sin f(x)) = f'(x)\cos(f(x))$

$\dfrac{d}{dx}(\cos f(x)) = -f'(x)\sin(f(x))$

$\dfrac{d}{dx}(\tan f(x)) = f'(x)\sec^2(f(x))$

1 Given that a is a positive constant, sketch the graph of $y = |3x - a|$, indicating clearly in terms of a where the graph crosses or touches the coordinate axes.

Solve the inequality $|3x - a| < x$.

AQA (AEB) 1996

(Not from the live examinations for the current specification.)

2 The function f is defined for all real values of x by

$$f(x) = |2x - 3| - 1.$$

a Sketch the graph $y = f(x)$. Indicate the coordinates of the points where the graph crosses the x-axis and the coordinates of the point where the graph crosses the y-axis.

b State the range of f.

c Find the values of x for which $f(x) = x$.

AQA Jan 2002

3 The function f is defined for real values of x by

$$f(x) = 3 - |2x - 1|.$$

a i Sketch the graph of $y = f(x)$. Indicate the coordinates of the points where the graph crosses the coordinate axes.

  ii Hence show that the equation $f(x) = 4$ has no real roots.

b State the range of f.

c By finding the values of x for which $f(x) = x$, solve the inequality

$$f(x) < x.$$

AQA Jan 2003

4 **a** Sketch on the same diagram, the graphs of

$$y = |2x + 3| \text{ and } y = 2x^2 - 9,$$

stating the coordinates of any points where the graphs meet the axes.

b Deduce the number of roots of the equation

$$|2x + 3| = 2x^2 - 9.$$

Determine the value of each of these roots.

AQA Jan 2001

5 The diagram shows a sketch of the curve with equation $y = \sin 2x$ for $-\dfrac{\pi}{2} \leqslant x \leqslant \dfrac{\pi}{2}$.

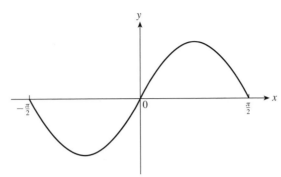

   a Find, in radians, the values of x in the interval $-\dfrac{\pi}{2} \leqslant x \leqslant \dfrac{\pi}{2}$ for which $\sin 2x = -\tfrac{1}{5}$.

   b **i** Draw on the same diagram sketches of the graphs with equations $y = |x|$ and $y = |\sin 2x|$ for $-\dfrac{\pi}{2} \leqslant x \leqslant \dfrac{\pi}{2}$.

      ii Hence state the number of times the graph of the curve with equation

$$y = |\sin 2x| - |x|$$

intersects the x-axis in the interval $-\dfrac{\pi}{2} \leqslant x \leqslant \dfrac{\pi}{2}$.

AQA June 2003

6 Describe, in each of the following cases, a single transformation which maps the graph of $y = e^x$ onto the graph of the function given.

   a $y = e^{3x}$

   b $y = e^{x-3}$

   c $y = \ln x$

AQA June 2001

7 **a** Describe a sequence of geometrical transformations that map the graph of $y = \cot x$ onto the graph of $y = 4 + \cot 2x$.

   b **i** Find the gradient of the curve with equation $y = 4 + \cot 2x$ at the point P with x-coordinate $\dfrac{\pi}{8}$.

      ii Find the equation of the tangent to the curve with equation $y = 4 + \cot 2x$ at the point P.

8 a Show that the equation

$$\tan^2 \theta + \sec \theta = 11$$

can be written as

$$x^2 + x - 12 = 0$$

where $x = \sec \theta$.

b Hence solve the equation

$$\tan^2 \theta + \sec \theta = 11$$

giving all solutions to the nearest 0.1° in the interval $0° < \theta < 360°$.

AQA June 2003

9 *No credit will be given for numerical answers without supporting working.*

Solve the equation

$$4 \cot^2 \theta + 12 \operatorname{cosec} \theta + 1 = 0$$

giving all values of θ to the nearest degree in the interval $0° \leqslant \theta \leqslant 360°$.

AQA (AEB) 1996

(Not from the live examinations for the current specification.)

10 Show, using the derivatives of $\sin x$ and $\cos x$, that the derivative of $\tan x$ is $\sec^2 x$.

11 The function f is defined for all $x > 0$ by $f(x) = 4 \ln x - \dfrac{1}{x}$.

a Find the derivative $f'(x)$.

b Explain why f is an increasing function.

c State, giving a reason, whether f^{-1} exists.

AQA Jan 2001

12 It is given that

$$y = \ln x - 2x^2 + 3, \qquad x > 0.$$

a Find $\dfrac{dy}{dx}$.

b Verify that y has a stationary value when $x = \frac{1}{2}$.

c Find the value of $\dfrac{d^2 y}{dx^2}$ when $x = \frac{1}{2}$.

d Hence determine whether this stationary value is a maximum or a minimum.

AQA Jan 2003

13

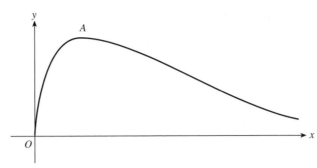

The diagram shows part of the curve with equation $y = 4\sqrt{x}e^{-4x}$.

Find the x-coordinate of the maximum point of the curve, A.

14 A curve has equation

$$y = x^2 - 3x + \ln x + 2, \qquad x > 0.$$

a **i** Find $\dfrac{dy}{dx}$.

ii Hence show that the gradient of the curve at the point where $x = 2$ is $\frac{3}{2}$.

b **i** Show that the x-coordinates of the stationary points of the curve satisfy the equation

$$2x^2 - 3x + 1 = 0.$$

ii Hence find the x-coordinates of each of the stationary points.

iii Find $\dfrac{d^2y}{dx^2}$.

iv Find the value of $\dfrac{d^2y}{dx^2}$ at each of the stationary points.

v Hence show that the y-coordinate of the maximum point is

$$\tfrac{3}{4} - \ln 2.$$

AQA May 2002

15 a i Draw on the same diagram sketches of the graphs with equations

$$y = x - 2 \text{ and } y = 2\ln x \text{ for } x > 0.$$

ii Hence state the number of roots of the equation

$$x - 2 = 2\ln x, \qquad x > 0.$$

b The curve, C, with equation

$$y = x - 2 - 2\ln x, \qquad x > 0$$

has only one stationary point.

i Find $\dfrac{dy}{dx}$.

ii Show that the y-coordinate of the stationary point is $-\ln 4$.

iii Find $\dfrac{d^2y}{dx^2}$.

iv Hence show that the stationary point is a minimum.

c The vertical lines $x = 6$ and $x = 7$ meet the curve C at points P and Q respectively.

i Show that the y-coordinate of P is $4 - \ln 36$.

ii The area of the trapezium bounded by the lines PQ, $x = 6$, $x = 7$ and the x-axis is A square units. Show that

$$A = \frac{p}{2} - \ln q$$

stating the values of the positive integers p and q.

AQA Nov 2002

16 A curve has equation $y = (x^2 + 5x + 4)\cos 3x$.

a Find $\dfrac{dy}{dx}$.

b Find the equation of the tangent to the curve at the point where $x = 0$.

AQA June 2002

Integration

Integration is a very powerful tool with applications as diverse as determining the exact surface area of an object such as a sports car, to finding the chance that an electrical component will not fail during the first five years of its use.

Earlier in the course you learnt how to integrate functions by applying differentiation in reverse and how to apply this to find areas under curves. This chapter extends this work to consider some specific techniques for finding integrals and also how to use integration to find volumes of certain solids.

After working through this chapter you should

- be able to apply integration to finding areas and volumes
- be able to integrate e^x, $\dfrac{1}{x}$ and related functions
- understand how to use integration by recognition in simple situations
- be able to use integration by substitution to find both definite and indefinite integrals
- recognise integrals of the form $\displaystyle\int \dfrac{f'(x)}{f(x)} \, dx$ and obtain their solution.

5.1 Integration as the limit of a sum

It can be shown that the area enclosed by the curve $y = f(x)$, the x-axis and the ordinates $x = a$ and $x = b$ can be found using

$$\frac{dA}{dx} = y \Rightarrow A = \int_a^b y \, dx \begin{cases} +\text{ve for area above } x\text{-axis} \\ -\text{ve for area below } x\text{-axis} \end{cases}$$

Refer to Section 11.4 of Book 1.

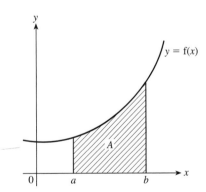

An alternative approach is to split the required area into strips, where a typical strip has width δx and area δA.

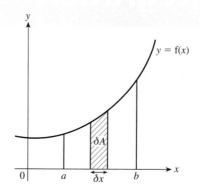

The total area is the sum of the areas of these strips, so

$$A = \sum_{x=a}^{x=b} \delta A$$

If P is the point (x, y) on the curve $y = f(x)$, then the area of a strip is approximately equal to the area of a rectangle of height y and width δx, so $\delta A \approx y\,\delta x$.

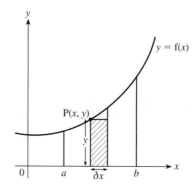

Note: The more rectangles the area is split into, the smaller will be their width, δx, and the more accurate the estimate of the area.

An approximate value for A is given by the sum of the areas of the rectangles:

$$A = \sum_{x=a}^{x=b} \delta A \approx \sum_{x=a}^{x=b} y\,\delta x$$

As a greater number of strips is taken, $\delta x \to 0$. In the limit, the **exact area**, A, will be given.

$$A = \lim_{\delta x \to 0} \sum_{x=a}^{x=b} y\,\delta x = \int_a^b y\,dx$$

In a similar way, it can also be shown that the area enclosed by the curve $y = f(x)$, the y-axis and the lines $y = c$ and $y = d$ can be found using

$$\frac{dA}{dy} = x \Rightarrow A = \int_c^d x\,dy$$

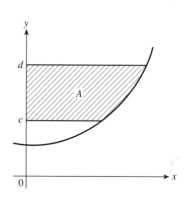

Note The word integration implies the putting together of parts to make up a whole. This fundamental aspect of the process is reinforced when finding the area under a curve by summing the areas of the rectangles. The symbol $\int$, an elongated S (for 'sum'), is a reminder that integration is essentially summation.

Note The specification does not require you to make use of the summation approach; $A = \int_a^b y\,dx$ and $A = \int_c^d x\,dy$ should usually be used straight away. However, the approach of summing the areas of rectangles can be useful in visualising the process, particularly when finding the area between two curves.

Example 1 The diagram shows the area enclosed between the curve $y = x^2 + 1$ and the line $y = 4x - 2$.

    a Find the coordinates of P and Q.

    b Find the shaded area.

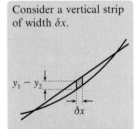

Solution **a** The line and curve intersect when

$$x^2 + 1 = 4x - 2$$

$$x^2 - 4x + 3 = 0$$

$$(x - 3)(x - 1) = 0$$

$$x = 3 \quad \text{or} \quad x = 1$$

When $x = 1$, $y = 2$, so P is $(1, 2)$.
When $x = 3$, $y = 10$, so Q is $(3, 10)$.

    b $\text{Area} \approx \displaystyle\lim_{\delta x \to 0} \sum_{x=1}^{x=3} (y_1 - y_2)\,\delta x$

$$= \int_1^3 (y_1 - y_2)\,dx$$

> Consider a vertical strip of width δx.
>
>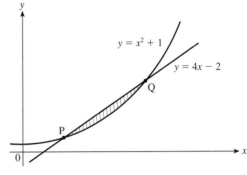
>
> This is approximately a rectangle with height $y_1 - y_2$. Area of strip $\approx (y_1 - y_2)\,\delta x$.

$$\text{Area} = \int_1^3 (4x - 3 - x^2)\,dx$$

$$= \left[2x^2 - 3x - \frac{x^3}{3} \right]_1^3$$

$$= 18 - 9 - 9 - (2 - 3 - \tfrac{1}{3})$$

$$= 1\tfrac{1}{3}$$

> Simplify $y_1 - y_2$, where $y_1 = 4x - 2$ and $y_2 = x^2 + 1$, so
> $$y_1 - y_2 = 4x - 2 - (x^2 + 1)$$
> $$= 4x - 3 - x^2$$

The area enclosed by the line and curve is $1\tfrac{1}{3}$ square units.

Note The same result would be obtained by finding $A_1 - A_2$ where A_1 is the area under $y = 4x - 2$ and A_2 is the area under $y = x^2 + 1$.

Example 2   The diagram shows the curve $y = \dfrac{16}{x^2}$.

P is the area between the curve, the x-axis and the lines $x = 2$ and $x = 8$. Q is the area between the curve, the y-axis and the lines $y = 4$ and $y = 9$. Find the ratio area P:area Q in simplest form.

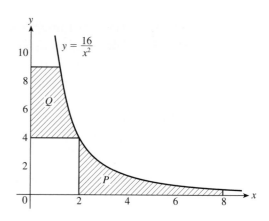

Solution   Area $P = \displaystyle\lim_{\delta x \to 0} \sum_{x=2}^{x=8} y\,\delta x$

$\qquad\qquad = \displaystyle\int_2^8 y\,\mathrm{d}x$

$\qquad\qquad = \displaystyle\int_2^8 \frac{16}{x^2}\,\mathrm{d}x$

$\qquad\qquad = 16\displaystyle\int_2^8 x^{-2}\,\mathrm{d}x$

$\qquad\qquad = 16\left[-x^{-1}\right]_2^8$

$\qquad\qquad = 16\left(-\tfrac{1}{8} - \left(-\tfrac{1}{2}\right)\right)$

$\qquad\qquad = 6$

Area $Q = \displaystyle\lim_{\delta y \to 0} \sum_{y=4}^{y=9} x\,\delta y$

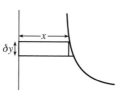

$\qquad\qquad = \displaystyle\int_4^9 x\,\mathrm{d}y$

$\qquad\qquad = \displaystyle\int_4^9 4y^{-\frac{1}{2}}\,\mathrm{d}y$

$\qquad\qquad = \left[\dfrac{4y^{\frac{1}{2}}}{\frac{1}{2}}\right]_4^9$

$\qquad\qquad = \left[8\sqrt{y}\right]_4^9$

$\qquad\qquad = 8(3 - 2)$

$\qquad\qquad = 8$

Area P:Area $Q = 6{:}8 = 3{:}4$

> Take a typical vertical strip and approximate to a rectangle with height y, width δx.

> Write in index form.

> Take a typical horizontal strip and approximate to a rectangle of length x, width δy.

> $y = \dfrac{16}{x^2} \Rightarrow x^2 = \dfrac{16}{y} \Rightarrow x = \dfrac{4}{y^{\frac{1}{2}}} = 4y^{-\frac{1}{2}}$

5.2 Solids of revolution

When a curve is rotated about a line, it traces out the surface of a solid known as a solid of revolution.

Rotation through 360°, or 2π radians, about the *x*-axis

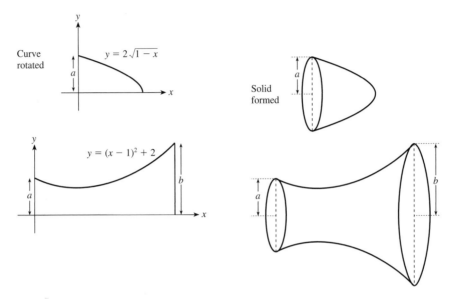

Rotation through 360°, or 2π radians, about the *y*-axis

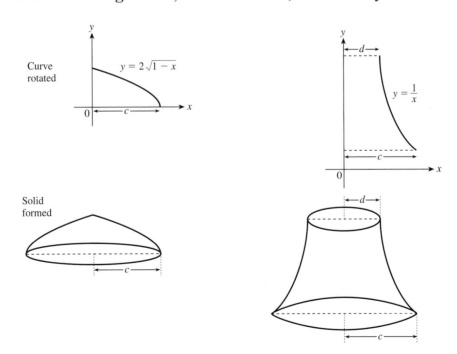

Note   When a curve is rotated, a hollow object is formed. If an area is rotated, a solid object is formed. Both are known as solids of revolution.

Volume of revolution

The volume of the solid formed by rotating a curve or an area is called the volume of revolution. It can be found by using the process of integration as the limit of a sum.

Consider the solid formed by rotating the area enclosed by the curve $y = f(x)$, the x-axis and the lines $x = a$ and $x = b$ through $360°$ about the x-axis.

To find the volume, imagine the solid split into slices.

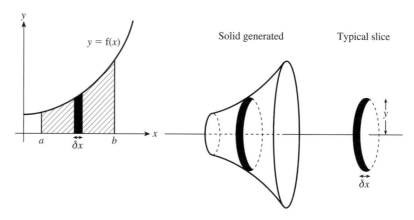

Let the volume of a typical slice be δV.

The slice is approximately a cylinder with radius y and thickness δx, so $\delta V \approx \pi y^2 \delta x$.

The total volume, V, is the sum of the volumes of the slices.

$$V = \sum_{x=a}^{x=b} \delta V \approx \sum_{x=a}^{x=b} \pi y^2 \delta x$$

Remember: volume of a cylinder, radius r, height h is given by $V = \pi r^2 h$.

The more slices that are taken, the smaller the value of δx and the more accurate the estimate of the volume. The **exact volume**, V, is given by the limit as $\delta x \to 0$:

$$V = \lim_{\delta x \to 0} \sum_{x=a}^{x=b} \pi y^2 \delta x = \int_a^b \pi y^2 \, dx$$

In a similar way, to find the volume of the solid formed by rotating the area enclosed by the curve $y = f(x)$, the y-axis and the lines $y = c$ and $y = d$ through $360°$ about the y-axis, imagine the solid split into slices horizontally.

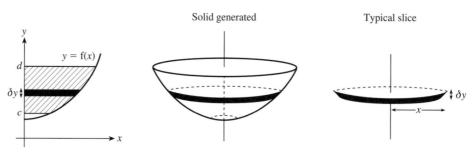

A typical slice, with volume δV, is approximately a cylinder of radius x and thickness δy. So $\delta V \approx \pi x^2 \, \delta y$.

Total volume $V = \lim\limits_{\delta y \to 0} \sum\limits_{y=c}^{y=d} \pi x^2 \, \delta y = \int_c^d \pi x^2 \, \mathrm{d}y$

These results are summarised as follows:

> If the area bounded by the curve $y = f(x)$, the x-axis, $x = a$ and $x = b$, is rotated through $360°$, or 2π radians, about the x-axis, the volume V of the solid generated is given by
>
> $$V = \int_a^b \pi y^2 \, \mathrm{d}x$$
>
> If the area bounded by the curve $y = f(x)$, the y-axis, $y = c$ and $y = d$ is rotated through $360°$ about the y-axis, the volume V of the solid generated is given by
>
> $$V = \int_c^d \pi x^2 \, \mathrm{d}y$$

Note   When calculating a volume, the constant π can be taken outside the integral sign. The volume is often given as a multiple of π.

Example 3   Find the volume of the solid formed when the area bounded by $y = x^3$, the x-axis, $x = 2$ and $x = 3$, is rotated through $360°$ about the x-axis.

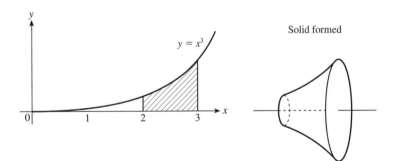

Solution

$$V = \int_a^b \pi y^2 \, \mathrm{d}x$$

Take π outside the integral sign and substitute $y = x^3$, so $y^2 = (x^3)^2 = x^6$.

$$= \pi \int_2^3 (x^3)^2 \, \mathrm{d}x$$

$$= \pi \int_2^3 x^6 \, \mathrm{d}x$$

$$= \pi \left[\frac{x^7}{7} \right]_2^3$$

$$= \frac{\pi}{7}(3^7 - 2^7)$$

Leave the answer as a multiple of π.

$$= 294\tfrac{1}{7}\pi \text{ cubic units}$$

Example 4   A vase is formed by rotating the area bounded by $y = \sqrt{x - 1}$, the y-axis, $y = 0$ and $y = 1$ through $360°$ about the y-axis. Calculate its volume.

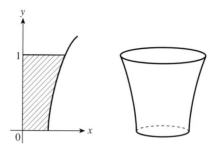

Solution

$$V = \int_c^d \pi x^2 \, dy$$

$$= \pi \int_0^1 (y^2 + 1)^2 \, dy$$

$$= \pi \int_0^1 (y^4 + 2y^2 + 1) \, dy$$

$$= \pi \left[\frac{y^5}{5} + \frac{2y^3}{3} + y \right]_0^1$$

$$= \pi \left(\frac{1^5}{5} + \frac{2 \times 1^3}{3} + 1 - 0 \right)$$

$$= 1\tfrac{13}{15} \pi$$

> Substitute for x^2 where $y = \sqrt{x - 1}$, so $y^2 = x - 1$, i.e. $x = y^2 + 1$ and $x^2 = (y^2 + 1)^2$.

> Leave the answer as a multiple of π.

The volume of the vase is $1\tfrac{13}{15}\pi$ cubic units.

Using calculus, some familiar volumes used in earlier work can be derived, as illustrated in Examples 5 and 6.

Example 5   When the area enclosed by the line $y = \dfrac{r}{h}x$, the x-axis and $x = h$, is rotated through $360°$ about the x-axis, a cone is formed with base radius r and height h.

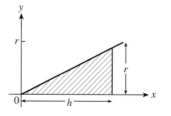

Solid generated

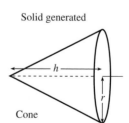

Cone

Show that the volume of the cone is given by $V = \tfrac{1}{3}\pi r^2 h$.

Solution

$$V = \int_a^b \pi y^2 \, dx$$

$$= \int_0^h \pi \frac{r^2}{h^2} x^2 \, dx$$

$$= \frac{\pi r^2}{h^2} \int_0^h x^2 \, dx$$

$$= \frac{\pi r^2}{h^2} \left[\frac{x^3}{3} \right]_0^h$$

$$= \frac{\pi r^2}{h^2} \left(\frac{h^3}{3} \right)$$

$$= \tfrac{1}{3} \pi r^2 h$$

The volume of a cone is $\tfrac{1}{3}\pi r^2 h$.

Example 6   This diagram shows the area enclosed by the curve $y = \sqrt{r^2 - x^2}$ and the x-axis.

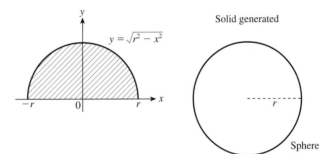

When this area is rotated through $360°$ about the x-axis, a sphere with radius r is obtained.

Show that V, the volume of the sphere, is given by $V = \tfrac{4}{3}\pi r^3$.

Solution

$$V = \int_a^b \pi y^2 \, dx$$

$$= \pi \int_{-r}^r (r^2 - x^2) \, dx$$

$$= 2\pi \int_0^r (r^2 - x^2) \, dx$$

$$= 2\pi \left[r^2 x - \frac{x^3}{3} \right]_0^r$$

$$= 2\pi \left(r^3 - \frac{r^3}{3} \right)$$

$$= \tfrac{4}{3} \pi r^3$$

> $y = \sqrt{r^2 - x^2}$ so $y^2 = r^2 - x^2$

> $r^2 - x^2$ is an even function so use $\int_{-a}^a f(x) \, dx = 2\int_0^a f(x) \, dx$

The volume of a sphere of radius r is $\tfrac{4}{3}\pi r^3$.

Example 7   The diagram shows part of the curves $y = x^2$ and $y = \sqrt{x}$.

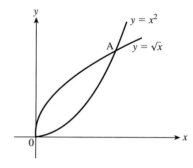

a   Find the coordinates of A.

b   Calculate the volume generated when the area enclosed by the curves is rotated through 360° about the y-axis.

Solution   **a**   The lines intersect when

$$x^2 = \sqrt{x}$$
$$x^4 = x$$
$$x^4 - x = 0$$
$$x^3(x - 1) = 0$$
$$x = 0 \quad \text{or} \quad x = 1$$

When $x = 1$, $y = 1$, so A is $(1, 1)$.

b   The required volume is $V_1 - V_2$, where V_1 is the volume generated by rotating the curve $y = x^2$ about the y-axis and V_2 is the volume generated by rotating the curve $y = \sqrt{x}$ about the y-axis.

Volume generated by the curve $y = x^2$

$$V_1 = \int_0^1 \pi x^2 \, dy$$

$$= \int_0^1 \pi y \, dy$$

$$= \pi \left[\frac{y^2}{2} \right]_0^1$$

$$= \frac{\pi}{2}$$

Volume generated by the curve $y = \sqrt{x}$

$$V_2 = \int_0^1 \pi x^2 \, dy$$

$$= \int_0^1 \pi y^4 \, dy$$

$$= \pi \left[\frac{y^5}{5} \right]_0^1$$

$$= \frac{\pi}{5}$$

$$\boxed{y = \sqrt{x} \Rightarrow y^2 = x \Rightarrow y^4 = x^2}$$

So the required volume is

$$V_1 - V_2 = \frac{\pi}{2} - \frac{\pi}{5} = \frac{3\pi}{10}$$

Care must be taken to visualise correctly the solid being formed. This is illustrated in Example 8 in which two different solids are formed by rotating the same area about different lines.

Example 8 (Extension)

The diagram shows the curve $y = x^2 + 2$ and the line $y = 6$.

a Find the x coordinates of the points of intersection of the curve and the line.

b Find the volume generated when the segment cut off by $y = 6$ from the curve $y = x^2 + 2$ is rotated through $360°$ about the x-axis.

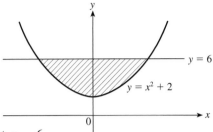

c Repeat part **b** for rotation through $360°$ about $y = 6$.

Solution

a $y = 6$ and $y = x^2 + 1$ intersect when

$$x^2 + 2 = 6$$
$$x^2 = 4$$
$$x = \pm 2$$

The line and curve intersect when $x = 2$ and $x = -2$.

b

Solid generated   Typical slice

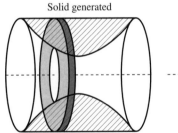

> Rotating about the x-axis forms this solid.
> *Note*: the area of the ring is *not* $\pi(y_1 - y_2)^2$. A typical slice of the solid is a ring with cross-section area $\delta A = \pi(y_1{}^2 - y_2{}^2)$, where $y_1 = 6$ and $y_2 = x^2 + 2$.

$$\delta A = \pi\left(6^2 - (x^2 + 2)^2\right) = \pi\left(36 - (x^4 + 4x^2 + 4)\right) = \pi(32 - x^4 - 4x^2)$$

Then $\quad \delta V \approx \pi(32 - x^4 - 4x^2)\delta x$

> Volume of a prism = area of cross-section × height

$$V = \sum_{x=a}^{x=b} \delta V \approx \sum_{x=-2}^{x=2} \pi(32 - x^4 - 4x^2)\,\delta x$$

$$V = \pi \int_{-2}^{2} (32 - x^4 - 4x^2)\,dx$$

> Even function: $\int_{-a}^{a} f(x)\,dx = 2\int_{0}^{a} f(x)\,dx$

$$= 2\pi \int_{0}^{2} (32 - x^4 - 4x^2)\,dx$$

$$= 2\pi \left[32x - \frac{x^5}{5} - \frac{4x^3}{3} \right]_{0}^{2}$$

$$= 2\pi \left(64 - \frac{32}{5} - \frac{32}{3} \right)$$

$$= 93\tfrac{13}{15}\pi$$

113

Alternatively: If preferred, the volume can be obtained by finding $V_1 - V_2$, where V_1 is the volume generated by rotating the line $y_1 = 6$ about the x-axis and V_2 is the volume generated by rotating the curve $y_2 = x^2 + 2$ about the x-axis.

Solid generated by the line $y_1 = 6$

$$V_1 = \int_{-2}^{2} \pi y_1{}^2 \, dx$$

$$= \int_{-2}^{2} \pi(6^2) \, dx$$

$$= \pi \int_{-2}^{2} 36 \, dx$$

$$= 2\pi \left[36x \right]_0^2$$

$$= 144\pi$$

> *Note*: the solid formed by the line $y_1 = 6$ is a cylinder, so the volume can be found directly using $V = \pi r^2 h = \pi \times 6^2 \times 4 = 144\pi$

Solid generated by the curve $y_2 = x^2 + 2$

$$V_2 = \int_{-2}^{2} \pi y_2{}^2 \, dx$$

$$= \int_{-2}^{2} \pi(x^2 + 2)^2 \, dx$$

$$= \pi \int_{-2}^{2} (x^4 + 4x^2 + 4) \, dx$$

$$= 2\pi \left[\frac{x^5}{5} + \frac{4x^3}{3} + 4x \right]_0^2$$

$$= 2\pi \left(\frac{32}{5} + \frac{32}{3} + 8 \right)$$

$$= 50 \tfrac{2}{15} \pi$$

So volume generated when segment is rotated about the x-axis is
$$V_1 - V_2 = 144\pi - 50\tfrac{2}{15}\pi = 93\tfrac{13}{15}\pi$$

c Solid generated   Typical slice

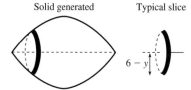

Consider an element of volume, δV.

$$\delta V \approx \pi(6 - y)^2 \, \delta x$$

$$= \pi(6 - (x^2 + 2))^2 \delta x$$

$$= \pi(4 - x^2)^2 \delta x$$

$$= \pi(16 - 8x^2 + x^4) \, \delta x$$

$$V = \sum_{x=a}^{x=b} \delta V \approx \sum_{x=-2}^{x=2} \pi(16 - 8x^2 + x^4)\delta x$$

> Rotating about $y = 6$ forms this solid.

> The slice is approximately a cylinder with radius $(6 - y)$ and height (thickness) δx.

$$V = \pi \int_{-2}^{2} (16 - 8x^2 + x^4)\,dx$$

$$= 2\pi \int_{0}^{2} (16 - 8x^2 + x^4)\,dx$$

$$= 2\pi \left[16x - \frac{8x^3}{3} + \frac{x^5}{5} \right]_{0}^{2}$$

$$= 2\pi \left(32 - \frac{64}{3} + \frac{32}{5} \right)$$

$$= 34\tfrac{2}{15}\,\pi$$

For an even function use
$\int_{-a}^{a} f(x)dx = 2\int_{0}^{a} f(x)dx$ to
make the calculation easier.

Exercise 5A

1 By considering typical strips, or otherwise, find these areas.

a

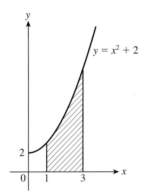

$y = x^2 + 2$

b

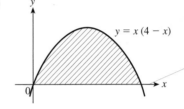

$y = x(4 - x)$

c

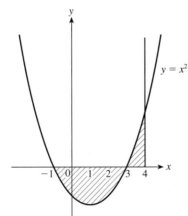

$y = x^2 - 2x - 3$

d

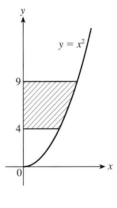

$y = x^2$

e

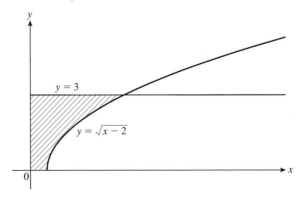

$y = 3$

$y = \sqrt{x - 2}$

f

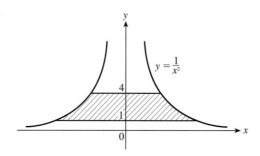

2 In each of the following find the coordinates of P and Q, and the area of the shaded region.

a

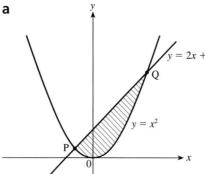

b

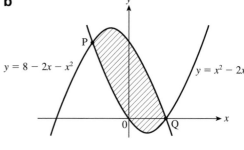

c

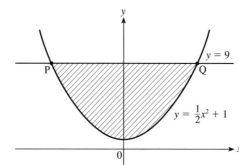

3 The curves $y = 4 - x^2$ and $y = x^2 + 2$ intersect at A and B.

   a Draw a sketch of the curves and calculate the coordinates of A and B.

   b Find the area enclosed by the two curves.

4 Find the volume of the solid formed when each shaded area is rotated through 360° about the *x*-axis.

a

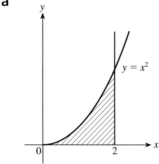

b

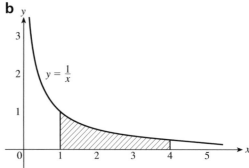

c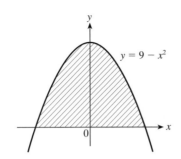

d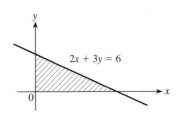

5   Find the volume of the solid formed when each shaded area is rotated through 360° about the y-axis.

a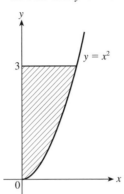

b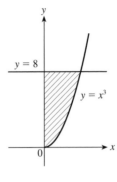

c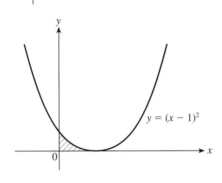

d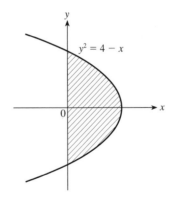

6   Find the volume of the solid formed when $y = 4x^2$, from $y = 1$ to $y = 2$, is rotated through 180° about the y-axis.

7   Find the volumes of the solids generated by rotating through 360° about the x-axis each of the areas bounded by these lines and curves. Draw a sketch each time.

    a $x + 2y - 12 = 0$, $x = 0$, $y = 0$ **b** $y = x^2 + 1$, $y = 0$, $x = 0$, $x = 1$

    c $y = \sqrt{x}$, $y = 0$, $x = 2$ **d** $y = x(x - 2)$, $y = 0$

8 The diagram shows part of the curves $y^2 = 8x$ and $y = x^2$.

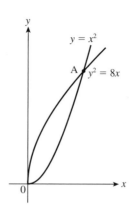

a Find the coordinates of A.

b Calculate the volume generated when the area enclosed by the curves is rotated through $360°$ about the x-axis.

c Calculate the volume generated when the area enclosed by the curves is rotated through $360°$ about the y-axis.

9 The curve $y = x^2 + 4$ meets the y-axis at A, and B is the point on the curve where $x = 2$.

a Find the area between the arc of the curve AB, the axes and the line $x = 2$.

b If this area is rotated through $360°$ about the x-axis, show that the volume swept out is approximately 188 cubic units.

10 Find the volumes of the solids generated by rotating through $360°$ about the y-axis each of the areas bounded by these lines and curves. Draw a sketch each time.

a $y = 2x - 1$, $y = 2$, $x = 0$ **b** $y = \sqrt{2x - 4}$, $y = 0$, $y = 2$, $x = 0$

c $y = x^2$, $x = 0$, $y = 1$, $y = 4$ **d** $y = \dfrac{12}{x}$, $x = 0$, $y = 1$, $y = 3$

11 By rotating the area enclosed by $y = \dfrac{h}{r}x$, the y-axis and $y = h$, through $360°$ about the y-axis, derive the volume of a cone with height h cm and base radius r cm.

12 a Sketch the curve $x^2 + y^2 = 9$ for values of x between 0 and 3 and shade the area enclosed by this portion of the curve and the y-axis.

b Find the volume of the solid generated when the shaded area is rotated through $180°$ about the y-axis.

c Describe the solid generated.

5.3 Integration of e^x and $\frac{1}{x}$

In Chapter 4 the derivatives of e^x and $\ln x$ were obtained, leading to the results:

$$\frac{d}{dx}(e^x) = e^x \quad \text{and} \quad \frac{d}{dx}(\ln x) = \frac{1}{x}$$

Applying the process of differentiation in reverse, it follows that:

$$\int e^x dx = e^x + c \quad \text{and} \quad \int \frac{1}{x} dx = \ln x + c$$

So $\frac{1}{x}$, i.e. x^{-1}, the only power of x which could not be integrated earlier, can now be integrated using the function $\ln x$.

Example 9 Find the area enclosed by the curve $y = e^x$, the y-axis and the tangent to $y = e^x$ at $(1, e)$.

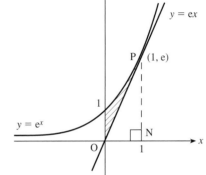

$$y = e^x$$

$$\Rightarrow \qquad \frac{dy}{dx} = e^x$$

At $(1, e)$ $\dfrac{dy}{dx} = e$

$\therefore$ tangent has gradient e and passes through $(1, e)$.

$\therefore$ equation of tangent is $y - e = e(x - 1)$ Using $y - y_1 = m(x - x_1)$

i.e. the equation of the tangent is $y = ex$. So the tangent passes through O.

$$\text{Area required} = \int_0^1 e^x \, dx - \text{area} \triangle \text{OPN}$$

$$= \left[e^x \right]_0^1 - \frac{e \times 1}{2} \qquad \qquad PN = e, \ ON = 1$$

$$= e^1 - e^0 - \frac{e}{2} \qquad \qquad \textit{Remember}: e^0 = 1$$

$$= \frac{e}{2} - 1$$

Limits of the definite integral $\displaystyle\int_a^b \frac{1}{x}\,dx$

If a and b are both positive

$$\int_a^b \frac{1}{x}\,dx = \Big[\ln x\Big]_a^b = \ln b - \ln a = \ln\frac{b}{a}$$

> Since a and b are positive, $\ln a$ and $\ln b$ can be evaluated.

If a and b are both negative, a problem arises since $\ln x$ is not defined for $x < 0$.

Remember   The domain of $\ln x$ is $x \in \mathbb{R}$, $x > 0$.

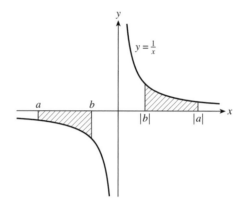

By the symmetry of the graph, it can be seen that, for a and b negative,

$$\int_a^b \frac{1}{x}\,dx = \Big[\ln |x|\Big]_a^b = \ln |b| - \ln |a| = \ln\frac{b}{a}$$

> *Note*: $\ln |b| < \ln |a|$. The integral is $-$ve because the area is below the x-axis.

If a and b have different signs the integral

$$\int_a^b \frac{1}{x}\,dx$$

cannot be evaluated because $\dfrac{1}{x}$ is not defined for $x = 0$.

➤
> **The definite integral**
>
> $$\int_a^b \frac{1}{x}\,dx$$
>
> **can be evaluated if *both* limits have the same sign, i.e. the discontinuity ($x = 0$) does not lie between a and b. Then**
>
> $$\int_a^b \frac{1}{x}\,dx = \Big[\ln |x|\Big]_a^b$$

Example 10   Find $\displaystyle\int_2^3 \frac{1}{x}\,\mathrm{d}x$ and $\displaystyle\int_{-2}^{-1} \frac{1}{x}\,\mathrm{d}x$

and explain why $\displaystyle\int_{-1}^2 \frac{1}{x}\,\mathrm{d}x$

cannot be evaluated.

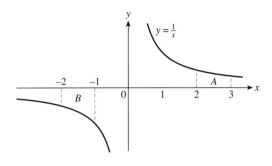

$$\int_2^3 \frac{1}{x}\,\mathrm{d}x = \Big[\ln|x|\Big]_2^3$$

> This integral represents area A. The limits of integration are both +ve.

$$= \ln 3 - \ln 2$$

> $\log a - \log b = \log\frac{a}{b}$

$$= \ln\tfrac{3}{2}$$

$$\int_{-2}^{-1} \frac{1}{x}\,\mathrm{d}x = \Big[\ln|x|\Big]_{-2}^{-1}$$

> This integral represents area B. The limits of integration are both −ve.

$$= \ln|-1| - \ln|-2|$$

$$= \ln 1 - \ln 2$$

> *Remember*: $\ln 1 = 0$

$$= -\ln 2$$

> Integral is −ve because the area is below x-axis.

The graph of $y = \dfrac{1}{x}$ is undefined at $x = 0$.

So the area between $x = -1$ and $x = 2$ is not defined. Therefore

$$\int_{-1}^2 \frac{1}{x}\,\mathrm{d}x$$

> Limits of integration have different signs, so the integral cannot be evaluated.

cannot be evaluated.

Exercise 5B

1   Evaluate

   a $2\displaystyle\int_0^6 e^x\,\mathrm{d}x$ **b** $\displaystyle\int_{2.5}^{7.5} \frac{1}{x}\,\mathrm{d}x$ **c** $\displaystyle\int_1^e \frac{1}{x}\,\mathrm{d}x$

2   Find the area enclosed by the curve $y = e^x$, the axes and $x = 3$.

3   Find the area enclosed by the curve $y = \dfrac{1}{x}$, the x-axis, $x = 2$ and $x = 4$.

4   Find the area enclosed by the curve $y = e^x - x$, the x-axis, $x = -1$ and $x = 1$.

5   Find the area enclosed by the curve $y = \dfrac{1}{x}$, the x-axis, $x = -6$ and $x = -1$.

6   **a**   Show that the curve $y = \dfrac{1}{x} + 4x^2$ has a minimum point and find its coordinates.

     b   Find the area enclosed by the curve, the x-axis, $x = \tfrac{1}{2}$ and $x = 2$.

5.4 Integration by recognition

In previous work, integrals have been found by applying the process of differentiation in reverse. Here are some simple examples.

$$\frac{d}{dx}(x^{n+1}) = (n+1)x^n \;\Rightarrow\; \int x^n\,dx = \frac{x^{n+1}}{n+1} + c,\, n \neq -1$$

$$\frac{d}{dx}(ke^x) = ke^x \qquad \Rightarrow \int ke^x\,dx = ke^x + c$$

Note   The function to be integrated is called the **integrand**.

When the integrand is more complicated it may still be possible to use this technique, but often other strategies are needed. The secret lies in knowing which line of approach to take. In this section, knowledge of when and how to apply some of the methods is built up and consolidated.

Note   Graphical calculators and some scientific calculators will perform definite integrals numerically; this can be used as a useful check of results.

The **integration by recognition** technique is based on recognising that the integrand is the result of differentiating a particular function, usually a composite function.

Remember   The chain rule for composite functions:

$$\frac{d}{dx}g(f(x)) = g'(f(x))f'(x) \quad \text{(Section 4.1)}$$

Applying the chain rule in reverse:

➤
$$\boxed{\int \mathbf{g'(f(x))f'(x)}\,\mathbf{dx} = \mathbf{g(f(x))} + c}$$

This looks very complicated but, with practice, some integrals can be recognised straight away.

Note   Sometimes this is written

$$\frac{d}{dx}f(g(x)) = f'(g(x))g'(x) \Rightarrow \int f'(g(x))g'(x)\,dx = f(g(x)) + c$$

Integrals of the type $\int k(f(x))^n f'(x)\,dx$

Consider first when $f(x)$ is a linear function of the form $f(x) = ax + b$.

Example 11

These examples illustrate a pattern in the differentials and integrals.

a

$$\frac{d}{dx}(3x-2)^5 = 15(3x-2)^4$$

Use the chain rule.

so $\int 15(3x-2)^4\,dx = (3x-2)^5 + c$

$$\int (3x-2)^4\,dx = \frac{1}{15}(3x-2)^5 + c$$

An adjustment factor of $\frac{1}{15}$ is needed.

$$\int 2(3x-2)^4\,dx = \frac{2}{15}(3x-2)^5 + c$$

b

$$\frac{d}{dx}(6-2x)^8 = -16(6-2x)^7$$

Use the chain rule.

so $\int -16(6-2x)^7\,dx = (6-2x)^8 + c$

$$\int (6-2x)^7\,dx = -\frac{1}{16}(6-2x)^8 + c$$

An adjustment factor of $-\frac{1}{16}$ is needed.

$$\int 5(6-2x)^7\,dx = -\frac{5}{16}(6-2x)^8 + c$$

Remember Using the chain rule

$$\frac{d}{dx}(ax+b)^{n+1} = (n+1)(ax+b)^n \times a$$
$$= a(n+1)(ax+b)^n$$

Applying this in reverse:

➤

> **For $n \neq -1$**
>
> $$\int (ax+b)^n\,dx = \frac{1}{a(n+1)}(ax+b)^{n+1} + c$$
>
> **If the integrand is multiplied by a constant k, then, for $n \neq -1$**
>
> $$\int k(ax+b)^n\,dx = \frac{k}{a(n+1)}(ax+b)^{n+1} + c$$

Restriction on n

Remember   Provided that $n \neq -1$

$$\int x^n \, dx = \frac{x^{n+1}}{n+1} + c$$

If $n = -1$ the formula would give

$$\frac{x^0}{0}$$

which is undefined. In a similar way, the case when $n = -1$ is excluded when integrating $(ax + b)^n$, since

$$\frac{(ax + b)^0}{0}$$

is undefined. When $n = -1$, the integral is

$$\int \frac{1}{ax + b} \, dx$$

and this is investigated in Section 5.6.

Example 12   *Integrals of the type*

$$\int k(ax + b)^n \, dx \quad for \; n \neq -1$$

are easy to recognise. Look for a linear function, raised to a power and multiplied only by a constant, as in these examples.

a $\displaystyle\int (3x + 2)^4 \, dx = \frac{1}{3 \times 5}(3x + 2)^5 + c$

$$= \frac{1}{15}(3x + 2)^5 + c$$

b $\displaystyle\int \frac{1}{(7x + 3)^3} \, dx = \int (7x + 3)^{-3} \, dx$ | Write in index form. |

$$= \frac{1}{7 \times (-2)}(7x + 3)^{-2} + c$$

$$= -\frac{1}{14}(7x + 3)^{-2} + c$$

$$= -\frac{1}{14(7x + 3)^2} + c$$

c $\displaystyle\int 5\sqrt{3x + 4} \, dx = 5 \int (3x + 4)^{\frac{1}{2}} \, dx$ | Write in index form. |

$$= \frac{5}{3 \times \frac{3}{2}}(3x + 4)^{\frac{3}{2}} + c$$

$$= \frac{10}{9}(3x + 4)^{\frac{3}{2}} + c$$

Note It is advisable to check by differentiating to verify that the correct numerical adjustment factor has been calculated. For example, to check part **c** find

$$\frac{d}{dx}\left(\frac{10}{9}(3x+4)^{\frac{3}{2}}+c\right) = \frac{10}{9} \times \frac{3}{2}(3x+4)^{\frac{1}{2}} \times 3 = 5(3x+4)^{\frac{1}{2}}$$

Example 13 Find the equation of the curve which passes through $(5, 6)$ and for which $\dfrac{dy}{dx} = \dfrac{1}{\sqrt{5+4x}}$.

$$\frac{dy}{dx} = \frac{1}{\sqrt{5+4x}}$$

> To find y, integrate with respect to x.

$$y = \int \frac{1}{\sqrt{5+4x}}\, dx$$

> Write in index form so that it is easier to recognise the standard format of $\int (ax+b)^n\, dx$.

$$= \int (5+4x)^{-\frac{1}{2}}\, dx$$

$$= \frac{1}{4 \times \frac{1}{2}}(5+4x)^{\frac{1}{2}} + c$$

> Check: $\dfrac{d}{dx}\left(\frac{1}{2}(5+4x)^{\frac{1}{2}}+c\right)$
> $= \frac{1}{2} \times \frac{1}{2}(5+4x)^{-\frac{1}{2}} \times 4 = (5+4x)^{-\frac{1}{2}}$

$$= \tfrac{1}{2}\sqrt{5+4x} + c$$

When $x = 5$, $y = 6$

> Find c using the fact that the curve goes through $(5, 6)$.

$$\therefore \quad 6 = \tfrac{1}{2}\sqrt{5+20} + c$$

$$= \tfrac{5}{2} + c$$

$$c = \tfrac{7}{2}$$

Equation of curve is $y = \frac{1}{2}\sqrt{5+4x} + \frac{7}{2}$.

Example 14 Evaluate $\displaystyle\int_1^2 (2x-1)^3\, dx$.

$$\int_1^2 (2x-1)^3\, dx = \left[\tfrac{1}{8}(2x-1)^4\right]_1^2$$

> $\dfrac{1}{a(n+1)} = \dfrac{1}{2 \times 4} = \dfrac{1}{8}$

$$= \tfrac{1}{8}\left[(2x-1)^4\right]_1^2$$

> It is a good idea to take out $\frac{1}{8}$ as a factor before substituting the limits.

$$= \tfrac{1}{8}(3^4 - 1)$$

$$= 10$$

Example 15 *This example extends the theory to the more general case for any function* f(x).

a Consider $\dfrac{d}{dx}(3x^2+6)^4 = 4 \times (3x^2+6)^3 \times 6x$

$$= 24x(3x^2+6)^3$$

It follows that

$$\int x(3x^2+6)^3\, dx = \tfrac{1}{24}(3x^2+6)^4 + c$$

> A numerical adjustment factor of $\frac{1}{24}$ is needed.

Any multiple of this integral can be found, for example

$$\int 5x(3x^2 + 6)^3 \, dx = \tfrac{5}{24}(3x^2 + 6)^4 + c$$

A numerical adjustment factor of $\tfrac{5}{24}$ is needed.

b Consider $\dfrac{d}{dx}(x^3 + 3x)^5 = 5(x^3 + 3x)^4 \times (3x^2 + 3)$

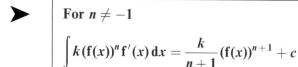

Note: $(3x^2 + 3) = 3(x^2 + 1)$

$$= 15(x^2 + 1)(x^3 + 3x)^4$$

It follows that

$$\int (x^2 + 1)(x^3 + 3x)^4 \, dx = \tfrac{1}{15}(x^3 + 3x)^5 + c$$

A numerical adjustment factor of $\tfrac{1}{15}$ is needed.

In general

> **For $n \neq -1$**
>
> $$\int k(f(x))^n f'(x) \, dx = \frac{k}{n+1}(f(x))^{n+1} + c$$

Note: Look for the integral of a function raised to a power and multiplied only by the derivative (or a multiple of the derivative) of the function.

In practice, the numerical adjustment factor is best found by trial. For example, to find

$$\int x(5x^2 + 3)^6 \, dx$$

- Notice that x is a multiple of the derivative of the function in the bracket (it is $\tfrac{1}{10}$ of it).
- Increase the power of the bracket by 1 and make a guess that the answer is a multiple of $(5x^2 + 3)^7$.
- Check by differentiating:

$$\frac{d}{dx}(5x^2 + 3)^7 = 7(5x^2 + 3)^6 \times 10x = 70x(5x^2 + 3)^6$$

- Adjust using the numerical factor $\tfrac{1}{70}$.

$$\therefore \quad \int x(5x^2 + 3)^6 \, dx = \frac{1}{70}(5x^2 + 3)^7 + c$$

Note The adjustment factor must be a number; it *cannot* be a function of x. For example, in

$$\int x(2x + 1)^5 \, dx$$

the derivative of the bracket is 2, whereas there is a factor of x in the integrand. So

$$\int x(2x + 1)^5 \, dx$$

cannot be done by recognition. Instead, the strategy of applying a substitution can be used and this is discussed in Section 5.5.

Example 16 **a** Find $\int 7x^2(x^3+1)^5\,dx$.

b Evaluate $\int_0^1 \dfrac{x}{\sqrt{(x^2+1)}}\,dx$.

Solution **a** $\int 7x^2(x^3+1)^5\,dx$ $\dfrac{d}{dx}(x^3+1)^6$

$= \frac{7}{18}(x^3+1)^6 + c$ $= 6(x^3+1)^5 \times 3x^2$

$= 18x^2(x^3+1)^5$

> The derivative of the bracket is $3x^2$. Since x^2 is a factor of the integrand, try $\dfrac{d}{dx}(x^3+1)^6$.

> An adjustment factor of $\frac{7}{18}$ is needed.

b $\int_0^1 \dfrac{x}{\sqrt{(x^2+1)}}\,dx$ $\dfrac{d}{dx}(x^2+1)^{\frac12}$

$= \int_0^1 x(x^2+1)^{-\frac12}\,dx$ $= \frac12 (x^2+1)^{-\frac12} \times 2x$

$= \left[\sqrt{x^2+1}\right]_0^1$ $= x(x^2+1)^{-\frac12}$

$= \sqrt{2} - 1$

> Write in index form.

> The derivative of the bracket is $2x$ and x is a factor of the integrand, so try $\dfrac{d}{dx}(x^2+1)^{\frac12}$.

> No adjustment factor is needed.

Integrals of the type $\int kf'(x)e^{f(x)}\,dx$

Consider first when $f(x)$ is a linear function of the form $f(x) = ax + b$.

Remember $\dfrac{d}{dx}e^{ax+b} = ae^{ax+b}$ (Section 4.2, page 69)

Applying the chain rule in reverse:

➤
$$\int e^{ax+b}\,dx = \frac{1}{a}e^{ax+b} + c$$

Also note that

$$\int ke^{ax+b}\,dx = \frac{k}{a}e^{ax+b} + c$$

> Look for the integral of an exponential function multiplied only by a constant, when the exponent is linear.

For example:

$\int e^{2x+5}\,dx = \dfrac{1}{2}e^{2x+5} + c$ $\int e^{3-4x}\,dx = -\dfrac{1}{4}e^{3-4x} + c$

$\int 6e^{-3x}\,dx = -\dfrac{6}{3}e^{-3x} + c = -2e^{-3x} + c$ $\int 4e^{6x-5}\,dx = \dfrac{4}{6}e^{6x-5} + c = \dfrac{2}{3}e^{6x-5} + c$

Example 17 The sketch shows the graph of $y = e^{1 - \frac{1}{2}x}$.

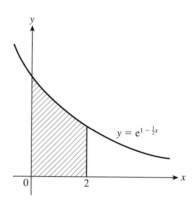

a Find the exact area enclosed between the axes and the line $x = 2$.

b Find the volume of the solid generated when the area is rotated through $360°$ about the x-axis.

Solution **a** $A = \displaystyle\int_0^2 y \, dx$

State the formula to be used.

$\qquad = \displaystyle\int_0^2 e^{1 - \frac{1}{2}x} \, dx$

Use $\displaystyle\int e^{ax+b} dx = \frac{1}{a} e^{ax+b} + c$, with $a = -\frac{1}{2}$.

$\qquad = \left[-2e^{1 - \frac{1}{2}x} \right]_0^2$

Take care with $\dfrac{1}{a} = \dfrac{1}{-\frac{1}{2}} = -2$.

$\qquad = -2 \left[e^{1 - \frac{1}{2}x} \right]_0^2$

$\qquad = -2(e^0 - e^1)$

Remember: $e^0 = 1$, $e^1 = e$

$\qquad = 2(e - 1)$

This is the exact answer. Leave it in terms of e.

b $V = \displaystyle\int_0^2 \pi y^2 \, dx$

$y = e^{1 - \frac{1}{2}x}$ so $y^2 = (e^{1 - \frac{1}{2}x})^2 = e^{2-x}$

$\qquad = \pi \displaystyle\int_0^2 e^{2 - x} \, dx$

$\qquad = -\pi \left[e^{2 - x} \right]_0^2$

$\qquad = -\pi(1 - e^2)$

$\qquad = \pi(e^2 - 1)$

In general, for any function f(x),

$$\frac{\mathrm{d}}{\mathrm{d}x}k\mathrm{e}^{\mathrm{f}(x)} = k\mathrm{f}'(x)\,\mathrm{e}^{\mathrm{f}(x)}$$

It follows that

$$\int k\mathrm{f}'(x)\,\mathrm{e}^{\mathrm{f}(x)}\,\mathrm{d}x = k\mathrm{e}^{\mathrm{f}(x)} + c$$

> Look for the integral of an exponential function multiplied only by the derivative of the exponent, or a multiple of it.

Often a numerical adjustment factor is required. This is best found by considering the derivative of the exponent, for example:

$$\int x\mathrm{e}^{3x^2}\,\mathrm{d}x = \tfrac{1}{6}\mathrm{e}^{3x^2} + c$$

> Notice $\frac{\mathrm{d}}{\mathrm{d}x}(3x^2) = 6x$. The adjustment factor is $\tfrac{1}{6}$.

$$\int (x+1)\,\mathrm{e}^{x^2+2x}\,\mathrm{d}x = \tfrac{1}{2}\mathrm{e}^{x^2+2x} + c$$

> Notice $\frac{\mathrm{d}}{\mathrm{d}x}(x^2+2x) = 2x+2 = 2(x+1)$.
> The adjustment factor is $\tfrac{1}{2}$.

Note An integral such as $\int x^2\mathrm{e}^{4x+2}\,\mathrm{d}x$ cannot be evaluated by this method because the derivative of the exponent is 4, but, in the integrand, the exponential function is multiplied by x^2. The technique required is integration by parts and this is described in Section 6.2.

Exercise 5C

1 Find

    a $\displaystyle\int (3x+1)^4\,\mathrm{d}x$ **b** $\displaystyle\int (2+5x)^3\,\mathrm{d}x$ **c** $\displaystyle\int (1-2x)^5\,\mathrm{d}x$

    d $\displaystyle\int (5-\tfrac{1}{2}x)^6\,\mathrm{d}x$ **e** $\displaystyle\int \sqrt{2+x}\,\mathrm{d}x$ **f** $\displaystyle\int (4x+1)^{\frac{1}{2}}\,\mathrm{d}x$

    g $\displaystyle\int \frac{1}{(2x-1)^4}\,\mathrm{d}x$ **h** $\displaystyle\int (2x-1)^{-2}\,\mathrm{d}x$ **i** $\displaystyle\int \frac{1}{\sqrt{3x-1}}\,\mathrm{d}x$

    j $\displaystyle\int \frac{1}{4(x+3)^2}\,\mathrm{d}x$ **k** $\displaystyle\int 4(3+2x)^{-2}\,\mathrm{d}x$ **l** $\displaystyle\int \frac{1}{(1-2x)^{\frac{3}{2}}}\,\mathrm{d}x$

2 Evaluate

    a $\displaystyle\int_0^1 (4x+1)^4\,\mathrm{d}x$ **b** $\displaystyle\int_0^3 \sqrt{2x+1}\,\mathrm{d}x$ **c** $\displaystyle\int_2^5 \frac{1}{(x-1)^3}\,\mathrm{d}x$

3 Find the equation of the curve which passes through $(0, 7)$ and for which

$$\frac{\mathrm{d}y}{\mathrm{d}x} = (x+2)^3$$

4 A curve has gradient function $k(2x-1)^2$ and passes through the origin and the point $(1, 2)$. Find the equation of the curve.

5 Find these shaded areas.

a

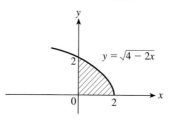

$y = \sqrt{4 - 2x}$

b

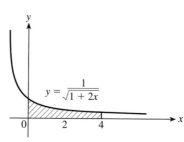

$y = \dfrac{1}{\sqrt{1 + 2x}}$

c

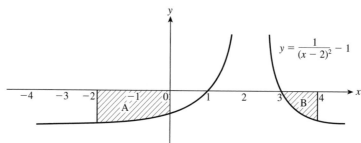

$y = \dfrac{1}{(x - 2)^2} - 1$

6 This diagram shows part of the graph of

$$y = \frac{1}{x + 1}$$

Find the volume of the solid generated when the shaded area is rotated through $360°$ about the x-axis.

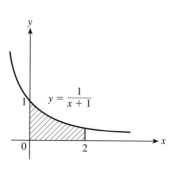

$y = \dfrac{1}{x + 1}$

7 Find, using the method of recognition,

a $\displaystyle\int x(3x^2 + 1)^4 \, dx$

b $\displaystyle\int x^2(2 - x^3) \, dx$

c $\displaystyle\int 2x(x^2 - 5)^3 \, dx$

d $\displaystyle\int 2x\sqrt{x^2 + 6} \, dx$

e $\displaystyle\int \frac{x}{(1 + x^2)^2} \, dx$

f $\displaystyle\int (2x - 3)(x^2 - 3x + 7)^3 \, dx$

g $\displaystyle\int \frac{x}{\sqrt{3 + x^2}} \, dx$

h $\displaystyle\int \frac{x + 1}{(x^2 + 2x + 3)^4} \, dx$

i $\displaystyle\int \frac{x^2 + 2}{\sqrt{x^3 + 6x}} \, dx$

8 Find

a $\displaystyle\int e^{3x + 1} \, dx$

b $\displaystyle\int e^{1 - 4x} \, dx$

c $\displaystyle\int \tfrac{1}{2} e^{2x + 3} \, dx$

d $\displaystyle\int e^{2x} \times e^{3x + 1} \, dx$

e $\displaystyle\int \frac{e^{5x + 2}}{e^x} \, dx$

f $\displaystyle\int \frac{1}{e^{x + 1}} \, dx$

9 Show that

a $\displaystyle\int_0^1 e^{2x-1}\,dx = \frac{e^2-1}{2e}$

b $\displaystyle\int_0^\infty 4e^{1-4x}\,dx = e$

10 This diagram shows the curve $y = e^{1-2x}$.

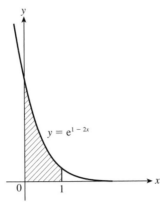

a Find the area of the shaded region.

b Find the volume of the solid generated when the shaded area is rotated through $360°$ about the x-axis.

11 The displacement of a particle from O, t seconds after leaving O, is s m. When $t = 4$, $s = 3$. Given that $\dfrac{ds}{dt} = e^{2-0.5t}$, find s when $t = 2$.

12 This diagram shows the curve $y = \sqrt{1+2x}$. The curve meets the x-axis at A and the y-axis at B. The normal to the curve at B meets the x-axis at C.

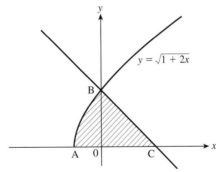

a Find the coordinates of A and B.

b Find the equation of the normal to the curve at B.

c Find the coordinates of C.

d Find the area, shown shaded in the diagram, which is enclosed by the curve, the normal and the x-axis.

e Find the volume of the solid generated when this area is rotated through $360°$ about the x-axis.

13 **a** Sketch the curve $y = \sqrt{2x+4}$.

b Find the points of intersection of the curve $y = \sqrt{2x+4}$ and the line $y = x+2$.

c Show that the area enclosed by the curve $y = \sqrt{2x+4}$ and the line $y = x+2$ is $\frac{2}{3}$ square units.

14 Find by recognition

   a $\displaystyle\int 3xe^{x^2}\,dx$
      b $\displaystyle\int 3x^2 e^{-x^3}\,dx$
      c $\displaystyle\int \frac{x}{e^{x^2}}\,dx$

15 a Find the point of intersection of the curves $y = e^{3x-1}$ and $y = e^x$.

   b Find the area bounded by the curves and the y-axis, giving your answer correct to 3 significant figures.

16 a Sketch the curve $y = (x+2)^2$ and the line $y = 4$.

   b Find the volume of the solid generated when the area enclosed by the line and the curve is rotated through $360°$ about the line $y = 4$.

   c Find the volume of the solid generated when the same arc is rotated through $360°$ about the x-axis.

17 a Find $\dfrac{d}{dx}(2^x)$.

   b Find **i** $\displaystyle\int 2^x\,dx$ **ii** $\displaystyle\int 3^{2x-1}\,dx$

5.5   Integration by substitution

On page 126, it was noted that $\displaystyle\int x(2x+1)^5\,dx$ cannot be solved by recognition.

A laborious approach is to multiply out all the brackets and then integrate term by term. A more efficient approach, however, is to use the method of **integration by substitution**, or **change of variable**. This is a very flexible method; it can be used even when it is not possible to expand the function as a finite number of terms.

Note   All integrals that can be done by recognition can also be done by applying a suitable substitution.

The method of substitution involves transforming an integral with respect to one variable, say x, into an integral with respect to a related variable, say u.

In general, let $f(x)$ be a function of x and let

$$y = \int f(x)\,dx$$

Then

$$\frac{dy}{dx} = f(x)$$

If u is a function of x, then, by the chain rule

$$\frac{dy}{du} = \frac{dy}{dx} \times \frac{dx}{du}$$

i.e. $\dfrac{dy}{du} = f(x) \times \dfrac{dx}{du}$

Integrating with respect to u gives

$$y = \int f(x)\frac{dx}{du}\,du$$

This gives the result used in the method known as substitution or change of variable:

➤ $$\boxed{\int f(x)\,dx = \int f(x)\frac{dx}{du}\,du}$$

Example 18 *In this example, the integration part is relatively straightforward, but the algebraic manipulation needs to be done with care. The neatest answer is obtained by dealing with the algebraic fractions and then taking out factors as soon as possible, preferably before substituting back for u to obtain the answer in terms of x.*

Find $\int x(2x+1)^5\,dx$ using the substitution $u = 2x+1$.

$\int x(2x+1)^5\,dx$

Let $u = 2x+1$ Define the substitution.

$x = \frac{1}{2}(u-1)$

$= \int x(2x+1)^5\frac{dx}{du}\,du$

$\dfrac{dx}{du} = \dfrac{1}{2}$

$= \int \frac{1}{2}(u-1)u^5\frac{1}{2}\,du$ The substitution results in an integral in u that is easy to find.

$= \frac{1}{4}\int(u^6 - u^5)\,du$ Take out the factor of $\frac{1}{4}$ and integrate with respect to u.

$= \frac{1}{4}\left(\frac{u^7}{7} - \frac{u^6}{6}\right) + c$ Subtract the fractions.

$= \frac{1}{4}\left(\frac{6u^7 - 7u^6}{42}\right) + c$ Take out a factor of u^6.

$= \frac{1}{168}u^6(6u - 7) + c$ Give the answer in terms of x by substituting $u = 2x+1$.

$= \frac{1}{168}(2x+1)^6(6(2x+1) - 7) + c$

$= \frac{1}{168}(12x-1)(2x-1)^6 + c$

Example 19 *When applying a substitution the idea is to arrive at an integral that can be found by applying standard techniques.*
In this example, this is achieved by using either $u = \sqrt{3x-1}$ or $u = 3x-1$. Whichever method is used, great care must be taken with the algebra manipulation to obtain the answer in its simplified form.

Find $\int x\sqrt{3x-1}\,dx$.

Solution   *Method 1*: using $u = \sqrt{3x-1}$

$$\int x\sqrt{3x-1}\,dx$$

Let $u = \sqrt{3x-1}$ — Define the substitution.

$$= \int x\sqrt{3x-1}\,\frac{dx}{du}\,du$$

Then $x = \frac{1}{3}(u^2+1)$

$$= \int \frac{1}{3}(u^2+1)u\frac{2u}{3}\,du$$

$$\frac{dx}{du} = \frac{2u}{3}$$

$$= \frac{2}{9}\int(u^4+u^2)\,du$$

Integrate with respect to u.

$$= \frac{2}{9}\left(\frac{u^5}{5}+\frac{u^3}{3}\right)+c$$

Add the fractions.

$$= \frac{2}{9}\left(\frac{3u^5+5u^3}{15}\right)+c$$

Factorise here.

$$= \frac{2}{135}u^3(3u^2+5)+c$$

Substitute $\sqrt{3x-1}$ for u.

$$= \frac{2}{135}(3x-1)^{\frac{3}{2}}(3(3x-1)+5)+c$$

$u^3 = \left(\sqrt{3x-1}\right)^3 = (3x-1)^{\frac{3}{2}}$

$$= \frac{2}{135}(9x+2)(3x-1)^{\frac{3}{2}}+c$$

Method 2: using $u = 3x-1$

$$\int x\sqrt{3x-1}\,dx$$

Let $u = 3x-1$

$$= \int x\sqrt{3x-1}\,\frac{dx}{du}\,du$$

Then $x = \frac{1}{3}(u+1)$

$$= \int \frac{1}{3}(u+1)u^{\frac{1}{2}}\frac{1}{3}\,du$$

$$\frac{dx}{du} = \frac{1}{3}$$

$$= \frac{1}{9}\int\left(u^{\frac{3}{2}}+u^{\frac{1}{2}}\right)\,du$$

Integrate with respect to u.

$$= \frac{1}{9}\left(\frac{u^{\frac{5}{2}}}{\frac{5}{2}}+\frac{u^{\frac{3}{2}}}{\frac{3}{2}}\right)+c$$

Take care with the algebra.

$$= \frac{1}{9}\left(\frac{2u^{\frac{5}{2}}}{5}+\frac{2u^{\frac{3}{2}}}{3}\right)+c$$

Take out factors and add the fractions.

$$= \frac{2}{9}u^{\frac{3}{2}}\left(\frac{3u+5}{15}\right)+c$$

Now substitute $3x-1$ for u.

$$= \frac{2}{135}(3x-1)^{\frac{3}{2}}(3(3x-1)+5)+c$$

$$= \frac{2}{135}(9x+2)(3x-1)^{\frac{3}{2}}+c$$

Example 20   *The integrals in this example were done by recognition in Section 5.4 but they can also be done by applying a suitable substitution.*

Using a method of substitution, find

a $\displaystyle\int e^{2x+5}dx$ **b** $\displaystyle\int (3x+2)^4\,dx$ **c** $\displaystyle\int x(3x^2+6)^3\,dx$

Solution **a** $\displaystyle\int e^{2x+5}\,dx$

$\displaystyle = \int e^{2x+5}\frac{dx}{du}\,du$

$\displaystyle = \int e^u \times \tfrac{1}{2}\,du$

$\displaystyle = \tfrac{1}{2}\int e^u\,du$

$\displaystyle = \tfrac{1}{2}e^u + c$

$\displaystyle = \tfrac{1}{2}e^{2x+5} + c$

Let $u = 2x+5$

Then $x = \tfrac{1}{2}(u-5)$

$\displaystyle \frac{dx}{du} = \frac{1}{2}$

> Integrate with respect to u.

> Now substitute $2x+5$ for u.

b $\displaystyle\int (3x+2)^4\,dx$

$\displaystyle = \int (3x+2)^4\frac{dx}{du}\,du$

$\displaystyle = \int u^4 \times \tfrac{1}{3}\,du$

$\displaystyle = \int \tfrac{1}{3}u^4\,du$

$\displaystyle = \tfrac{1}{15}u^5 + c$

$\displaystyle = \tfrac{1}{15}(3x+2)^5 + c$

Let $u = 3x+2$

Then $x = \dfrac{1}{3}(u-2)$

$\displaystyle \frac{dx}{du} = \frac{1}{3}$

> Integrate with respect to u.

> Now substitute $3x+2$ for u.

c $\displaystyle\int x(3x^2+6)^3\,dx$

$\displaystyle = \int (3x^2+6)^3 x\frac{dx}{du}\,du$

$\displaystyle = \int u^3\frac{1}{6}\,du$

$\displaystyle = \frac{1}{6}\int u^3\,du$

$\displaystyle = \frac{1}{24}u^4 + c$

$\displaystyle = \frac{1}{24}(3x^2+6)^4 + c$

Let $u = 3x^2 + 6$

$\displaystyle \frac{du}{dx} = 6x$

$\displaystyle \frac{dx}{du} = \frac{1}{6x}$

$\displaystyle x\frac{dx}{du} = \frac{1}{6}$

> Rather than writing the substitution in terms of x, it is easier to find $\dfrac{du}{dx}$, and use
> $$\frac{dx}{du} = \frac{1}{\frac{du}{dx}}$$

> $x\dfrac{dx}{du}$ can now be substituted into the integral.

Definite integrals and changing the limits

The method of substitution involving changing the variable can also be applied to definite integrals. Although it is possible to deal with the limits when the final expression in x has been found, it is usually more convenient to change the limits to those of the new variable. This often makes calculations easier to perform and avoids some of the algebraic manipulation.

Example 21   This diagram shows a sketch of the curve $y = x(x-2)^4$.

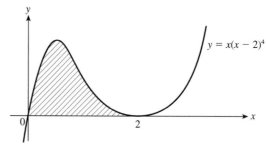

Find the shaded area.

Solution $\text{Area} = \displaystyle\int_a^b y\,dx$

> Remember the method for finding area under curve.

$= \displaystyle\int_0^2 x(x-2)^4\,dx$ Let $u = x - 2$

> Use a substitution.

$= \displaystyle\int_{x=0}^{x=2} x(x-2)^4\frac{dx}{du}\,du$ then $x = u + 2$

$\dfrac{dx}{du} = 1$

$= \displaystyle\int_{-2}^{0} (u+2)u^4 \times 1\,du$ Limits:

x	0	2
u	-2	0

$= \displaystyle\int_{-2}^{0} (u^5 + 2u^4)\,du$

> Change the x limits to u limits:
> When $x = 0$, $u = 0 - 2 = -2$
> When $x = 2$, $u = 2 - 2 = 0$
> If the x limits are changed to u limits then there is no need to simplify or to substitute back for x.

$= \left[\dfrac{u^6}{6} + \dfrac{2u^5}{5}\right]_{-2}^{0}$

$= 0 - \left(\dfrac{(-2)^6}{6} + \dfrac{2(-2)^5}{5}\right)$

$= 2\frac{2}{15}$

The shaded area is $2\frac{2}{15}$ square units.

Exercise 5D

1   Using the change of variable $u = 3x - 2$, show that

$$\int x(3x-2)^6\,dx = \tfrac{1}{504}(3x-2)^7(21x+2) + c$$

2   **a**   Using $u = \sqrt{2x+1}$, show that

$$\int x\sqrt{2x+1}\,dx = \tfrac{1}{15}(2x+1)^{\frac{3}{2}}(3x-1) + c$$

b   Now verify the result by using the substitution $u = 2x + 1$.

3 Find these integrals, using the given change of variable.

a $\displaystyle\int 3x\sqrt{4x-1}\,dx \qquad u=\sqrt{4x-1}$

b $\displaystyle\int x\sqrt{5x+2}\,dx \qquad u=5x+2$

c $\displaystyle\int x(2x-1)^6\,dx \qquad u=2x-1$

d $\displaystyle\int \frac{x}{\sqrt{x-2}}\,dx \qquad u=\sqrt{x-2}$

e $\displaystyle\int (x+2)(x-1)^4\,dx \qquad u=x-1$

f $\displaystyle\int (x-2)^5(x+3)^2\,dx \qquad u=x-2$

g $\displaystyle\int \frac{x(x-4)}{(x-2)^2}\,dx \qquad u=x-2$

h $\displaystyle\int \frac{x-1}{\sqrt{2x+3}}\,dx \qquad u=\sqrt{2x+3}$

4 Show that

$$\int_{0.5}^{3} x\sqrt{2x+3}\,dx = 11.6$$

to three significant figures.

5 Evaluate these definite integrals by using a suitable change of variable.

a $\displaystyle\int_{2}^{3} x\sqrt{x-2}\,dx$ **b** $\displaystyle\int_{0}^{1} x(x-1)^4\,dx$ **c** $\displaystyle\int_{1}^{2} \frac{x}{\sqrt{2x-1}}\,dx$

d $\displaystyle\int_{1}^{2} (2x-1)(x-2)^3\,dx$ **e** $\displaystyle\int_{-\frac{3}{8}}^{0} \frac{x+3}{\sqrt{2x+1}}\,dx$

6 This diagram shows the graph of $y=x(x-3)^4$.

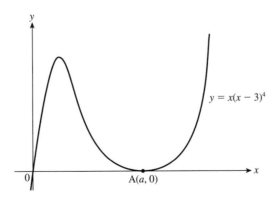

a The curve touches the x-axis at A $(a,0)$. Find the value of a.

b Find the area enclosed by the curve and the x-axis.

7 The diagram shows the graph of $y = x\sqrt{3-x}$.

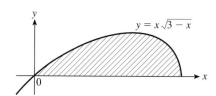

Find the shaded area, giving your answer correct to 2 decimal places.

8 In each part, both integrals can be found by using the method of substitution. One of the integrals, however, can be done by recognition of the derivative of a composite function. Say which it is and find both integrals.

a **i** $\displaystyle\int x\sqrt{3x-4}\,dx$ **ii** $\displaystyle\int x\sqrt{3x^2-4}\,dx$

b **i** $\displaystyle\int x(x^2+5)^6\,dx$ **ii** $\displaystyle\int x(x+5)^6\,dx$

c **i** $\displaystyle\int \frac{x}{\sqrt{x-1}}\,dx$ **ii** $\displaystyle\int \frac{x}{\sqrt{x^2-1}}\,dx$

9 By using the substitution $u = 8 + e^x$, or otherwise, find

$$\int_0^{\ln 8} \frac{e^x}{\sqrt{8+e^x}}\,dx$$

5.6   Integrals of the type $\displaystyle\int \frac{kf'(x)}{f(x)}\,dx$

First consider the linear function $f(x) = ax + b$ and integrals of the form

$$\int \frac{k}{ax+b}\,dx$$

Note the pattern in these differentials and integrals.

$$\frac{d}{dx}\ln(x+3) = \frac{1}{x+3}$$

so $$\int \frac{1}{x+3}\,dx = \ln(x+3) + c$$

$$\frac{d}{dx}\ln(2x-1) = \frac{2}{2x-1}$$

so $\displaystyle\int \frac{1}{2x - 1}\,dx = \frac{1}{2}\ln(2x - 1) + c$

A numerical adjustment of $\frac{1}{2}$ is needed.

$$\frac{d}{dx}\ln(3x + 5) = \frac{3}{3x + 5}$$

so $\displaystyle\int \frac{4}{3x + 5}\,dx = \frac{4}{3}\ln(3x + 5) + c$

A numerical adjustment of $\frac{4}{3}$ is needed.

Note To obtain a general rule, remember that

$$\frac{d}{dx}\ln(ax + b) = \frac{a}{ax + b}$$

and apply the chain rule in reverse.

➤
$$\int \frac{1}{ax + b}\,dx = \frac{1}{a}\ln(ax + b) + c$$

It also follows that

$$\int \frac{k}{ax + b}\,dx = \frac{k}{a}\ln(ax + b) + c$$

Look for an integrand in which the numerator is a constant and the denominator is a linear function of x.

Example 22 Find **a** $\displaystyle\int \frac{2}{3x + 5}\,dx$ **b** $\displaystyle\int \frac{1}{4 - 5x}\,dx$

Solution **a** $\displaystyle\int \frac{2}{3x + 5}\,dx = \frac{2}{3}\ln(3x + 5) + c$

Recognise integral of the type $\displaystyle\int \frac{k}{ax + b}\,dx$.

b $\displaystyle\int \frac{1}{4 - 5x}\,dx = \frac{1}{(-5)}\ln(4 - 5x) + c$

Take care with signs.

$$= -\frac{1}{5}\ln(4 - 5x) + c$$

The definite integral $\displaystyle\int_{x_1}^{x_2} \frac{1}{ax + b}\,dx$

As was seen earlier, care must be taken when evaluating definite integrals of this type. This is illustrated by considering the curve

$$y = \frac{1}{2x - 3}$$

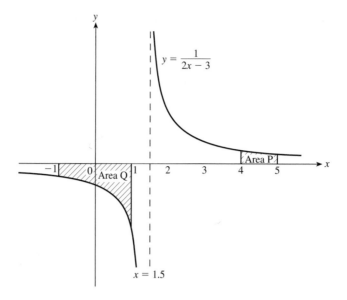

Note: The function does not exist when $2x - 3 = 0$, i.e. when $x = 1.5$.

The area between the curve, the x-axis, $x = x_1$ and $x = x_2$ is given by

$$\int_{x_1}^{x_2} y \, dx = \int_{x_1}^{x_2} \frac{1}{2x - 3} \, dx$$

However, this integral can be evaluated only if *both limits* are greater than 1.5 or if *both limits* are smaller than 1.5, i.e. provided that the discontinuity, when $x = 1.5$, does not lie between the two limits.

Consider the areas P and Q shown shaded in the diagram.

To find area P, between $x = 4$ and $x = 5$:

$$\int_4^5 y \, dx = \int_4^5 \frac{1}{2x - 3} \, dx$$
$$= \frac{1}{2} \Big[\ln(2x - 3) \Big]_4^5$$
$$= \frac{1}{2} (\ln 7 - \ln 5)$$
$$= \frac{1}{2} \ln(1.4)$$
$$= 0.17 \, (2 \text{ s.f.})$$

Use $\ln a - \ln b = \ln \dfrac{a}{b}$

So, the area P is 0.17 square units.

When this method is used to find the area Q, between $x = -1$ and $x = 1$, there is a problem when the limits are substituted, because

$$\frac{1}{2} \Big[\ln(2x - 3) \Big]_{-1}^{1}$$

gives

$$\frac{1}{2} (\ln(-1) - \ln(-5))$$

but $\ln(-1)$ and $\ln(-5)$ do not exist.

The area, however, can be seen in the diagram and does exist.

Instead, $\ln|2x - 3|$ must be considered.

$$\int_{-1}^{1} y \, dx = \int_{-1}^{1} \frac{1}{2x - 3} \, dx$$

$$= \frac{1}{2} \left[\ln|2x - 3| \right]_{-1}^{1} \qquad \text{Put in the modulus signs.}$$

$$= \frac{1}{2} (\ln|-1| - \ln|-5|) \qquad \ln|-a| = a$$

$$= \frac{1}{2} (\ln 1 - \ln 5) \qquad \ln 1 = 0$$

$$= -\frac{1}{2} \ln 5$$

$$= -0.80 \ (2 \text{ s.f.})$$

As expected, the integration gives a negative value, because the area lies below the x-axis.

So, the area Q is 0.80 square units.

Note $\int_{0}^{2} \frac{1}{2x - 3} \, dx$ is meaningless, because the curve is undefined when $x = 1.5$ and this integral cannot be evaluated.

In general

$$\int_{x_1}^{x_2} \frac{1}{ax + b} \, dx = \frac{1}{a} \left[\ln|ax + b| \right]_{x_1}^{x_2}$$

$$= \frac{1}{a} (\ln|ax_2 + b| - \ln|ax_1 + b|)$$

Note: $\frac{1}{ax + b}$ is not defined when $ax + b = 0$, and so the integral can be evaluated only if $ax_1 + b$ and $ax_2 + b$ have the same sign.

Integrals of the type $\int \dfrac{cx + d}{ax + b} \, dx$

Example 23 *This example demonstrates two approaches: either use a substitution or divide the numerator by the denominator.*

Find $\int \dfrac{x + 1}{x - 1} \, dx$.

Solution   *Using a substitution*

$$\int \frac{x+1}{x-1} \, dx$$

Let $u = x - 1$

Rearrange to give x and find $\dfrac{dx}{du}$.

$$x = u + 1$$

$$= \int \frac{x+1}{x-1} \frac{dx}{du} \, du$$

$$\frac{dx}{du} = 1$$

$$= \int \frac{u+1+1}{u} \, du$$

$$= \int \left(1 + \frac{2}{u}\right) du$$

Integrate with respect to u.

$$= u + 2\ln u + c$$

$$= x - 1 + 2\ln(x-1) + c$$

$$= x + 2\ln(x-1) + d$$

Writing d for $c - 1$.

Alternatively: dividing the numerator by the denominator

$$\frac{x+1}{x-1} = \frac{x-1+2}{x-1}$$

$$= 1 + \frac{2}{x-1}$$

So $$\int \frac{x+1}{x-1} \, dx = \int \left(1 + \frac{2}{x-1}\right) dx$$

$$= x + 2\ln(x-1) + c$$

The general case of $\int \dfrac{f'(x)}{f(x)} \, dx$ when f(x) is any function

Remember $\dfrac{d}{dx} k\ln(f(x)) = \dfrac{kf'(x)}{f(x)}$

See p. 73.

It follows that

➤ $$\boxed{\int \frac{kf'(x)}{f(x)} \, dx = k\ln(f(x)) + c}$$

Look for an integrand in which the numerator is the derivative (or multiple of the derivative) of the denominator. Sometimes a numerical adjustment factor is needed and this is best found by inspection.

Example 24   Find

a $\displaystyle\int \frac{5x^2}{x^3+1} \, dx$ **b** $\displaystyle\int \frac{2x-3}{3x^2-9x+4} \, dx$ **c** $\displaystyle\int \frac{e^x}{2+e^x} \, dx$

Solution **a** $\displaystyle\int \frac{5x^2}{x^3+1}\, dx$

$\dfrac{d}{dx}(x^3+1)=3x^2$, so the numerator is a multiple of the derivative of the denominator.

$\qquad = \dfrac{5}{3}\ln(x^3+1)+c$

A numerical adjustment factor of $\frac{5}{3}$ is needed.

b $\displaystyle\int \frac{2x-3}{3x^2-9x+4}\, dx$

$\dfrac{d}{dx}(3x^2-9x+4)=6x-9=3(2x-3)=3\times$ numerator

$\qquad = \dfrac{1}{3}\ln(3x^2-9x+4)+c$

A numerical adjustment factor of $\frac{1}{3}$ is needed.

c $\displaystyle\int \frac{e^x}{2+e^x}\, dx$

$\dfrac{d}{dx}(2+e^x)=e^x$ so the numerator is the derivative of the denominator.

$\qquad = \ln(2+e^x)+c$

Example 25   The area enclosed by the curve

$$y=\sqrt{\frac{x}{x^2+1}}$$

the x-axis and the line $x=10$ is rotated through $360°$ about the x-axis.

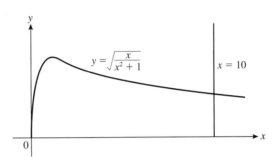

Find the volume of the solid generated, giving the answer correct to 3 significant figures.

Solution $\displaystyle V=\int_0^{10}\pi y^2\, dx$

$y=\sqrt{\dfrac{x}{x^2+1}}\Rightarrow y^2=\dfrac{x}{x^2+1}$

$\qquad = \pi\displaystyle\int_0^{10}\frac{x}{x^2+1}\, dx$

Recognise type $\displaystyle\int \frac{kf'(x)}{f(x)}\, dx$ with $f(x)=x^2+1,\ f'(x)=2x$.

$\qquad = \tfrac{1}{2}\pi\Big[\ln(x^2+1)\Big]_0^{10}$

$\qquad = \tfrac{1}{2}\pi\ln 101$

The modulus signs are not needed, because x^2+1 is always positive.

$\qquad = 7.25$ cubic units (3 s.f.)

1 Find these integrals.

    a $\displaystyle\int \frac{1}{3x-2}\,dx$ **b** $\displaystyle\int \frac{3}{2+5x}\,dx$

    c $\displaystyle\int \frac{1}{4-2x}\,dx$ **d** $\displaystyle\int \frac{4}{5(3-x)}\,dx$

2 Evaluate

    a $\displaystyle\int_1^3 \frac{1}{3x}\,dx$ **b** $\displaystyle\int_1^2 \frac{1}{2x+3}\,dx$ **c** $\displaystyle\int_1^3 (2+0.5x)^{-1}\,dx$

3 a Sketch the curve $y = \dfrac{1}{x-4}$.

    b Which of these integrals *cannot* be evaluated?

      i $\displaystyle\int_1^2 \frac{1}{x-4}\,dx$ **ii** $\displaystyle\int_0^5 \frac{1}{x-4}\,dx$ **iii** $\displaystyle\int_5^6 \frac{1}{x-4}\,dx$ **iv** $\displaystyle\int_4^5 \frac{1}{x-4}\,dx$

    c Evaluate the integrals in part **b** that *are* possible.

4 a Sketch the curve

        $y = \dfrac{1}{3x-2}$

    b Find the area enclosed by the curve $y = \dfrac{1}{3x-2}$, the x-axis and

      i the lines $x = 3$ and $x = 4$
      ii the lines $x = -1$ and $x = 0$.

5 This diagram shows the curve $y = \dfrac{1}{\sqrt{2x-1}}$.

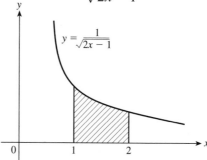

    The volume of the solid generated when the area enclosed by the curve, the x-axis and the lines $x = 1$ and $x = 2$ is rotated through $360°$ about the x-axis is $a \ln b$. Find the values of a and b.

6 Find these integrals.

    a $\displaystyle\int \frac{3x}{x^2-1}\,dx$ **b** $\displaystyle\int \frac{x}{1-x^2}\,dx$ **c** $\displaystyle\int \frac{2x+1}{x^2+x-2}\,dx$

    d $\displaystyle\int \frac{2x-3}{3x^2-9x+4}\,dx$ **e** $\displaystyle\int \frac{e^{2x}}{4+2e^{2x}}\,dx$ **f** $\displaystyle\int 3e^x(e^x+2)^{-1}\,dx$

7 Find these integrals.

a $\displaystyle\int \frac{x}{x+2}\,dx$ **b** $\displaystyle\int \frac{3x}{2x+3}\,dx$ **c** $\displaystyle\int \frac{2x}{3-x}\,dx$

8 Evaluate

a $\displaystyle\int_4^6 \frac{x}{x-2}\,dx$ **b** $\displaystyle\int_{-7}^{-5} \frac{x+1}{x+3}\,dx$ **c** $\displaystyle\int_{-0.5}^0 \frac{2-x}{x-1}\,dx$

9 This diagram shows the curve $y = \dfrac{x}{x^2+2}$.

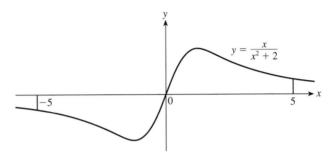

The area enclosed by the curve, the x-axis and the lines $x = -5$ and $x = 5$ is $\ln a$. Find the value of a.

10 Evaluate

$$\int_0^3 \frac{2x-1}{x^2-x+1}\,dx$$

Exercise 5F (Review)

1 Integrate with respect to x

a e^x **b** $8e^{4x}$ **c** $-e^{-x}$ **d** $\frac{1}{3}e^{-4x}$ **e** $5e^{-5x}$ **f** $-3e^{\frac{3x}{2}}$

2 Integrate with respect to x

a $\dfrac{1}{x}$ **b** $\dfrac{4}{x}$ **c** $\dfrac{1}{2x}$ **d** $\dfrac{1}{3x}$ **e** $-\dfrac{2}{3x}$ **f** $-\dfrac{5}{x}$

3 Find the area bounded by $y = e^{3x} - \dfrac{1}{x}$, the x-axis, $x = \frac{1}{2}$ and $x = 1$.

4 Find the area enclosed by the curve $y = e^{-2x}$, the axes and $x = 2$.

5 a Find the coordinates of any points where $y = e^{-x} - 1$ cuts the axes.

   b Find the area bounded by the curve $y = e^{-x} - 1$, the x-axis and $x = 4$.

6 a Show that the equation of the tangent to $y = e^{-3x}$ at the point $(0, 1)$, is $y = 1 - 3x$.

   b Find the area bounded by $y = e^{-3x}$, the axes and $x = -1$.

7 Find these integrals.

a $\displaystyle\int (3x+2)^3\,dx$ b $\displaystyle\int (2x+3)^2\,dx$ c $\displaystyle\int \frac{1}{(3x-4)^2}\,dx$

d $\displaystyle\int \sqrt{5x-2}\,dx$ e $\displaystyle\int e^{3x-7}\,dx$ f $\displaystyle\int 2e^{2-4x}\,dx$

8 Find

a $\displaystyle\int \frac{1}{\sqrt{3x+2}}\,dx$ b $\displaystyle\int \frac{1}{\sqrt[3]{4x-2}}\,dx$

9 Evaluate

a $\displaystyle\int_0^1 (2x+1)^5\,dx$ b $\displaystyle\int_{-2}^{-1} \frac{1}{x-4}\,dx$ c $\displaystyle\int_2^3 \frac{2}{4-3x}\,dx$

d $\displaystyle\int_0^2 e^{2x+1}\,dx$ e $\displaystyle\int_{-1}^2 3e^{1-x}\,dx$ f $\displaystyle\int_1^2 (4-x)^{-2}\,dx$

10 Find the area under the curve

$$y = \frac{1}{3-x}$$

from $x = 0$ to $x = 1$.

11 The area between the curve $y = e^{0.5x}$, $x = 0$, $x = 2$ and the x-axis is rotated through $360°$ about the x-axis. Find the volume of the solid formed.

12 Find the volume of revolution when the portion of the curve

$$y = \frac{2}{\sqrt{x}}$$

from $y = 0.5$ to $y = 2$ is rotated through $360°$ about the y-axis.

13 This diagram shows part of the curve $y = \dfrac{8}{x}$ and the line $x + y = 6$.

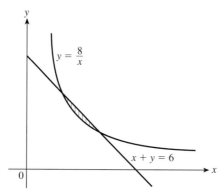

a Find the points of intersection of the line and the curve.

b Find the area enclosed by the line and the curve.

c Find the volume of the solid generated when this area is rotated through $360°$ about the x-axis.

14 The diagram shows part of the curves $y = x^3$ and $y = 2x^2$.

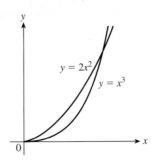

Find the volume of revolution when the area enclosed by the curves is rotated through 360° about the y-axis.

15 Given that $y = e^x + e^{2-x}$, find

    a the intercept of the curve on the y-axis

    b any stationary points of the curve and determine their nature

    c the area bounded by the curve, the axes and $x = 1$

16 Find these integrals, using the given substitution.

    a $\displaystyle\int x(3x + 1)^5 \, dx \quad u = 3x + 1$

    b $\displaystyle\int x\sqrt{5x - 2} \, dx \quad u = 5x - 2$

    c $\displaystyle\int \frac{y(y - 8)}{(y - 4)^2} \, dy \quad u = y - 4$

17 Find these integrals, either by direct recognition or by using a suitable substitution.

    a $\displaystyle\int x\sqrt{x^2 + 3} \, dx$ **b** $\displaystyle\int \frac{x}{\sqrt{2 + 3x^2}} \, dx$ **c** $\displaystyle\int 4xe^{-x^2} \, dx$

18 Evaluate

$$\int_{-1}^{0} \frac{x + 1}{x^2 + 2x - 3} \, dx$$

19 By using the substitution $u = \ln x$, or otherwise, find

$$\int_{e}^{e^4} \frac{1}{x\sqrt{\ln x}} \, dx$$

There is a 'Test yourself' exercise (Ty5) and an 'Extension exercise' (Ext5) on the CD-ROM.

➤ Key points

Area as the limit of a sum

The area enclosed by the curve $y = f(x)$, the x-axis and the lines $x = a$ and $x = b$ is given by

$$A = \lim_{\delta x \to 0} \sum_{x=a}^{x=b} y\delta x = \int_a^b y\,dx$$

→ $+$ve for area above x-axis

↘ $-$ve for area below x-axis

The area enclosed by the curve $y = f(x)$, the y-axis and the lines $y = c$ and $y = d$ is given by

$$A = \lim_{\delta y \to 0} \sum_{y=c}^{y=d} x\delta y = \int_c^d x\,dy$$

→ $+$ve for area to right of y-axis

↘ $-$ve for area to left of y-axis

Volumes of revolution

If the area bounded by the curve $y = f(x)$, the x-axis, $x = a$ and $x = b$, is rotated through $360°$, or 2π radians, about the x-axis, the volume V of the solid generated is given by

$$V = \int_a^b \pi y^2\,dx$$

If the area bounded by the curve $y = f(x)$, the y-axis, $y = c$ and $y = d$ is rotated through $360°$, or 2π radians, about the y-axis, the volume V of the solid generated is given by

$$V = \int_c^d \pi x^2\,dy$$

Integration by recognition

$$\int k(ax+b)^n dx = \frac{k}{a(n+1)}(ax+b)^{n+1} + c \qquad n \neq -1$$

$$\int k(f(x))^n f'(x)\,dx = \frac{k}{n+1}(f(x))^{n+1} + c$$

$$\int e^{ax+b}\,dx = \frac{1}{a}e^{ax+b} + c$$

$$\int k f'(x)e^{f(x)}\,dx = k e^{f(x)} + c$$

Integration by substitution (change of variable)

$$\int f(x)\,dx = \int f(x)\frac{dx}{du}\,du$$

Integrals resulting in a logarithmic function

$$\int \frac{1}{ax+b}\,dx = \frac{1}{a}\ln(ax+b) + c$$

The definite integral

$$\int_a^b \frac{1}{x}\,dx$$

can be evaluated if *both* limits have the same sign, i.e. the discontinuity ($x = 0$) does not lie between a and b. Then

$$\int_a^b \frac{1}{x}\,dx = \left[\ln|x|\right]_a^b$$

$$\int_{x_1}^{x_2} \frac{1}{ax+b}\,dx = \frac{1}{a}\left[\ln|ax+b|\right]_{x_1}^{x_2} = \frac{1}{a}(\ln|ax_2+b| - \ln|ax_1+b|)$$

$$\int \frac{k f'(x)}{f(x)}\,dx = k\ln(f(x)) + c$$

6 Further integration

This chapter considers other techniques of integration which you will need for the course.

Integrals of trigonometric functions are particularly important when considering oscillatory motion such as that of a vibrating string. Integration by parts is a technique required in areas as diverse as finding the centre of mass of an object to predicting the mean lifetime of a lightbulb.

After working through this chapter you should be able to

■ *integrate* $\sin x$, $\cos x$ *and other standard trigonometric functions*

■ *use integration by parts to find both definite and indefinite integrals*

■ *evaluate integrals using inverse trigonometrical functions.*

6.1 Integration of trigonometric functions

Integrals of $\sin x$, $\cos x$ and $\sec^2 x$

Standard trigonometric integrals can be obtained by applying the reverse process to the differentiation results obtained in Section 4.4.

Remember that x is measured in radians.

$$\frac{d}{dx}(\sin x) = \cos x \Rightarrow \int \cos x \, dx = \sin x + c$$

$$\frac{d}{dx}(\cos x) = -\sin x \Rightarrow \int \sin x \, dx = -\cos x + c$$

$$\frac{d}{dx}(\tan x) = \sec^2 x \Rightarrow \int \sec^2 x \, dx = \tan x + c$$

Also, using the chain rule:

$$\frac{d}{dx}\sin(ax+b) = a\cos(ax+b) \Rightarrow \int \cos(ax+b)\, dx = \frac{1}{a}\sin(ax+b) + c$$

$$\frac{d}{dx}\cos(ax+b) = -a\sin(ax+b) \Rightarrow \int \sin(ax+b)\, dx = -\frac{1}{a}\cos(ax+b) + c$$

$$\frac{d}{dx}\tan(ax+b) = a\sec^2(ax+b) \Rightarrow \int \sec^2(ax+b)\, dx = \frac{1}{a}\tan(ax+b) + c$$

For example

$$\int \cos 5x \, dx = \frac{1}{5} \sin 5x + c$$

$$\int \sin \tfrac{1}{2} x \, dx = -2 \cos \tfrac{1}{2} x + c$$

$$\int 2 \cos\left(3x + \frac{\pi}{2}\right) dx = \frac{2}{3} \sin\left(3x + \frac{\pi}{2}\right) + c$$

$$\int \sec^2(4x + 1) \, dx = \frac{1}{4} \tan(4x + 1) + c$$

$$\int \frac{1}{2\cos^2 x} \, dx = \int \tfrac{1}{2} \sec^2 x \, dx = \tfrac{1}{2} \tan x + c$$

> *Care must be taken when evaluating definite integrals especially when dealing with areas below the x-axis.*

Example 1 Find the area enclosed by the curve $y = \sin x$, the x-axis and the line $x = \frac{\pi}{3}$.

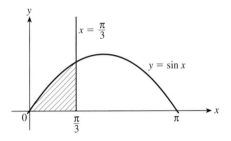

Solution $A = \displaystyle\int_0^{\frac{\pi}{3}} y \, dx$

Remember that x is measured in radians.

$$= \int_0^{\frac{\pi}{3}} \sin x \, dx$$

$$= \Big[-\cos x \Big]_0^{\frac{\pi}{3}}$$

$$= -\cos\frac{\pi}{3} - (-\cos 0)$$

Remember: $\cos 0 = 1$

$$= -0.5 + 1$$

$$= 0.5$$

Example 2 Evaluate $\displaystyle\int_0^{\frac{\pi}{6}} (6 \cos 2x) \, dx$.

$$\int_0^{\frac{\pi}{6}} (6 \cos 2x) \, dx = \Big[3 \sin 2x \Big]_0^{\frac{\pi}{6}}$$

The factor 3 can be taken out before applying the limits.

$$= 3\left(\sin\frac{\pi}{3} - \sin 0 \right)$$

$\sin\dfrac{\pi}{3} = \dfrac{\sqrt{3}}{2}$

$$= \frac{3\sqrt{3}}{2}$$

$\sin 0 = 0$

Example 3   Find the area enclosed by $y = \sin 2x$ and $y = 3 \sin 2x$ between $x = 0$ and $x = \pi$.

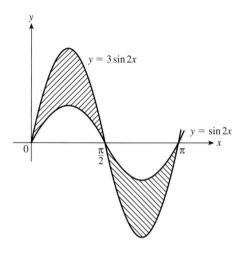

Solution   From the symmetry of the graph, it can be seen that the required area can be found by doubling the value of the area between $x = 0$ and $x = \dfrac{\pi}{2}$.

Remember   The area between the curves $y = \mathrm{f}(x)$ and $y = \mathrm{g}(x)$ is

$$\int_a^b (\mathrm{f}(x) - \mathrm{g}(x)) \, \mathrm{d}x$$

Let A be the area between $x = 0$ and $x = \dfrac{\pi}{2}$.

$$A = \int_0^{\frac{\pi}{2}} (3 \sin 2x - \sin 2x) \, \mathrm{d}x$$

$$= \int_0^{\frac{\pi}{2}} (2 \sin 2x) \, \mathrm{d}x$$

$$= \left[-\cos 2x \right]_0^{\frac{\pi}{2}}$$

$$= (-\cos \pi - (-\cos 0))$$

$$= 1 + 1$$

$$= 2$$

> If possible, simplify the expression before integrating.

> $\displaystyle \int 2 \sin 2x \, \mathrm{d}x = 2 \times \left(-\frac{1}{2} \cos 2x \right)$

> Take care when putting in the limits and remember that $\cos \pi = -1$ and $\cos 0 = 1$.

> Required area $= 2 \times A$

So, the required area is 4 square units.

Example 4 Find the volume generated when the portion of the curve $y = \tan x$ between the x-axis, $x = 0$ and $x = \dfrac{\pi}{4}$ is rotated through 2π radians about the x-axis.

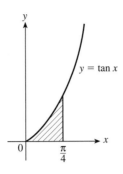

$$V = \int_a^b \pi y^2 \, dx$$

$$= \pi \int_0^{\frac{\pi}{4}} \tan^2 x \, dx$$

$$= \pi \int_0^{\frac{\pi}{4}} (\sec^2 x - 1) \, dx$$

$$= \pi \left[\tan x - x \right]_0^{\frac{\pi}{4}}$$

$$= \pi \left(1 - \frac{\pi}{4} \right)$$

Remember the formula for volume of revolution.

Use $\tan^2 x = \sec^2 x - 1$

Recognise a standard trigonometric integral.

The volume is $\pi \left(1 - \dfrac{\pi}{4} \right)$ cubic units.

Displacement, velocity and acceleration

Velocity v is defined as the rate of change of displacement, s, from a fixed origin, with respect to time, t, i.e. $v = \dfrac{ds}{dt}$. This implies that

$$s = \int v \, dt$$

Acceleration, a, is defined as the rate of change of velocity with respect to time, so

$$a = \frac{dv}{dt}$$

This implies that

$$v = \int a \, dt$$

Note Also $a = \dfrac{d}{dt} \left(\dfrac{ds}{dt} \right) = \dfrac{d^2 s}{dt^2}$.

Example 5   A particle moves in a straight line such that its velocity in $\mathrm{m\,s}^{-1}$, t s after passing a fixed point O is given by $v = 3\cos t - 2\sin t$.

a   Find its displacement from O after $\frac{1}{2}\pi$ s.

b   Find the velocity of the particle at this instant.

Solution   **a** $v = \dfrac{ds}{dt} \Rightarrow s = \displaystyle\int v\,dt$

> Integrate the velocity to obtain the displacement.

$$= \int (3\cos t - 2\sin t)\,dt$$

$$= 3\sin t + 2\cos t + c$$

When $t = 0$, $s = 0$

> To find c, substitute the conditions given in the question.

$$\therefore \qquad 0 = 2 + c$$

$$c = -2$$

so $s = 3\sin t + 2\cos t - 2$

> State the relationship between s and t.

When $t = \frac{1}{2}\pi$, $s = 3 + 0 - 2 = 1$

The displacement after $\frac{1}{2}\pi$ s is 1m.

b   When $t = \frac{1}{2}\pi$,

$v = 3\cos\frac{1}{2}\pi - 2\sin\frac{1}{2}\pi = -2$

so the velocity is $-2\,\mathrm{m\,s}^{-1}$.

Integration of related trigonometric functions

Applying the chain rule in reverse, it is possible to recognise integrals relating to functions of $\sin x$, $\cos x$ and $\sec^2 x$:

$$\frac{d}{dx}\sin(f(x)) = f'(x)\cos(f(x)) \Rightarrow \int f'(x)\cos(f(x))\,dx = \sin(f(x)) + c$$

$$\frac{d}{dx}\cos(f(x)) = -f'(x)\sin(f(x)) \Rightarrow \int f'(x)\sin(f(x))\,dx = -\cos(f(x)) + c$$

$$\frac{d}{dx}\tan(f(x)) = f'(x)\sec^2(f(x)) \Rightarrow \int f'(x)\sec^2(f(x))\,dx = \tan(f(x)) + c$$

Note   Integrals of this type can also be done by using a substitution, though time is saved if the integral is recognised directly.

Example 6 Find these integrals, first by direct recognition, and then by using a suitable substitution.

a $\displaystyle\int 2x\cos(x^2+3)\,\mathrm{d}x$ **b** $\displaystyle\int x^2\sec^2(x^3)\,\mathrm{d}x$

Solution **a** *By recognition*

$$\int 2x\cos(x^2+3)\,\mathrm{d}x = \sin(x^2+3)+c$$

Recognise directly, with $f(x)=x^2+3$, $f'(x)=2x$

Using a substitution

$$\int 2x\cos(x^2+3)\,\mathrm{d}x$$

$$=\int 2x\cos(x^2+3)\frac{\mathrm{d}x}{\mathrm{d}u}\,\mathrm{d}u$$

$$=\int 2\cos(x^2+3)\times x\frac{\mathrm{d}x}{\mathrm{d}u}\,\mathrm{d}u$$

$$=\int 2\cos u \times \tfrac{1}{2}\,\mathrm{d}u$$

$$=\sin u + c$$

$$=\sin(x^2+3)+c$$

Let $u=x^2+3$

$$\frac{\mathrm{d}u}{\mathrm{d}x}=2x$$

$$\therefore\ \frac{\mathrm{d}x}{\mathrm{d}u}=\frac{1}{2x}$$

$$x\frac{\mathrm{d}x}{\mathrm{d}u}=\tfrac{1}{2}$$

Use $\dfrac{\mathrm{d}x}{\mathrm{d}u}=\dfrac{1}{\frac{\mathrm{d}u}{\mathrm{d}x}}$

b *By recognition*

$$\int x^2\sec^2(x^3)\,\mathrm{d}x = \frac{1}{3}\tan(x^3)+c$$

$f(x)=x^3$, $f'(x)=3x^2$

Using a substitution

$$\int x^2\sec^2(x^3)\,\mathrm{d}x$$

$$=\int x^2\sec^2(x^3)\frac{\mathrm{d}x}{\mathrm{d}u}\,\mathrm{d}u$$

$$=\int \sec^2(x^3)\times x^2\frac{\mathrm{d}x}{\mathrm{d}u}\,\mathrm{d}u$$

$$=\int \sec^2 u \times \tfrac{1}{3}\,\mathrm{d}u$$

$$=\tfrac{1}{3}\tan u + c$$

$$=\tfrac{1}{3}\tan(x^3)+c$$

Let $u=x^3$

$$\frac{\mathrm{d}u}{\mathrm{d}x}=3x^2$$

$$\therefore\ \frac{\mathrm{d}x}{\mathrm{d}u}=\frac{1}{3x^2}$$

$$x^2\frac{\mathrm{d}x}{\mathrm{d}u}=\tfrac{1}{3}$$

155

Example 7 Find these integrals, first by direct recognition, and then using a suitable substitution.

    a $\displaystyle\int \cos x e^{\sin x}\, dx$ **b** $\displaystyle\int \frac{\sin x}{1-\cos x}\, dx$

Solution   **a** *By recognition*

$$\int \cos x e^{\sin x}\, dx = e^{\sin x} + c$$

> *Remember:* $\displaystyle\int f'(x) e^{f(x)}\, dx = e^{f(x)} + c$
> with $f(x) = \sin x$ and $f'(x) = \cos x$

Using a substitution

$$\int \cos x e^{\sin x}\, dx$$

$$= \int \cos x e^{\sin x}\frac{dx}{du}\, du$$

$$= \int e^{\sin x}\cos x \frac{dx}{du}\, du$$

$$= \int e^{u}\, du$$

$$= e^{u} + c$$

$$= e^{\sin x} + c$$

Let $u = \sin x$

$$\frac{du}{dx} = \cos x$$

$$\therefore \quad \frac{dx}{du} = \frac{1}{\cos x}$$

$$\cos x \frac{dx}{du} = 1$$

b *By recognition*

$$\int \frac{\sin x}{1-\cos x}\, dx = \ln(1 - \cos x) + c$$

> *Remember:* $\displaystyle\int \frac{f'(x)}{f(x)}\, dx = \ln(f(x)) + c$
> with $f(x) = 1 - \cos x$ and $f'(x) = \sin x$

Using a substitution

$$\int \frac{\sin x}{1-\cos x}\, dx$$

$$= \int \frac{\sin x}{1-\cos x}\frac{dx}{du}\, du$$

$$= \int \frac{1}{1-\cos x} \times \sin x \frac{dx}{du}\, du$$

$$= \int \frac{1}{u}\, du$$

$$= \ln u + c$$

$$= \ln(1 - \cos x) + c$$

Let $u = 1 - \cos x$

$$\frac{du}{dx} = \sin x$$

$$\therefore \quad \frac{dx}{du} = \frac{1}{\sin x}$$

$$\sin x \frac{dx}{du} = 1$$

> $\displaystyle\int \frac{1}{u}\, du = \ln u + c$

Recognition of other standard trigonometric integrals

Using the derivatives of $\sec x$, $\operatorname{cosec} x$ and $\cot x$ (Section 4.4) and applying the process in reverse:

$$\frac{\mathrm{d}}{\mathrm{d}x}(\sec x) = \sec x \tan x \qquad \Rightarrow \int \sec x \tan x \,\mathrm{d}x = \sec x + c$$

$$\frac{\mathrm{d}}{\mathrm{d}x}(\operatorname{cosec} x) = -\operatorname{cosec} x \cot x \Rightarrow \int \operatorname{cosec} x \cot x \,\mathrm{d}x = -\operatorname{cosec} x + c$$

$$\frac{\mathrm{d}}{\mathrm{d}x}(\cot x) = -\operatorname{cosec}^2 x \qquad \Rightarrow \int \operatorname{cosec}^2 x \,\mathrm{d}x = -\cot x + c$$

Sometimes these standard integrals appear in disguise.

Example 8   **a** $\displaystyle\int \frac{\cos 2x}{\sin^2 2x}\,\mathrm{d}x = \int \frac{1}{\sin 2x} \times \frac{\cos 2x}{\sin 2x}\,\mathrm{d}x$

> This integrand can be written as one of the standard integrals that can be recognised.

$$= \int \operatorname{cosec} 2x \cot 2x \,\mathrm{d}x$$

$$= -\tfrac{1}{2}\operatorname{cosec} 2x + c$$

> An adjustment factor of $\tfrac{1}{2}$ is needed.

b $\displaystyle\int \cot^2 x \,\mathrm{d}x = \int (\operatorname{cosec}^2 x - 1)\,\mathrm{d}x$

$$= -\cot x - x + c$$

> Use the trigonometric identity $1 + \cot^2 x = \operatorname{cosec}^2 x$, and the integral of $\operatorname{cosec}^2 x$.

Integrals of $\tan x$ and $\cot x$

$$\tan x = \frac{\sin x}{\cos x}$$

$$\int \tan x \,\mathrm{d}x = \int \frac{\sin x}{\cos x}\,\mathrm{d}x$$

> Use $\displaystyle\int \frac{k\mathrm{f}'(x)}{\mathrm{f}(x)}\,\mathrm{d}x = k\ln \mathrm{f}(x) + c$ with $\mathrm{f}(x) = \cos x$, $\mathrm{f}'(x) = -\sin x$ and $k = -1$

$$= -\ln(\cos x) + c$$

$$= \ln((\cos x)^{-1}) + c$$

> $-\ln a = \ln(a^{-1})$

$$= \ln\left(\frac{1}{\cos x}\right) + c$$

> $(\cos x)^{-1} = \dfrac{1}{\cos x}$

$$= \ln(\sec x) + c$$

➤ $$\boxed{\int \tan x \,\mathrm{d}x = -\ln(\cos x) + c = \ln(\sec x) + c}$$

In a similar way, it can be shown that

➤ $$\boxed{\int \cot x \,\mathrm{d}x = \int \frac{\cos x}{\sin x}\,\mathrm{d}x = \ln(\sin x) + c}$$

Example 9 $\int (\tan 3x + \cot 5x)\,dx = \frac{1}{3}\ln(\sec 3x) + \frac{1}{5}\ln(\sin 5x) + c$

Exercise 6A

1 Find these integrals.

a $\displaystyle\int \sin 3x\,dx$

b $\displaystyle\int \cos 3x\,dx$

c $\displaystyle\int 2\sin 4x\,dx$

d $\displaystyle\int 2\cos 2x\,dx$

e $\displaystyle\int -\tfrac{1}{2}\sin 6x\,dx$

f $\displaystyle\int 6\cos 4x\,dx$

g $\displaystyle\int \sin(2x+1)\,dx$

h $\displaystyle\int 3\cos(2x-1)\,dx$

i $\displaystyle\int \tfrac{2}{3}\sin(\tfrac{1}{2}x)\,dx$

2 Find these integrals.

a $\displaystyle\int \sec^2 2x\,dx$

b $\displaystyle\int 3\sec^2\left(x-\frac{\pi}{4}\right)dx$

c $\displaystyle\int (1+\tan^2 x)\,dx$

d $\displaystyle\int \tan^2(\tfrac{1}{2}x)\,dx$

e $\displaystyle\int \frac{1}{\cos^2 4x}\,dx$

f $\displaystyle\int \frac{1}{(1-\sin^2 x)}\,dx$

3 Evaluate

a $\displaystyle\int_{-\frac{\pi}{4}}^{\frac{\pi}{4}} \cos 2x\,dx$

b $\displaystyle\int_{0}^{\frac{\pi}{3}} 2\sin 3x\,dx$

c $\displaystyle\int_{-\frac{\pi}{3}}^{\frac{\pi}{6}} \sec^2 x\,dx$

4 a Evaluate $\displaystyle\int_{0}^{2\pi} \sin x\,dx$.

b Explain the answer with the aid of a diagram.

5 Find the area enclosed by the curve $y = \sin 2x$, the x-axis and $x = \dfrac{\pi}{4}$.

6 a Sketch the curve $y = 1 + \cos x$ from $x = -\pi$ to $x = \pi$.

b Find the area enclosed by the curve and the x-axis between these limits.

7 a Find the maximum value of $y = 2\sin x - x$ which is given by a value of x between 0 and $\frac{1}{2}\pi$.

b Sketch the graph of y for $0 \leqslant x \leqslant \frac{1}{2}\pi$.

c Find the area enclosed by the curve, the x-axis and the line $x = \frac{1}{2}\pi$.

8 Find the volume of the solid generated when the area enclosed by the curve $y = \sec x$, the lines $y = 2$, $x = -\dfrac{\pi}{3}$ and $x = \dfrac{\pi}{3}$, is rotated through 2π radians about the x-axis.

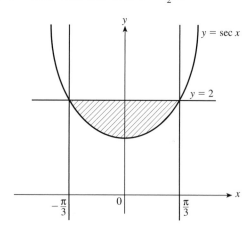

9 This diagram shows part of the curves of $y = \sqrt{\cos 2x}$ and $y = \sqrt{\cos x}$.

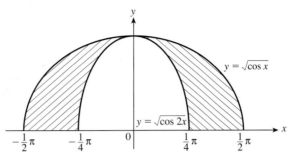

a The area bounded by $y = \sqrt{\cos x}$ and the x-axis between $x = -\frac{1}{2}\pi$ and $x = \frac{1}{2}\pi$ is rotated through 2π radians about the x-axis and the volume of the solid generated is V_1. Find V_1.

b The area bounded by $y = \sqrt{\cos 2x}$ and the x-axis between $x = -\frac{1}{4}\pi$ and $x = \frac{1}{4}\pi$ is rotated through 2π radians about the x-axis and the volume of the solid generated is V_2. Find V_2.

c Hence find the volume generated when the shaded area is rotated through $360°$ about the x-axis.

10 This diagram shows part of the graphs of $y = \frac{1}{2}\sec^2 x$ and $y = \tan^2 x$. The curves intersect at $\left(\frac{\pi}{4}, 1\right)$ and $\left(-\frac{\pi}{4}, 1\right)$.

Find the area enclosed by the two curves for values of x between $-\frac{\pi}{4}$ and $\frac{\pi}{4}$.

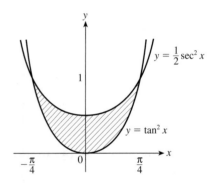

11 This diagram shows the curves $y = \sin 2x$ and $y = 3\sin x$, for values of x such that $0 \leqslant x \leqslant 2\pi$. Find the shaded area.

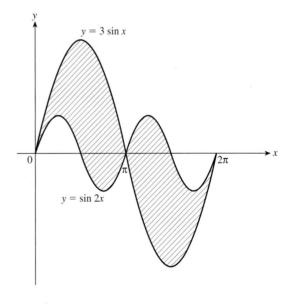

12 Find these integrals.

a $\displaystyle\int \sec 3x \tan 3x \, dx$

b $\displaystyle\int -\cot\left(\tfrac{1}{2}x\right) \operatorname{cosec}\left(\tfrac{1}{2}x\right) dx$

c $\displaystyle\int \frac{2}{\sin^2 3x} \, dx$

d $\displaystyle\int \frac{\operatorname{cosec} x}{\tan x} \, dx$

13 Find these integrals.

a $\displaystyle\int \frac{1}{\cos^2 x - 1} \, dx$

b $\displaystyle\int \frac{\sin^2 x}{\cos^2 x} \, dx$

c $\displaystyle\int \frac{\cos^2 x}{\sin^2 x} \, dx$

d $\displaystyle\int \frac{1}{\sin^2 2x} \, dx$

14 Find these integrals, either by direct recognition or by using the given substitution.

a $\displaystyle\int 4x \cos\left(x^2\right) dx \qquad u = x^2$

b $\displaystyle\int \left(\frac{1}{\sqrt{x}} \cos \sqrt{x}\right) dx \qquad u = \sqrt{x}$

c $\displaystyle\int \frac{e^{\tan x}}{\cos^2 x} \, dx \qquad u = \tan x$

d $\displaystyle\int \frac{3\cos x}{2 + \sin x} \, dx \qquad u = 2 + \sin x$

e $\displaystyle\int \cos 2x \, e^{-\sin 2x} \, dx \qquad u = \sin 2x$

15 Find these integrals.

a $\displaystyle\int \tan 3x \, dx$

b $\displaystyle\int \cot 3x \, dx$

c $\displaystyle\int \cos x \operatorname{cosec} x \, dx$

d $\displaystyle\int \sin 2x \sec 2x \, dx$

e $\displaystyle\int \frac{\cos 4x}{\sin 4x} \, dx$

16 a Sketch the curve $y = \tan x$ between $x = 0$ and $x = \dfrac{\pi}{2}$.

b Show that the area enclosed by the curve $y = \tan x$, the axes and the line $x = \dfrac{\pi}{3}$ is $\ln 2$.

c Find the volume of the solid generated when the area is rotated through 2π radians about the x-axis.

6.2 Integration by parts

When the integrand is a product of two functions, i.e. in the form

$$\int f(x) \times g(x)\, dx$$

there are various possible lines of approach.

It may be possible to integrate by simplifying the integrand, for example

$$\int 4x^2(3-2x)\, dx = \int (12x^2 - 8x^3)\, dx = 4x^3 - 2x^4 + c$$

The integral may be one that can be done by recognition, or using a substitution for example

$$\int 7x^2(x^3+1)^5\, dx = \frac{7}{18}(x^3+1)^6 + c \qquad \text{(Recognition – page 127)}$$

$$\int xe^{3x^2}\, dx = \frac{1}{6}e^{3x^2} + c \qquad \text{(Recognition – page 129)}$$

$$\int x\sqrt{(3x-1)}\, dx = \frac{2}{135}(9x+2)(3x-1)^{\frac{3}{2}} + c \qquad \text{(Substitution – page 133)}$$

$$\int 2x\cos(x^2+3)\, dx = \sin(x^2+3) + c \qquad \text{(Recognition – page 155)}$$

$$\int \cos x\, e^{\sin x}\, dx = e^{\sin x} + c \qquad \text{(Recognition – page 156)}$$

Remember Integrals that can be done by recognition can also be done by substitution.
There are some integrals, however, that *cannot* be done by one of the above strategies, e.g. integrals such as

$$\int x\cos x\, dx \qquad \text{and} \qquad \int x^2 e^{4x}\, dx$$

The technique of integration by parts may often be successfully applied. The method is based on the idea of differentiating a product.

If u and v are functions of x,

$$\frac{d}{dx}(uv) = u\frac{dv}{dx} + v\frac{du}{dx}$$

Integrating each side with respect to x

$$uv = \int u\frac{dv}{dx}\, dx + \int v\frac{du}{dx}\, dx$$

Rearranging:

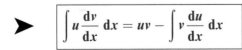

$$\boxed{\int u\frac{dv}{dx}\, dx = uv - \int v\frac{du}{dx}\, dx}$$

Note The method of integration by parts can only be attempted if the function chosen as $\frac{dv}{dx}$ can be integrated.

Example 10 Use integration by parts to find $\int 3x\cos x\,dx$.

Define u and $\dfrac{dv}{dx}$.

Consider $\quad\displaystyle\int \underset{u}{\overset{}{3x}}\,\underset{\tfrac{dv}{dx}}{\overset{}{\cos x}}\,dx$

$u = 3x \Rightarrow \dfrac{du}{dx} = 3$

Differentiate u to get $\dfrac{du}{dx}$.

$\dfrac{dv}{dx} = \cos x \Rightarrow v = \sin x$

Integrate $\dfrac{dv}{dx}$ to get v.

$$\int u\frac{dv}{dx}\,dx = uv - \int v\frac{du}{dx}\,dx$$

Use integration by parts formula.

$$\int 3x\cos x\,dx = 3x \times \sin x - \int (\sin x \times 3)\,dx$$

$$= 3x\sin x - \int 3\sin x\,dx$$

It is possible to find $\int 3\sin x\,dx$.

$$= 3x\sin x + 3\cos x + c$$

Choice of u and $\dfrac{dv}{dx}$

When integrating by parts, the aim is to obtain the integral

$$\int v\frac{du}{dx}\,dx$$

that is easier to tackle than the original integral. With some integrals, this is not possible; with others, the choice of functions for u and $\dfrac{dv}{dx}$ is crucial.

Consider what happens in Example 10 when u and $\dfrac{dv}{dx}$ are chosen the other way round.

Consider $\quad\displaystyle\int \underset{u}{\overset{}{\cos x}} \times \underset{\tfrac{dv}{dx}}{\overset{}{3x}}\,dx$

$u = \cos x \Rightarrow \dfrac{du}{dx} = -\sin x$

$\dfrac{dv}{dx} = 3x \Rightarrow v = \tfrac{3}{2}x^2$

$$\int u\frac{dv}{dx}\,dx = uv - \int v\frac{du}{dx}\,dx$$

$$\int \cos x \times 3x\,dx = \cos x \times \tfrac{3}{2}x^2 - \int \tfrac{3}{2}x^2(-\sin x)\,dx$$

$$= \tfrac{3}{2}x^2\cos x + \int \tfrac{3}{2}x^2\sin x\,dx$$

This integral is more difficult to integrate than the original one.

So the choice of $u = \cos x$ and $\dfrac{dv}{dx} = 3x$ is not helpful.

If one of the functions is a polynomial in x (e.g. x, x^2, x^3, etc.), then it is often useful to substitute u for the polynomial, since the power of x will reduce when $\dfrac{du}{dx}$ is found.

If the polynomial is taken as $\dfrac{dv}{dx}$, then the power of x will increase when it is integrated to find v, and this can make the related integration difficult, if not impossible, to perform.

There are, however, exceptions to this strategy, as shown in Example 11.

Example 11   Find $\displaystyle\int x^3 \ln x\,dx$.

> The integrand is the product of two functions, x^3 and $\ln x$. If $\ln x$ is chosen as $\dfrac{dv}{dx}$, this leads to $v = \displaystyle\int \ln x\,dx$ which presents difficulties. So instead try $u = \ln x$ and $\dfrac{dv}{dx} = x^3$.

Consider $\displaystyle\int \overset{\overset{u}{\downarrow}}{\ln x} \times \overset{\overset{\frac{dv}{dx}}{\downarrow}}{x^3}\,dx$

$u = \ln x \Rightarrow \dfrac{du}{dx} = \dfrac{1}{x}$

$\dfrac{dv}{dx} = x^3 \Rightarrow v = \dfrac{1}{4}x^4$

$\displaystyle\int u\dfrac{dv}{dx}\,dx = uv - \int v\dfrac{du}{dx}\,dx$

$\displaystyle\int \ln x \times x^3\,dx = \ln x \times \dfrac{1}{4}x^4 - \int \dfrac{1}{4}x^4 \times \dfrac{1}{x}\,dx$

> Simplify the second integral. Take out the factor of $\frac{1}{4}$.

$= \dfrac{1}{4}x^4 \ln x - \dfrac{1}{4}\displaystyle\int x^3\,dx$

> The function is now easy to integrate.

$= \dfrac{1}{4}x^4 \ln x - \dfrac{1}{16}x^4 + c$

$= \dfrac{1}{16}x^4(4\ln x - 1) + c$

Special integral: $\displaystyle\int \ln x\,dx$

The method used in Example 11 can be adapted to find $\displaystyle\int \ln x\,dx$.

> Write $\displaystyle\int \ln x\,dx$ as $\displaystyle\int \ln x \times 1\,dx$ and take $u = \ln x$ and $\dfrac{dv}{dx} = 1$.

Example 12 $\displaystyle\int \ln x\,dx = \int \ln x \times 1\,dx$

$u = \ln x \Rightarrow \dfrac{du}{dx} = \dfrac{1}{x}$

$= \ln x \times x - \displaystyle\int x \times \dfrac{1}{x}\,dx$

$\dfrac{dv}{dx} = 1 \Rightarrow v = x$

$= x\ln x - \displaystyle\int 1\,dx$

$= x\ln x - x + c$

Example 13

To find some integrals, it may be necessary to apply the integration by parts procedure more than once.

Find $\int x^2 e^{4x}\,dx$.

Solution   Consider $\int \underset{\underset{u}{}}{x^2} \underset{\underset{\frac{dv}{dx}}{}}{e^{4x}}\,dx$

$u = x^2 \Rightarrow \dfrac{du}{dx} = 2x$

$\dfrac{dv}{dx} = e^{4x} \Rightarrow v = \tfrac{1}{4}e^{4x}$

> Define u and $\dfrac{dv}{dx}$.
> Take as u the polynomial in x.

$\int u\dfrac{dv}{dx}\,dx = uv - \int v\dfrac{du}{dx}\,dx$

$\int x^2 e^{4x}\,dx = x^2\left(\tfrac{1}{4}e^{4x}\right) - \int \tfrac{1}{4}e^{4x} \times 2x\,dx$

$\qquad\qquad = \tfrac{1}{4}x^2 e^{4x} - \tfrac{1}{2}\int xe^{4x}\,dx \qquad ①$

> Take out the factor of $\tfrac{1}{2}$ to simplify the integral.

> Use integration by parts again to find $\int xe^{4x}\,dx$. Redefine u and $\dfrac{dv}{dx}$.

Consider $\int \underset{\underset{u}{}}{x} \underset{\underset{\frac{dv}{dx}}{}}{e^{4x}}\,dx$

$u = x \Rightarrow \dfrac{du}{dx} = 1$

$\dfrac{dv}{dx} = e^{4x} \Rightarrow v = \tfrac{1}{4}e^{4x}$

$\int u\dfrac{dv}{dx}\,dx = uv - \int v\dfrac{du}{dx}\,dx$

$\int xe^{4x}\,dx = x\left(\tfrac{1}{4}e^{4x}\right) - \int \tfrac{1}{4}xe^{4x} \times 1\,dx$

$\qquad\qquad = \tfrac{1}{4}xe^{4x} - \tfrac{1}{4}\int e^{4x}\,dx$

$\qquad\qquad = \tfrac{1}{4}xe^{4x} - \tfrac{1}{16}e^{4x}$

> Substitute this into original working ①.

So $\int x^2 e^{4x}\,dx = \tfrac{1}{4}x^2 e^{4x} - \tfrac{1}{2}\left(\tfrac{1}{4}xe^{4x} - \tfrac{1}{16}e^{4x}\right) + c$

$\qquad\qquad = e^{4x}\left(\tfrac{1}{4}x^2 - \tfrac{1}{8}x + \tfrac{1}{32}\right) + c$

$\qquad\qquad = \tfrac{1}{32}e^{4x}(8x^2 - 4x + 1) + c$

> Remember to include the arbitrary constant.

Definite integration using integration by parts

$$\int_a^b u\dfrac{dv}{dx}\,dx = \left[uv\right]_a^b - \int_a^b v\dfrac{du}{dx}\,dx$$

Example 14   This diagram shows the graph of $y = x \sin x$ between $x = -3\pi$ and $x = 3\pi$.

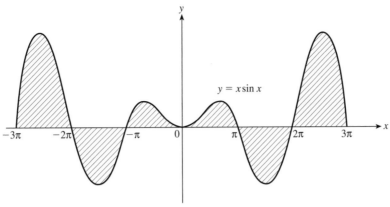

$y = x \sin x$

Note: The curve is symmetrical about the y-axis.

Find the area enclosed by the curve and the x-axis between these limits.

Solution   Since some of the curve lies below the x-axis and some above, the area must be found in stages:

- From $x = 0$ to $x = \pi$
- From $x = \pi$ to $x = 2\pi$
- From $x = 2\pi$ to $x = 3\pi$

Find the sum of these areas and double it to find the required area.

The area from $x = 0$ to $x = \pi$ is given by

$$\int_0^\pi y\,\mathrm{d}x = \int_0^\pi x \sin x\,\mathrm{d}x \qquad\qquad u = x \Rightarrow \frac{\mathrm{d}u}{\mathrm{d}x} = 1$$

Define u and $\dfrac{\mathrm{d}v}{\mathrm{d}x}$.

$$\frac{\mathrm{d}v}{\mathrm{d}x} = \sin x \Rightarrow v = -\cos x$$

$$\int_0^\pi x \sin x\,\mathrm{d}x = \Big[x(-\cos x)\Big]_0^\pi - \int_0^\pi (-\cos x)\,1\,\mathrm{d}x$$

Work out the limits as soon as possible.

$$= \pi(-\cos \pi) - 0 + \int_0^\pi \cos x\,\mathrm{d}x$$

$$= \pi + \Big[\sin x\Big]_0^\pi$$

$$= \pi + \sin \pi - \sin 0$$

$$= \pi$$

So, the area from $x = 0$ to $x = \pi$ is π.

The area from $x = \pi$ to $x = 2\pi$ is given by

$$\int_\pi^{2\pi} x \sin x\,\mathrm{d}x$$

$$= \Big[x(-\cos x)\Big]_\pi^{2\pi} + \Big[\sin x\Big]_\pi^{2\pi}$$

Work out these values very carefully, remembering that $\cos 2\pi = 1$, $\cos \pi = -1$ and $\sin 2\pi = 0$.

$$= 2\pi(-\cos 2\pi) - \pi(-\cos \pi) + \sin 2\pi - \sin \pi$$

$$= -2\pi - \pi + 0$$

$$= -3\pi$$

So, the area from $x = \pi$ to $x = 2\pi$ is 3π.

This integral gives a negative value because the curve is below the x-axis.

The area from $x = 2\pi$ to $x = 3\pi$ is given by

$$\int_{2\pi}^{3\pi} x \sin x \, dx$$

$$= \Big[x(-\cos x) \Big]_{2\pi}^{3\pi} + \Big[\sin x \Big]_{2\pi}^{3\pi}$$

$$= 3\pi(-\cos 3\pi) - 2\pi(-\cos 2\pi) + 0 \qquad \boxed{\cos 3\pi = -1 \text{ and } \sin 3\pi = 0.}$$

$$= 3\pi + 2\pi$$

$$= 5\pi$$

So, the area from $x = 2\pi$ to $x = 3\pi$ is 5π. $\boxed{\text{There is a pattern emerging. Does it continue?}}$

The area from $x = -3\pi$ to $x = 3\pi$ is

$$2(\pi + 3\pi + 5\pi) = 18\pi$$

Therefore, the required area is 18π square units.

Extension

Integrals of the type $\int e^{ax} \sin bx \, dx$ and $\int e^{ax} \cos bx \, dx$

A special technique is used to find these integrals.

Example 15

$$\int e^x \sin x \, dx$$

$\boxed{\text{For } \int e^x \sin x \, dx, \text{ let } u = e^x \text{ and } \dfrac{dv}{dx} = \sin x}$

$$= e^x(-\cos x) - \int (-\cos x) e^x \, dx$$

$$= -e^x \cos x + \int e^x \cos x \, dx$$

$\boxed{\text{For } \int e^x \cos x \, dx, \text{ let } u = e^x \text{ and } \dfrac{dv}{dx} = \cos x}$

$$= -e^x \cos x + \left(e^x \sin x - \int e^x \sin x \, dx \right)$$

$$= e^x(\sin x - \cos x) - \int e^x \sin x \, dx$$

$$\therefore \quad 2\int e^x \sin x \, dx = e^x(\sin x - \cos x)$$

$\boxed{\textit{Note: } \text{This integral is the same as the original one and so the two integrals can be collected together on the LHS.}}$

$$\int e^x \sin x \, dx = \tfrac{1}{2} e^x(\sin x - \cos x) + c$$

$\boxed{\text{Remember to include the arbitrary integration constant.}}$

Note   It is interesting to check this result by letting $u = \sin x$ and $\dfrac{dv}{dx} = e^x$ in the initial stage of working.

Exercise 6B

1 Find these integrals.

a $\displaystyle\int x\cos x\,\mathrm{d}x$ **b** $\displaystyle\int x\sin 2x\,\mathrm{d}x$ **c** $\displaystyle\int 3x\sin\left(x+\frac{\pi}{3}\right)\mathrm{d}x$

d $\displaystyle\int x\mathrm{e}^x\,\mathrm{d}x$ **e** $\displaystyle\int x\mathrm{e}^{-3x}\,\mathrm{d}x$ **f** $\displaystyle\int 2x\mathrm{e}^{3x+1}\,\mathrm{d}x$

2 Find $\displaystyle\int x(1+x)^6\,\mathrm{d}x$

    a using integration by parts   **b** using a suitable substitution.

3 Find $\displaystyle\int \frac{x}{2\mathrm{e}^x}\,\mathrm{d}x$.

4 Evaluate

    a $\displaystyle\int_0^{\frac{\pi}{3}} x\sin 3x\,\mathrm{d}x$ **b** $\displaystyle\int_0^1 (2x+1)\,\mathrm{e}^x\,\mathrm{d}x$

5 This diagram shows the curve $y = x\cos x$ for x between -2π and 2π.

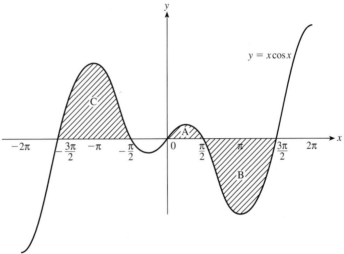

Find areas A, B and C.

6 Find

    a $\displaystyle\int x\sec^2 x\,\mathrm{d}x$ **b** $\displaystyle\int x\tan^2 x\,\mathrm{d}x$

    c $\displaystyle\int x\operatorname{cosec}^2 x\,\mathrm{d}x$ **d** $\displaystyle\int x\cot^2 x\,\mathrm{d}x$

7 Find these integrals.

    a $\displaystyle\int x^2\cos x\,\mathrm{d}x$ **b** $\displaystyle\int x^2\sin 2x\,\mathrm{d}x$

    c $\displaystyle\int x^2\mathrm{e}^{-3x}\,\mathrm{d}x$ **d** $\displaystyle\int x^3\mathrm{e}^x\,\mathrm{d}x$

8 a The curve $y = 2xe^x$ has a minimum point at P.

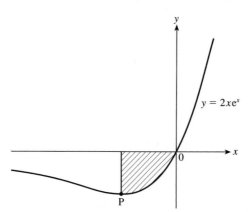

Find the coordinates of P.

b Find the area enclosed by the curve $y = 2xe^x$, the x-axis and the vertical line through P, shown shaded in the diagram.

c This area is rotated through $360°$ about the x-axis. Find the volume of the solid generated.

9 Find

a $\int x \ln x \, dx$

b $\int x^2 \ln x \, dx$

c $\int x^4 \ln x \, dx$

d $\int \dfrac{\ln x}{x^3} \, dx$

10 Find

a $\int \ln 2x \, dx$

b $\int \ln (x^2) \, dx$

c $\int \ln\left(\sqrt{x-1}\right) dx$

11 a $\int e^x \cos x \, dx$

b $\int e^{2x} \sin 3x \, dx$

c $\int e^{3x} \cos 2x \, dx$

d $\int e^{4x} \sin 3x \, dx$

12 a Find

$$\frac{d}{dx}(e^{x^2})$$

and deduce

$$\int x^3 e^{x^2} \, dx$$

b Find $\int x^5 e^{x^3} \, dx$. **c** Find $\int xe^{-x^2} \, dx$. **d** Find $\int x^3 e^{-x^2} \, dx$.

13 **a** Find $\dfrac{\mathrm{d}}{\mathrm{d}x}(a^x)$ and $\displaystyle\int xa^x\,\mathrm{d}x$.

b Find $\dfrac{\mathrm{d}}{\mathrm{d}x}(\log_a x)$ and $\displaystyle\int x(\log_a x)\,\mathrm{d}x$.

c Find $\displaystyle\int (\log_a x)\,\mathrm{d}x$.

6.3 Integrals of the type $\displaystyle\int \dfrac{1}{\sqrt{a^2-x^2}}\,dx$ and $\displaystyle\int \dfrac{1}{a^2+x^2}\,dx$

Earlier in the course you were introduced to the inverse trigonometrical functions. This section shows how they arise in the solution of certain integrals by making use of the identities $1-\sin^2\theta=\cos^2\theta$ and $1+\tan^2\theta=\sec^2\theta$.

Example 16 Find $\displaystyle\int \dfrac{1}{\sqrt{1-x^2}}\,\mathrm{d}x$.

$\displaystyle\int \dfrac{1}{\sqrt{1-x^2}}\,\mathrm{d}x$ 　　　　Let $x=\sin u$ 　　　 Define the substitution.

$=\displaystyle\int \dfrac{1}{\sqrt{1-x^2}}\dfrac{\mathrm{d}x}{\mathrm{d}u}\,\mathrm{d}u$ 　　 $\dfrac{\mathrm{d}x}{\mathrm{d}u}=\cos u$

$=\displaystyle\int \dfrac{1}{\sqrt{1-\sin^2 u}}\cos u\,\mathrm{d}u$ 　　 $\sqrt{1-x^2}=\sqrt{1-\sin^2 u}=\sqrt{\cos^2 u}=\cos u$

$=\displaystyle\int \dfrac{1}{\cos u}\cos u\,\mathrm{d}u$

$=\displaystyle\int 1\,\mathrm{d}u$

$=u+c$ 　　　　 $x=\sin u\Rightarrow u=\sin^{-1}x$

$=\sin^{-1}x+c$

Example 17   Find $\displaystyle\int \frac{1}{4+x^2}\,dx$.

$$\int \frac{1}{4+x^2}\,dx \qquad\qquad \text{Let } x = 2\tan u \qquad\qquad \boxed{\text{Define the substitution.}}$$

$$= \int \frac{1}{4+x^2}\frac{dx}{du}\,du \qquad\qquad \frac{dx}{du} = 2\sec^2 u$$

$$= \int \frac{1}{4+4\tan^2 u}\,2\sec^2 u\,du \qquad\qquad \boxed{4+x^2 = 4+4\tan^2 u = 4(1+\tan^2 u) = 4\sec^2 u}$$

$$= \int \frac{1}{4\sec^2 u}\,2\sec^2 u\,du$$

$$= \int \frac{1}{2}\,du$$

$$= \frac{1}{2}u + c \qquad\qquad \boxed{x = 2\tan u \Rightarrow \frac{x}{2} = \tan u \Rightarrow u = \tan^{-1}\frac{x}{2}}$$

$$= \frac{1}{2}\tan^{-1}\frac{x}{2} + c$$

In general

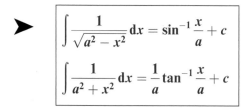

$$\int \frac{1}{\sqrt{a^2-x^2}}\,dx = \sin^{-1}\frac{x}{a} + c$$

$$\int \frac{1}{a^2+x^2}\,dx = \frac{1}{a}\tan^{-1}\frac{x}{a} + c$$

Both of these results are included in the formulae booklet for the course and can be quoted, as the following two examples show.

Example 18   Find $\displaystyle\int \frac{1}{8+x^2}\,dx$.

$$\int \frac{1}{8+x^2}\,dx = \frac{1}{\sqrt{8}}\tan^{-1}\frac{x}{\sqrt{8}} + c \qquad \text{Use the standard result with } a = \sqrt{8}.$$

$$= \frac{\sqrt{2}}{4}\tan^{-1}\frac{\sqrt{2}x}{4} + c \qquad \frac{1}{\sqrt{8}} = \frac{1}{2\sqrt{2}} = \frac{\sqrt{2}}{4}$$

Example 19   Find $\displaystyle\int_0^{\frac{3\sqrt{3}}{2}} \frac{1}{\sqrt{9-x^2}}\,dx$.

$$\int_0^{\frac{3\sqrt{3}}{2}} \frac{1}{\sqrt{9-x^2}}\,dx = \left[\sin^{-1}\frac{x}{3}\right]_0^{\frac{3\sqrt{3}}{2}}$$

> Use the standard result with $a = 3$.

$$= \sin^{-1}\frac{\sqrt{3}}{2} - \sin^{-1}0$$

> Using the special triangles, find the angle u in the PV range, $-\dfrac{\pi}{2} \leqslant u \leqslant \dfrac{\pi}{2}$, whose sine is $\dfrac{\sqrt{3}}{2}$.

$$= \frac{\pi}{3} - 0$$

$$= \frac{\pi}{3}$$

Exercise 6C

1 Find

a $\displaystyle\int \frac{1}{\sqrt{25-x^2}}\,dx$

b $\displaystyle\int \frac{1}{\sqrt{3-x^2}}\,dx$

c $\displaystyle\int \frac{1}{\sqrt{64-x^2}}\,dx$

d $\displaystyle\int \frac{1}{\sqrt{27-x^2}}\,dx$

2 Find

a $\displaystyle\int \frac{1}{25+x^2}\,dx$

b $\displaystyle\int \frac{1}{7+x^2}\,dx$

c $\displaystyle\int \frac{1}{32+x^2}\,dx$

d $\displaystyle\int \frac{1}{36+x^2}\,dx$

3 Evaluate the following, leaving π in your answers.

a $\displaystyle\int_1^{\sqrt{3}} \frac{1}{1+x^2}\,dx$

b $\displaystyle\int_0^2 \frac{1}{\sqrt{16-x^2}}\,dx$

c $\displaystyle\int_{-\sqrt{3}}^{\sqrt{3}} \frac{1}{\sqrt{4-x^2}}\,dx$

d $\displaystyle\int_0^{\sqrt{3}} \frac{1}{9+x^2}\,dx$

e $\displaystyle\int_0^{\sqrt{5}} \frac{1}{5+x^2}\,dx$

f $\displaystyle\int_{\sqrt{3}}^{2\sqrt{3}} \frac{1}{\sqrt{12-x^2}}\,dx$

4 a Find $\displaystyle\int \frac{1}{\sqrt{36-x^2}}\,dx$ using the substitutions

　　i $x = 6\sin u$　　　　**ii** $x = 6\cos u$

b Find $\displaystyle\int_3^6 \frac{1}{\sqrt{36-x^2}}\,dx$ using the substitutions

　　i $x = 6\sin u$　　　　**ii** $x = 6\cos u$

1 Find these integrals, either by direct recognition or by using a suitable substitution.

  a $\displaystyle\int x^2 \sin(x^3)\,dx$ **b** $\displaystyle\int \sin x \cos^6 x\,dx$ **c** $\displaystyle\int \tan^6 x \sec^2 x\,dx$

2 Find

  a $\displaystyle\int \cos 3x\,dx$ **b** $\displaystyle\int \frac{\sec 2x}{\cos 2x}\,dx$ **c** $\displaystyle\int 4\sin 2x \cos 2x\,dx$

3 Show that the area of the region enclosed by the curves $y = \sec^2 x$ and $y = \tan^2 x$ between $x = -\dfrac{\pi}{6}$ and $x = \dfrac{\pi}{6}$ is $\dfrac{\pi}{3}$.

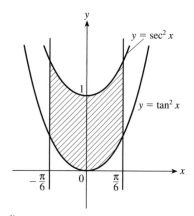

4 This diagram shows part of the curves $y = \cos x$ and $y = \sin x$.

  a Find the coordinates of B.

  b Find the coordinates of P, the point of intersection of the curves shown on the diagram.

  c Show that the area of the shaded region is $2 - \sqrt{2}$.

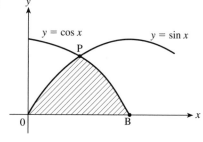

5 Show that $\displaystyle\int_{\frac{\pi}{6}}^{\frac{\pi}{3}} \frac{1}{\cos^2 x}\,dx = \frac{2\sqrt{3}}{3}$

6 This diagram shows part of the curve $y = \cot x$.

  a Find the area enclosed by the curve, the x-axis and the lines $x = \dfrac{\pi}{3}$ and $x = \dfrac{\pi}{2}$.

  b Find the volume of the solid generated when this area is rotated through $360°$ about the x-axis.

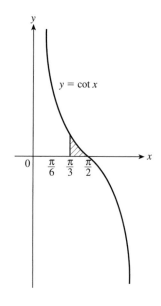

7 Find the volume of the solid generated when the portion of the curve $y = \sin^{-1} x$ from $y = 0$ to $y = \pi$ is rotated through $360°$ about the y-axis.

8 Find

a $\displaystyle\int 3x \cos 2x \, dx$ b $\displaystyle\int 3x e^{-4x} \, dx$

c $\displaystyle\int x^6 \ln x \, dx$ d $\displaystyle\int x^3 \cos x \, dx$

9 Find

a $\displaystyle\int \frac{1}{\sqrt{49 - x^2}} \, dx$ b $\displaystyle\int \frac{1}{\sqrt{50 - x^2}} \, dx$

10 Find

a $\displaystyle\int \frac{1}{81 + x^2} \, dx$ b $\displaystyle\int \frac{1}{80 + x^2} \, dx$

11 Evaluate

a $\displaystyle\int_1^{\sqrt{3}} \frac{1}{3 + x^2} \, dx$ b $\displaystyle\int_{-\frac{\sqrt{3}}{2}}^{\frac{1}{2}} \frac{1}{\sqrt{1 - x^2}} \, dx$

There is a 'Test yourself' exercise (Ty6) and an 'Extension exercise' (Ext6) on the CD-ROM.

➤ Key points

Trigonometric functions

$$\int \cos x \, dx = \sin x + c \qquad \int \cos(ax+b) \, dx = \frac{1}{a}\sin(ax+b) + c$$

$$\int \sin x \, dx = -\cos x + c \qquad \int \sin(ax+b) \, dx = -\frac{1}{a}\cos(ax+b) + c$$

$$\int \sec^2 x \, dx = \tan x + c \qquad \int \sec^2(ax+b) \, dx = \frac{1}{a}\tan(ax+b) + c$$

$$\int f'(x)\cos(f(x)) \, dx = \sin(f(x)) + c$$

$$\int f'(x)\sin(f(x)) \, dx = -\cos(f(x)) + c$$

$$\int f'(x)\sec^2(f(x)) \, dx = \tan(f(x)) + c$$

$$\int \sec x \tan x \, dx = \sec x + c$$

$$\int \operatorname{cosec} x \cot x \, dx = -\operatorname{cosec} x + c$$

$$\int \operatorname{cosec}^2 x \, dx = -\cot x + c$$

$$\int \tan x \, dx = -\ln(\cos x) + c = \ln(\sec x) + c$$

$$\int \cot x \, dx = \ln(\sin x) + c$$

$$\int \frac{1}{\sqrt{a^2 - x^2}} \, dx = \sin^{-1}\frac{x}{a} + c$$

$$\int \frac{1}{a^2 + x^2} \, dx = \frac{1}{a}\tan^{-1}\frac{x}{a} + c$$

Integration by parts

$$\int u \frac{dv}{dx} \, dx = uv - \int v \frac{du}{dx} \, dx$$

$$\int_a^b u \frac{dv}{dx} \, dx = \left[uv \right]_a^b - \int_a^b v \frac{du}{dx} \, dx$$

Numerical methods

Previous chapters have dealt with various methods for finding the exact solutions of equations. However, many equations cannot be solved easily, or sometimes at all, by using these methods and alternative numerical methods are required. These methods will allow you to find approximate roots of equations to any degree of accuracy. You have also seen how the trapezium rule can be used to find approximate values for integrals. In this chapter you will learn about two other methods of numerical integration.

The extraordinary increase in computing power over recent years allows us to use numerical methods to solve equations, or evaluate integrals, to any required degree of accuracy with astonishing speed. It is now possible, for example, to find the value of a number such as π or e to 100, 200 or even many thousands of decimal places in only a fraction of a second – a task which less than a century ago would have taken many hours of calculation.

After working through this chapter you should be able to

- *find the roots of an equation by using graphical methods*
- *locate an interval containing a root of an equation by looking for a change of sign*
- *use a decimal search to reduce the size of an interval containing a root of an equation*
- *form, and then use, an iterative formula to produce a sequence which converges to a root of an equation*
- *use the mid-ordinate rule and Simpson's rule to find approximate values for integrals.*

It is a great help to have a graphical calculator or spreadsheet facility, or at least a calculator capable of using the 'last answer', when working through this chapter. If one is not available, it may be necessary to give answers to a lower degree of accuracy than asked for in the exercises.

7.1 Introduction to numerical methods

A significant part of pure mathematics is concerned with precision – with equations that have exact solutions, with integrals that can be expressed exactly, and so on. However, most equations do not have solutions that can be expressed exactly and most functions cannot be integrated precisely.

For example, the equation

$$2 \sin \theta - \theta = 0$$

and the integral

$$\int_1^3 e^{-x^2} dx$$

do not have exact solutions.

In 'real world' practical problems, numerical solutions to different degrees of accuracy are required depending on the situation. A measurement of the area of a field to the nearest square metre may be acceptable, whereas calculations for constructing a spacecraft would need a high degree of accuracy.

Numerical methods can be used to solve problems to whatever degree of accuracy is required.

Consider this problem: OACB is a sector of a circle, centre O, radius r, where angle $AOB = \theta$.

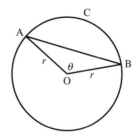

Assume θ is in radians unless stated otherwise.

Find θ if the area of segment ACB equals the area of $\triangle OAB$.

Area of segment ACB

$$= \text{Area of sector OACB} - \text{Area of } \triangle OAB$$

$$= \tfrac{1}{2}r^2\theta - \tfrac{1}{2}r^2\sin\theta$$

$$= \tfrac{1}{2}r^2(\theta - \sin\theta)$$

Area of sector $= \tfrac{1}{2}r^2\theta$
Area of $\triangle = \tfrac{1}{2}r^2\sin\theta$

But Area of segment ACB = Area of $\triangle OAB$

$$\therefore \quad \tfrac{1}{2}r^2(\theta - \sin\theta) = \tfrac{1}{2}r^2\sin\theta$$

Divide both sides by $\tfrac{1}{2}r^2$.

$$\theta - \sin\theta = \sin\theta$$

$$\theta - 2\sin\theta = 0$$

Can this equation be solved algebraically? There is no *algebraic* method of solving such an equation with terms in both θ and a trigonometrical ratio of θ.

Does this equation have a solution? When $\theta = 0$, the equation is satisfied. For small values of θ, area of $\triangle$ > area of segment. For AB approaching a diameter, i.e. θ approaching π, area of segment > area of $\triangle$. So, since the areas change continuously, there is some value of θ between 0 and π for which the areas are equal.

How can this equation be solved? Three approaches will be used:

■ Graphical solutions

■ Location of roots in an interval

■ Iteration

7.2 Graphical solutions

This section looks at solving equations f(x) = 0 and f(x) = g(x), graphically.

Intersection points on the x-axis

> **The roots of the equation f(x) = 0 are the values of x where the curve y = f(x) cuts the x-axis.**

So, to solve an equation such as

$$x - 2\sin x = 0$$

plot $y = x - 2\sin x$ for $0 < x < \pi$ either on graph paper or, preferably, on a graphical calculator or a computer.

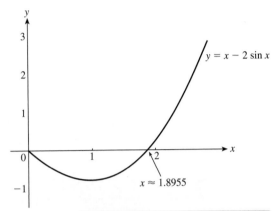

For $x < 1.8955$ (try $x = 1.89$), y is $-$ve
For $x > 1.8955$ (try $x = 1.9$), y is $+$ve

Find, as accurately as possible, where the graph cuts the x-axis, i.e. where $y = 0$. This value of x will be a solution of $x - 2\sin x = 0$.

Using the zoom and trace facilities on a graphical calculator x can be found correct to 4 decimal places as $x = 1.8955$.

Note: 1.8955 radians = $1.8955 \times \dfrac{180°}{\pi} = 108.6°$

Intersection points of two curves

> **An equation f(x) = g(x) can be solved by finding the x-coordinate(s) where the graphs of curve y = f(x) and curve y = g(x) intersect.**

So, an alternative method of solving $x - 2\sin x = 0$ graphically is to plot $y = x$ and $y = 2\sin x$ on the same axes and find their point of intersection for $0 < x < \pi$.

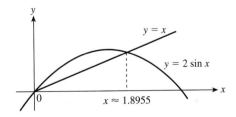

$y = x$ and $y = 2\sin x$ are simultaneous equations. The value of x at which the graphs intersect will make $x = 2\sin x$, i.e. $x - 2\sin x = 0$.

Change of sign

These diagrams show that, when the graph of a continuous function, $y = f(x)$, crosses the x-axis, there is a change of sign of the function.

Curves without a break such as $y = x^2$ and $y = x^3$ are called **continuous**. The graphs of $y = \frac{1}{x}$ and $y = \frac{1}{x^2}$ have breaks in the curve at $x = 0$. Such curves are said to be **discontinuous**. Discontinuous curves cannot be drawn without taking the pencil off the paper.

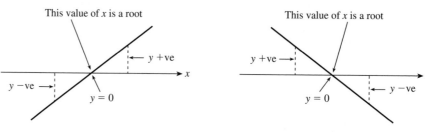

The function will be negative on one side of the crossing point and positive on the other. The crossing point corresponds to a root of the equation $y = 0$.

7.3 Location of roots in an interval

The second approach to solving $x - 2 \sin x = 0$ also uses the idea that the function has a change of sign, from $-$ve to $+$ve or from $+$ve to $-$ve, when the graph of the function crosses the x-axis.

If $y = f(x)$ then the roots (solutions) of $f(x) = 0$ are the values of x where the graph crosses the x-axis.

> **If two values x_1, x_2 can be found such that $x_1 < x_2$ and $f(x_1)$ and $f(x_2)$ have *different* signs, then $f(x) = 0$ has at least one root in (x_1, x_2), provided $f(x)$ is continuous in the interval (x_1, x_2).**

(x_1, x_2) is the open-ended interval $x_1 < x < x_2$

These two diagrams show a function $y = f(x)$ which has a change in sign for y in the interval (x_1, x_2), so $f(x) = 0$ has at least one root, α, in this interval.

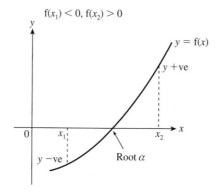

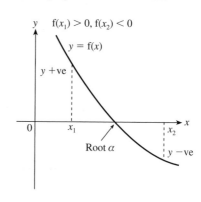

Possible problems with location of roots

When locating roots by looking for a change of sign in the function, drawing a sketch may help to avoid possible problems.

There could be *more than one root* of $f(x) = 0$ between x_1 and x_2.

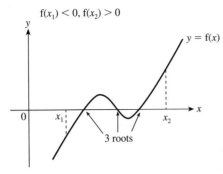

f$(x_1) < 0$, f$(x_2) > 0$

There is a change of sign and, in this case, there are 3 roots in the interval.

There could be a root *without* a change of sign.

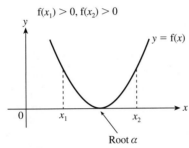

f$(x_1) > 0$, f$(x_2) > 0$

There is a root in the interval (x_1, x_2) but $f(x_1)$ and $f(x_2)$ both have the same sign.

There could be a change of sign but *no* root if $f(x)$ is discontinuous.

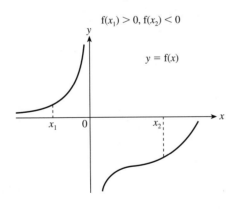

f$(x_1) > 0$, f$(x_2) < 0$

The graph does *not* cross the x-axis in this interval.

Methods of locating roots

Locating roots involves finding a small interval in which each root occurs. The greater the accuracy required, the smaller the interval needed. For example, if it is known that there is a root in the interval (1.91, 1.94) then the root is 1.9, correct to 1 decimal place. If greater accuracy is needed then a smaller interval is needed. If there is a root in the interval (1.9134, 1.9135) then the root is 1.913, correct to 3 decimal places.

Decimal search

Having located a root in an interval, say (1, 2), a traditional method of using a decimal search is to evaluate the function at increments of 0.1, i.e. 1.1, 1.2, 1.3, … until a change of sign is found. Having located a root in this reduced interval, say, (1.8, 1.9), the function is evaluated at increments of 0.01, i.e. 1.81, 1.82, 1.83, … until a change of sign is found. The process is repeated until the required accuracy is obtained. This approach involves many calculations but has the advantage that it can be programmed on a computer.

Example 1 Show that $x^3 - 3x^2 + 2 = 0$ has a root between $x = 2$ and $x = 3$.

Let $f(x) = x^3 - 3x^2 + 2$. Then
$$f(2) = 2^3 - 3 \times 2^2 + 2 = -2 < 0$$
$$f(3) = 3^3 - 3 \times 3^2 + 2 = 2 > 0$$
$f(x)$ is continuous, and $f(2) < 0$ and $f(3) > 0$.

So $x^3 - 3x^2 + 2 = 0$ has a root between $x = 2$ and $x = 3$.

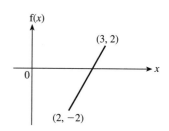

Example 2 *This example illustrates the decimal search method.*

Show that $x - 2 \sin x = 0$ has a root between $x = 1$ and $x = 2$ and find the root, to one decimal place.

Let $f(x) = x - 2 \sin x$. Then
$$f(1) = 1 - 2 \sin 1 = -0.68 \ldots < 0$$
$$f(2) = 2 - 2 \sin 2 = 0.18 \ldots > 0$$

> In a decimal search, having located the root in an interval, the interval is reduced by substituting different values of x. In each reduced interval, the function must still have a change of sign.

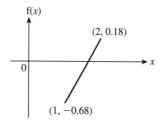

$f(x)$ is continuous, and $f(1) < 0$ and $f(2) > 0$.

So $f(x) = 0$ has a root in the interval (1, 2).

$$f(1.1) = -0.682 \ldots < 0$$
$$f(1.2) = -0.664 \ldots < 0$$
$$f(1.3) = -0.627 \ldots < 0$$
$$f(1.4) = -0.570 \ldots < 0$$
$$f(1.5) = -0.494 \ldots < 0$$
$$f(1.6) = -0.399 \ldots < 0$$
$$f(1.7) = -0.283 \ldots < 0$$
$$f(1.8) = -0.147 \ldots < 0$$
$$f(1.9) = 0.007 \ldots > 0$$

> This continuous function changes from −ve to +ve between 1 and 2. So $y = f(x)$ must cross the x-axis between $x = 1$ and $x = 2$.

> The dots indicate that the values have been truncated rather than rounded.

So the root is in the interval (1.8, 1.9).

$$f(1.81) = -0.133\ldots < 0$$
$$f(1.82) = -0.118\ldots < 0$$
$$f(1.83) = -0.103\ldots < 0$$
$$f(1.84) = -0.087\ldots < 0$$
$$f(1.85) = -0.072\ldots < 0$$

So the root is in the interval (1.85, 1.9).

> Both 1.85 and 1.9 equal 1.9, correct to 1 decimal place.

Hence $x = 1.9$, correct to one decimal place is the root of the equation $x - 2\sin x = 0$.

Note Most graphical calculators will generate a table. This is useful for carrying out a decimal search.

The CD-ROM (E7.3) shows how to use linear interpolation to obtain an approximate root of an equation.

Exercise 7A

1 a Given that $f(x) = x^3 - 5x + 1$, evaluate $f(x)$ for $x = -4, -3, -2, \ldots 2, 3, 4$.

  b Hence state the unit intervals in which the roots of $x^3 - 5x + 1 = 0$ lie.

  c Sketch the curve.

2 a Show that $x^3 - 4x + 2 = 0$ has roots in the intervals $(-3, -2)$, $(0, 1)$ and $(1, 2)$.

  b Using decimal searches, find these roots to one decimal place.

3 a By sketching the graph of $y = 3 + x^2 - x^3$ show that the equation $3 + x^2 - x^3 = 0$ has only one root.

  b Show that the root α is such that $1 < \alpha < 2$.

  c Find, by decimal search, the value of α to two decimal places.

4 a Sketch, on the same axes, $y = e^x$ and $y = 4 - x^2$.
     Use the sketch to deduce the number of roots of $e^x + x^2 - 4 = 0$.

  b Show that $e^x + x^2 - 4 = 0$ has roots in the intervals $(-2, -1)$ and $(1, 2)$.

  c Find, by decimal search, these roots to one decimal place.

5 a Sketch $y = \ln x$ and $y = e^{-x} + 1$ on the same axes and hence show that $\ln x = e^{-x} + 1$ has only one root.

  b Show that the root lies in the interval (2, 3).

  c Find, by a decimal search, the root to two decimal places.

6 a Show that $f(x) = \cos x - 2x$ has no stationary points.

  b Show that $\cos x - 2x = 0$ has only one root and that the root lies in the interval (0, 1).

  c Find the root to one decimal place.

7 The equation $2^x = x^3$ has two roots.

   a Show that the unit intervals in which the roots lie are (1, 2) and (9, 10). (Hint: put $f(x) = 2^x - x^3$)

   b Find the roots to one decimal place.

7.4 Iteration

Assume that the sequence defined by $x_{n+1} = 2 \sin x_n$ $(n \in \mathbb{Z}, n \geqslant 1)$ converges to a limit X, i.e. that $x_n \to X$ as $n \to \infty$. Then, for infinitely large n

$$X = 2 \sin X$$

or $X - 2 \sin X = 0$

So the limiting value X is a solution of $x - 2 \sin x = 0$.

Any equation can be rearranged to give $x = f(x)$, which can be written as an iterative formula

$$x_{n+1} = f(x_n)$$

For various rearrangements, the resulting sequence may, or may not, converge. If it does converge, the limiting value will be a root of the original equation.

To use the iterative formula a first value, x_1, must be chosen. This approximation to the root to be found can be chosen by sketching a graph and locating a root by looking for a change in sign. Example 3 illustrates this.

Example 3

Starting with $x_1 = 1$, use the iterative formula $x_{n+1} = 2 \sin x_n$ to find x_2, x_3, x_4 and x_5. Write down the first four decimal places of each term, keeping the accurate figure on the calculator for the following calculation.

Find the value to which this sequence converges to seven decimal places.

$x_1 = 1$

$x_2 = 2 \sin 1$

    $= 1.6829\ldots$

$x_3 = 2 \sin (1.6829)$

    $= 1.9874\ldots$

$x_4 = 1.8289\ldots$

$x_5 = 1.9337\ldots$

> The starting value, $x_1 = 1$ is given here.

> Check that the calculator is in radian mode.

> *Note*: This accurate result can be found in 20 seconds on a graphical calculator, or even more quickly in a spreadsheet.

Using the calculator $x_n \to 1.895\,494\,26\ldots$
So limit is $1.895\,4943$ correct to seven decimal places.

Note For iteration of $x_{n+1} = f(x_n)$ on a calculator the routine is
    x_1 ENTER (or EXE)
    f(ANS)
Each time ENTER (or EXE) is pressed the next term in the sequence will be displayed.

Example 4 *In this example, two possible rearrangements of the equation to be solved are considered.*

a Show that $x^3 + 1 = 4x$ has positive roots in the intervals $(0, 1)$ and $(1, 2)$.

b Show that two possible rearrangements of $x^3 - 4x + 1 = 0$ lead to the iterative formulae

$$x_{n+1} = \tfrac{1}{4}(x_n^3 + 1) \qquad \text{and} \qquad x_{n+1} = \sqrt[3]{4x_n - 1}$$

c Find the positive roots of $x^3 - 4x + 1 = 0$, to three decimal places, by putting $x_1 = 1$ in $x_{n+1} = \tfrac{1}{4}(x_n^3 + 1)$.

d Find the positive roots of $x^3 - 4x + 1 = 0$, to three decimal places, by putting $x_1 = 2$ in $x_{n+1} = \sqrt[3]{4x_n - 1}$.

e Show that putting $x_1 = 2$ in $x_{n+1} = \tfrac{1}{4}(x_n^3 + 1)$ leads to a divergent sequence.

Solution **a** $x^3 + 1 = 4x \Rightarrow x^3 - 4x + 1 = 0$

Let $f(x) = x^3 - 4x + 1$

> Always arrange with 0 on one side before locating roots and use function notation.

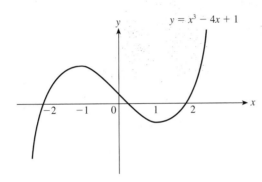

$y = x^3 - 4x + 1$

$f(0) = 1 > 0$

$f(1) = -2 < 0$

$f(2) = 1 > 0$

> A cubic equation has up to three real roots. This cubic cuts the x-axis in three distinct points, so there are three distinct real roots. The sketch shows two of them are positive.

$f(x)$ is continuous so a change in sign of $f(x)$ indicates a root of $f(x) = 0$

$\therefore$ $f(x) = 0$ has positive roots in the intervals $(0, 1)$ and $(1, 2)$.

b $x^3 - 4x + 1 = 0$

> Rearrange so that only $4x$ is on the LHS.

$4x = x^3 + 1$

> Multiply both sides by $\tfrac{1}{4}$.

$x = \tfrac{1}{4}(x^3 + 1)$

which gives the iterative formula

> Write $x = f(x)$ as $x_{n+1} = f(x_n)$.

$$x_{n+1} = \frac{x_n^3 + 1}{4}$$

Alternatively

$$x^3 - 4x + 1 = 0$$

> Rearrange so that only x^3 is on the LHS.

$$x^3 = 4x - 1$$

> Take cube root of both sides.

$$x = \sqrt[3]{4x - 1}$$

which gives the iterative formula

$$x_{n+1} = \sqrt[3]{4x_n - 1}$$

c $x_{n+1} = \frac{1}{4}(x_n^3 + 1)$

$x_1 = 1$

$x_2 = 0.5$

$x_3 = 0.28125\ldots$

$x_4 = 0.25556\ldots$

$x_5 = 0.25417\ldots$

$x_6 = 0.25410\ldots$

> Retain full accuracy on the calculator and write down more decimal places than are required in the answer.

> x_5 and x_6 both equal 0.254 correct to three decimal places, so no further terms need be calculated.

The sequence converges and the root is 0.254 correct to three decimal places.

Note   This does not constitute a proof that the root is 0.254 correct to three decimal places. To prove this it is necessary to calculate f(0.2535) and f(0.2545) and show that there is a change of sign.

d $x_{n+1} = \sqrt[3]{4x_n - 1}$

$x_1 = 2$

$x_2 = 1.91293\ldots$

$x_3 = 1.88066\ldots$

$x_4 = 1.86842\ldots$

$x_5 = 1.86373\ldots$

$x_6 = 1.86193\ldots$

$x_7 = 1.86123\ldots$

$x_8 = 1.86097\ldots$

> *Note*: A different value of x_1 is used in the alternative arrangement.

> *Note*: This sequence does not converge as quickly as the one in part **c**, but the terms are progressively closer so it does converge.

> x_7 and x_8 both equal 1.861 correct to three decimal places, so no further terms need be calculated.

The sequence converges and the root is 1.861 correct to three decimal places.

e $x_{n+1} = \dfrac{x_n^3 + 1}{4}$

$x_1 = 2$

$x_2 = 2.25$

$x_3 = 3.097\ldots$

$x_4 = 7.680\ldots$

$x_5 = 113.534\ldots$

> This sequence of terms illustrates how some rearrangements, with particular values of x_1, do not converge.

The sequence diverges.

Illustrating iteration graphically

Consider again the rearrangement $x = f(x)$ where $f(x) = \frac{1}{4}(x^3 + 1)$ (as in Example 4). The intersection of the graphs of the curve $y = \frac{1}{4}(x^3 + 1)$ and the line $y = x$ will give the roots of the equation $x = f(x)$, in this case $x = \frac{1}{4}(x^3 + 1)$ i.e. $x^3 - 4x + 1 = 0$.

On the graph start at x_1 on the x-axis and draw a *vertical* line to the curve. From that intersection, draw a *horizontal* line to the line $y = x$. Then, draw a vertical line to the curve, and a horizontal line to the line $y = x$ and so on.

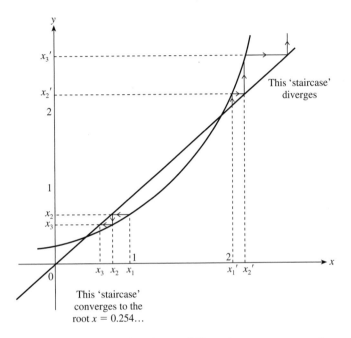

The effect of drawing vertical and horizontal lines is to generate the sequence $x_1, x_2, x_3, \ldots$

- The vertical line from $x = x_1$ meets the curve at $(x_1, f(x_1))$, i.e. (x_1, x_2) since $f(x_1) = x_2$.
- The horizontal line through (x_1, x_2) meets the line $y = x$ at (x_2, x_2).
- The vertical line through (x_2, x_2) meets the curve at $(x_2, f(x_2))$ i.e. (x_2, x_3) since $f(x_2) = x_3$.
- And so on.

Starting at $x_1 = 1$, the lines drawn form a 'staircase' approaching closer and closer to the intersection. The sequence converges to the root $x = 0.254\ldots$

Starting at $x_1' = 2$, the lines drawn form a 'staircase' which diverges. The root 1.861 *cannot* be found using the iterative formula $x_{n+1} = \frac{1}{4}(x_n^3 + 1)$, but can be found using an alternative iterative formula (see Example 4).

In some cases, the sequence approaches a root by a 'cobweb' rather than a 'staircase'.

Whether the sequence converges or not for a particular iterative formula, $x_{n+1} = f(x_n)$, depends on the gradient of $y = f(x)$ at the intersection with $y = x$.

There are four possible cases:

$|f'(x)| < 1 \Rightarrow$ staircase converges

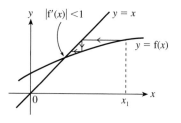

$|f'(x)| < 1 \Rightarrow$ cobweb converges

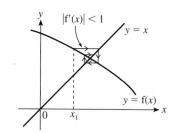

$|f'(x)| > 1 \Rightarrow$ staircase diverges

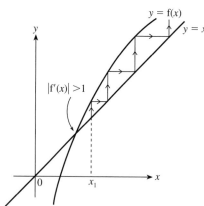

$|f'(x)| > 1 \Rightarrow$ cobweb diverges

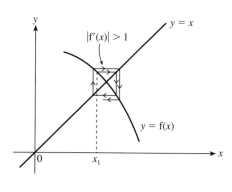

The condition for convergence is $|f'(x)| < 1$, i.e. $-1 < f'(x) < 1$ at the intersection with $y = x$.

The CD-ROM (E7.4) explains how iterative processes can be used in practical situations.

Exercise 7B

A calculator, preferably a graphical one, or spreadsheet facility, is needed for this exercise. If working without a graphical calculator or computer, give the answer to a lower degree of accuracy.

1 **a** Show that $x^3 - 9x + 1 = 0$ has a root between 0 and 1.

   b Show that $x^3 - 9x + 1 = 0$ has a root between 0.111 and 0.112. Hence state a root of $x^3 - 9x + 1 = 0$ to two decimal places.

2 a Show that the equation $x^2 - 5x + 1 = 0$ could be rearranged to give iterative formulae

$$x_{n+1} = \tfrac{1}{5}(x_n^2 + 1) \qquad \text{and} \qquad x_{n+1} = 5 - \frac{1}{x_n}$$

b Putting $x_1 = 0.2$ in $x_{n+1} = \tfrac{1}{5}(x_n^2 + 1)$ find x_2, x_3 and x_4.

Putting $x_1 = 4$ in $x_{n+1} = 5 - \dfrac{1}{x_n}$, find x_2, x_3 and x_4.

Hence write down two roots of the quadratic equation to three significant figures.

c Check the answers to part **b** using the formula for solving a quadratic equation.

3 a Use the iterative formula

$$x_{n+1} = \frac{1}{2}\left(x_n + \frac{12}{x_n}\right)$$

with $x_1 = 2$ to find x_2, x_3, x_4, x_5 and x_6.

b Compare x_6 with $\sqrt{12}$.

c By putting $x_{n+1} = x_n = X$ in the iterative formula in part **a** show that the sequence converges to $\sqrt{12}$.

4 a Show, by putting $x_{n+1} = x_n = x$, that if the sequence defined by

$$x_{n+1} = \frac{1}{2}\left(x_n + \frac{N}{x_n}\right)$$

converges, it will converge to $\sqrt{N}$.

b Hence find $\sqrt{127}$, to five decimal places, using the formula in part **a**. (Choose any value for x_1. Check the answer on a calculator.)

5 a Show that the equation $x^3 - 10x + 1 = 0$ can be rearranged in the form

$$x = \frac{x^3 + 1}{10}$$

b Use this rearrangement to form an iterative formula and use it to find, to four significant figures, the root which lies between 0 and 1.

6 a Show that $\theta = \sin\theta + 1$ has a root between 1 and 2.

b Use the iterative formula $\theta_{n+1} = \sin\theta_n + 1$ to find this root to seven decimal places.

c The sketch shows part of the graphs of $y = \sin\theta + 1$ and $y = \theta$. Draw a cobweb or staircase diagram to show how convergence takes place, indicating the positions of your successive approximations.

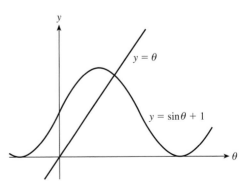

7 A circle with centre O has a chord AB such that $\angle AOB = \theta$ radians where $0 < \theta < \pi$. The area of the minor segment cut off by AB is one-sixth of the area of the circle.

  a Show that $\theta - \sin\theta = \dfrac{\pi}{3}$.

  b Show that the value of θ lies between 1 and 3.

  c Find θ to two decimal places.

8 a Show that $x^3 - 6x - 2 = 0$ has roots in the intervals $(-3, -2)$, $(-2, 0)$ and $(0, 3)$.

  b Show that three possible rearrangements of $x^3 - 6x - 2 = 0$ lead to the iterative formulae

  $$x_{n+1} = \frac{x_n^3 - 2}{6} \qquad x_{n+1} = \sqrt[3]{6x_n + 2} \qquad x_{n+1} = \frac{6x_n + 2}{x_n^2}$$

  c Find the roots of $x^3 - 6x - 2 = 0$, to seven decimal places, by

   i putting $x_1 = -2$ in $x_{n+1} = \dfrac{x_n^3 - 2}{6}$

   ii putting $x_1 = 0$ in $x_{n+1} = \sqrt[3]{6x_n + 2}$

   iii putting $x_1 = -2$ in $x_{n+1} = \dfrac{6x_n + 2}{x_n^2}$

9 In the diagram, O is the centre of the circle, radius r, and CD is a chord such that angle COD $= 2\theta$ radians. ABCD is a square.

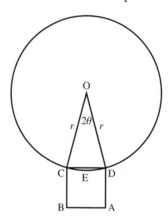

  a Given that the sector COD and the square ABCD have equal area, show that $\theta - 4\sin^2\theta = 0$.

  b Show that the equation in part **a** has one root, α, between 0.1 and 0.4 and another, β, between 2 and 2.5.

  c Show that the equation in part **a** can be rearranged as

  $$\sin\theta = \frac{\sqrt{\theta}}{2}$$

   Use an iterative process to find the root α to two decimal places.

  d By locating the root, β, in a suitable interval, find β, to one decimal place.

  e Draw a diagram to illustrate the solution corresponding to root β.

10 Assuming the sequences defined by the following iterative formulae converge, find, in the form $f(x) = 0$, the equation which each would solve.

a $\quad x_{n+1} = \dfrac{x_n^2 + 1}{4}$

b $\quad x_{n+1} = \dfrac{2x_n^3 + 10}{3x_n^2}$

c $\quad x_{n+1} = \dfrac{1}{5}\left(4x_n + \dfrac{50}{x_n^4}\right)$

d $\quad x_{n+1} = \dfrac{3x_n^3 - 5}{4x_n^2 + 1}$

e $\quad x_{n+1} = 1 + \dfrac{4x_n^2}{e^{x_n}}$

f $\quad x_{n+1} = \dfrac{\sin x_n - x_n + 2}{3 + x_n}$

11 Use the iterative formulae of Question 10 to solve, to four decimal places, each equation found. In each case, find the solution obtained by putting $x_1 = 1$.

12 a The equation $x^3 - 5x - 2 = 0$ has a root between 2 and 3. Use the iterative formula

$$x_{n+1} = \frac{2x_n^3 + 2}{3x_n^2 - 5}$$

starting with $x_1 = 2$, to find this root to three decimal places.

b Hence find, to two decimal places, a root of

$$2^{3x} - 5 \times 2^x - 2 = 0$$

13 a Show that $e^x - x = 4$ has a root between 1 and 2.

b Show that the iterative formula

$$x_{n+1} = \frac{e^{x_n}(x_n - 1) + 4}{e^{x_n} - 1}$$

leads to a solution of the equation in part **a**.

c Using $x_1 = 1$, find a root of $e^x - x = 4$, to six decimal places.

d Hence find, to four decimal places, a root of

$$e^{2\cos x} - 2\cos x = 4$$

14 The graphs of $y = x^3 + 2x - 2$ and $y = 5 + 3x$ intersect at the point P whose coordinates are (h, k).

a Show that h satisfies the equation

$$x^3 - x - 7 = 0$$

b Show that, by rearranging the equation in part **a** two possible iterative formulae are

$$x_{n+1} = x_n^3 - 7 \quad \text{and} \quad x_{n+1} = \sqrt[3]{x_n + 7}$$

c Using $x_1 = 2$ show that only one of the formulae gives a convergent sequence. Hence find the coordinates of P to four decimal places.

15 a Show that $x \ln x = 3$ has a root between 2.5 and 3.

b By using the iterative formula

$$x_{n+1} = \frac{3}{\ln x_n}$$

with $x_1 = 3$, find the root to six decimal places.

(Notice how slowly the sequence converges.)

7.5   Numerical integration

The area, A, 'under' a curve $y = \mathrm{f}(x)$ can be found *exactly* by integration.

$$A = \int_a^b \mathrm{f}(x)\,\mathrm{d}x$$

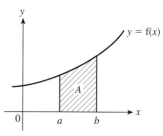

In Book 1 you met the trapezium rule and used it to find approximate values for areas.

Remember

> **Trapezium rule**
>
> $$\int_a^b \mathrm{f}(x)\,\mathrm{d}x \approx \frac{h}{2}[y_0 + y_n + 2(y_1 + y_2 + \cdots + y_{n-1})]$$

This section looks at two further methods of finding an approximate value for A, when $\mathrm{f}(x)$ *cannot* be integrated algebraically:

- The mid-ordinate rule
- Simpson's rule

The mid-ordinate rule

With the mid-ordinate rule, instead of approximating the area with trapeziums, the area is split into rectangles.

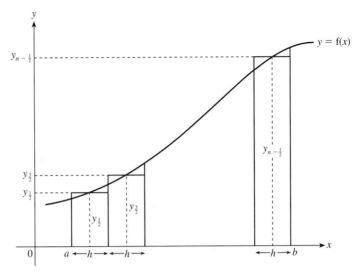

The area is approximated by n rectangles each of width h. The height of each rectangle is the y value half way along each rectangle, the mid-ordinate, denoted by $y_{\frac{1}{2}}, y_{\frac{3}{2}}, y_{\frac{5}{2}}, \cdots y_{n-\frac{1}{2}}$.

Summing the areas of the rectangles

$$A \approx h\left(y_{\frac{1}{2}} + y_{\frac{3}{2}} + y_{\frac{5}{2}} + \cdots + y_{n-\frac{1}{2}}\right)$$

> **Mid-ordinate rule**
>
> $$\int_a^b f(x)\,dx \approx h\left(y_{\frac{1}{2}} + y_{\frac{3}{2}} + y_{\frac{5}{2}} + \cdots + y_{n-\frac{1}{2}}\right)$$

Note A better approximation can be obtained by increasing the number of rectangles, i.e. by increasing n and reducing h.

Simpson's rule

With the trapezium and mid-ordinate rules, the area is split into vertical strips and the curve at the top of each strip is approximated by a straight line. With Simpson's rule, instead of approximating the curve by a straight line, it is approximated by a parabola. The area is split into n equal width strips; n must be *even* and therefore the number of ordinates *odd*.

> **Simpson's rule**
> **For even n**
>
> $$\int_a^b f(x)\,dx \approx \frac{h}{3}(y_0 + y_n + 4(y_1 + y_3 + \cdots + y_{n-1}) + 2(y_2 + y_4 + \cdots + y_{n-2}))$$

Simpson's rule can be stated as

$$\text{Integral} \approx \frac{h}{3} \times (\text{first} + \text{last} + 4 \times \text{odds} + 2 \times \text{evens})$$

The derivation of Simpson's rule can be constructed from Question 12 of Extension Exercise 7 on the CD-ROM.

Note   As with the previous methods, a better approximation can be obtained by increasing the number of strips.

Example 5   Estimate $\displaystyle\int_0^1 \cos x^2\,dx$ using

a   the mid-ordinate rule

b   Simpson's rule with five ordinates (four strips).

In each case, give the answer correct to three significant figures.

Solution   **a**   Let $y = \cos x^2$

*Note*: $\cos x^2 = \cos(x^2)$

n	0	1	2	3	4
x_n	0	0.25	0.5	0.75	1
Mid x		0.125	0.375	0.625	0.875	
Mid y		$\cos 0.125^2$	$\cos 0.375^2$	$\cos 0.625^2$	$\cos 0.875^2$	

These values are $y_{\frac{1}{2}}, y_{\frac{3}{2}}, y_{\frac{5}{2}}$ and $y_{\frac{7}{2}}$.

By the mid-ordinate rule

$$\int_0^1 \cos x^2\,dx \approx h(y_{\frac{1}{2}} + y_{\frac{3}{2}} + y_{\frac{5}{2}} + y_{\frac{7}{2}})$$

$h = 0.25$

$$\approx 0.25(\cos 0.125^2 + \cos 0.375^2 + \cos 0.625^2 + \cos 0.875^2)$$

Check calculator in radian mode.

$$\approx 0.909$$

So, the mid-ordinate rule with 4 intervals gives 0.909 as an approximation to the integral.

b   Let $y = \cos x^2$

There are 4 strips. For Simpson's rule n must be even.

n	0	1	2	3	4
x_n	0	0.25	0.5	0.75	1
y_n	$\cos 0$	$\cos 0.25^2$	$\cos 0.5^2$	$\cos 0.75^2$	$\cos 1^2$

These values are y_0, y_1, y_2, y_3 and y_4.

By Simpson's rule

$$\int_0^1 \cos x^2 \, dx \approx \frac{h}{3} \{y_0 + y_4 + 4(y_1 + y_3) + 2y_2\}$$

$$\approx \frac{0.25}{3} \left(\cos 0 + \cos 1 + 4(\cos 0.25^2 + \cos 0.75^2) + 2 \cos 0.5^2 \right)$$

$$\approx 0.905$$

So, Simpson's rule with 4 intervals gives 0.905 as an approximation to the integral.

Example 6   Given that $I = \displaystyle\int_0^{16} \sqrt{1 + x^2} \, dx$

   a   Find estimates for I using Simpson's rule with 2, 4 and 8 strips, giving your answers to four significant figures.

   b   Comment on your answers from part **a**.

Solution   **a**   Tabulating the values:

x values	y values
0	$y_0 = \sqrt{1}$
2	$y_1 = \sqrt{5}$
4	$y_2 = \sqrt{17}$
6	$y_3 = \sqrt{37}$
8	$y_4 = \sqrt{65}$
10	$y_5 = \sqrt{101}$
12	$y_6 = \sqrt{145}$
14	$y_7 = \sqrt{197}$
16	$y_8 = \sqrt{257}$

> It is easier to tabulate the values for 8 strips (9 ordinates) and then use this table for all three estimates.

With 2 strips:

$$I \approx \frac{8}{3} [y_0 + y_8 + 4y_4]$$

$$= \frac{8}{3} \times [1 + \sqrt{257} + 4 \times \sqrt{65}]$$

$$= 131.4$$

> The ordinates used are y_0, y_4, y_8 and so in this case y_4 is an *odd* ordinate.

With 4 strips:

$$I \approx \frac{4}{3} [y_0 + y_8 + 4(y_2 + y_6) + 2y_4]$$

$$= \frac{4}{3} \times [1 + \sqrt{257} + 4(\sqrt{17} + \sqrt{145}) + 2 \times \sqrt{65}]$$

$$= 130.4$$

> The ordinates used are y_0, y_2, y_4, y_6, y_8 and so in this case the *odd* ordinates are y_2 and y_6; y_4 is an *even* ordinate.

With 8 strips:

$$I \approx \frac{2}{3}[y_0 + y_8 + 4(y_1 + y_3 + y_5 + y_7) + 2(y_2 + y_4 + y_6)]$$

$$= \frac{2}{3} \times [1 + \sqrt{257} + 4(\sqrt{5} + \sqrt{37} + \sqrt{101} + \sqrt{197}) + 2 \times (\sqrt{17} + \sqrt{65} + \sqrt{145})]$$

$$= 130.1$$

b The estimates obtained using 2, 4 and 8 strips all give 130 correct to two significant figures suggesting that the value of I is 130 correct to two significant figures.

Exercise 7C

1 Giving your answers to three significant figures, estimate these integrals using first the mid-ordinate rule and then Simpson's rule.

a $\int_0^1 e^{x^2} \, dx$ using 4 strips

b $\int_0^1 \frac{1}{x + \cos x} \, dx$ using 2 strips

c $\int_1^2 \ln x^2 \, dx$ using 4 strips

d $\int_0^\pi \sin \sqrt{x} \, dx$ using 8 strips

2 Using 7 ordinates, compare the approximations to

$$\int_{0.1}^{0.7} x^{\cos x} \, dx$$

obtained by the mid-ordinate rule and Simpson's rule.
Give the answers to four significant figures.

3 Use Simpson's rule, with 11 ordinates, to find an approximation to the area bounded by $y = xe^{-x^2}$, the x-axis and $x = 1$.
Give the answer to four significant figures.

4 Use the mid-ordinate rule with 7 strips to estimate the volume obtained when

$$y = \frac{1}{\sqrt{x} + 1}$$

between $x = 0$ and $x = 1.4$, is rotated through 2π about the x-axis.
Give the answer to three significant figures.

5 The values of a function $f(x)$ are given in the table below.

x	1	2	3	4	5	6	7
$f(x)$	0.9	1.1	1.4	1.5	1.0	0.8	0.4

Find an approximate value for $\int_1^7 x^2 f(x) \, dx$ using Simpson's rule.

6 Given that $I = \int_0^1 e^{x^2+1} \, dx$

a Find estimates for I using the Simpson's rule with 2, 4 and 8 strips, giving your answers to three significant figures.

b Comment on the answers you obtained in part **a**.

Exercise 7D (Review)

1 a Show that the cubic equation $x^3 - 3x + 1 = 0$ can be arranged to give the iterative formulae

$$x_{n+1} = \frac{1}{3}(x_n^3 + 1) \quad \text{and} \quad x_{n+1} = \frac{3x_n - 1}{x_n^2}$$

b By putting $x_1 = 0.2$ show that only one of the iterative formulae defines a convergent sequence.

Hence find a root of $x^3 - 3x + 1 = 0$ to seven decimal places.

2 a Use the iterative formula

$$x_{n+1} = \frac{1 - x_n}{\cos x_n}$$

with $x_1 = 0.5$ to find the limit of the sequence $x_1, x_2, x_3, \ldots$ to two decimal places.

b Find the equation to which the limiting value of the sequence is a solution.

c With $x_1 = 0$ find x_2, x_3, x_4, x_5, x_6. Explain the results.

3 a Show that the equation $e^x - x = 2$ has one root between -2 and -1 and a second root between 1 and 2.

b Use the iterative formulae $x_{n+1} = e^{x_n} - 2$ and $x_{n+1} = \ln(x_n + 2)$ to find the two roots, to five decimal places.

c The sketches show parts of the graphs of

i $y = e^x - 2$ and $y = x$

ii $y = \ln(x + 2)$ and $y = x$

In each case, draw cobweb or staircase diagrams to show how convergence takes place, indicating the positions of your successive approximations.

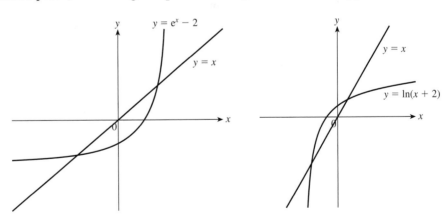

4 a Show that the equation $x^3 + 1 = 6x$ has a root between $x = 0$ and $x = 1$.

b Show that three possible arrangements of the equation in the form $x = f(x)$ are

$$x = \sqrt[3]{6x - 1} \qquad x = \frac{1}{6}(x^3 + 1) \qquad x = x^3 - 5x + 1$$

c Use each of the arrangements in part **b** to form an iterative formula of the form $x_{r+1} = f(x_r)$. Find which of the formulae provides a sequence which converges to the root between 0 and 1. Use this rearrangement to find the root to six significant figures.

5 The diagram shows a sketch of $y = e^x - 4x$. No scale is given.

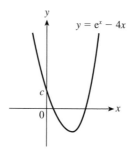

a State the value of c.

b By looking for changes in sign find the intervals of unit length in which the roots of $e^x - 4x = 0$ lie.

c Use decimal searches to find the roots, to two decimal places.

6 An industrial cooking bowl is formed by rotating about the y-axis from $y = 0$ to $y = 1$ the curve $y = e^{x^2} - 1$. The depth of the bowl is 1m.

a Show that the volume of the bowl is given by

$$\pi \int_0^1 \ln(y + 1) \, dy$$

b Use Simpson's rule and then the mid-ordinate rule, each with 5 ordinates, to find an approximation to the volume in m^3, to four significant figures.

7 Given that $f(x) = \ln(1 + \sin x)$, $0 \leqslant x \leqslant \pi$, find approximations to

a the area enclosed by the curve and x-axis

b the volume when the area in part **a** is rotated through 2π radians about the x-axis.

8 $f(x) = \sqrt[3]{1 + x^2}$

Find an approximation for

$$\int_0^{0.4} f(x) \, dx$$

giving the answer to four significant figures, using Simpson's rule with 9 ordinates.

There is an 'Extension exercise' (Ext7) on the CD-ROM.

➤ Key points

Solving equations

Graphically

- The roots of an equation $f(x) = 0$ are the values of x where the curve $y = f(x)$ cuts the x-axis.

- For an equation $f(x) = g(x)$ find the x-coordinate(s) where the graphs of curve $y = f(x)$ and curve $y = g(x)$ intersect.

Location of roots in an interval

If two values x_1, x_2 can be found such that $x_1 < x_2$ and $f(x_1)$ and $f(x_2)$ have *different* signs then $f(x) = 0$ has at least one root in (x_1, x_2), providing $f(x)$ is continuous in the interval (x_1, x_2).

Having located a root in an interval, use a **decimal search** to reduce the size of the interval.

Staircase and cobweb diagrams

These diagrams show how staircase and cobweb diagrams can be used to show the convergence, or divergence, of an iterative process.

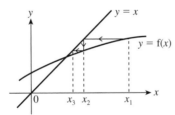

Staircase converges

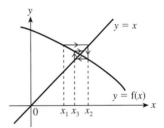

Cobweb converges

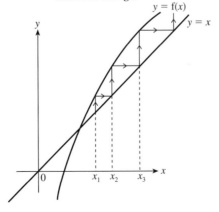

Staircase diverges

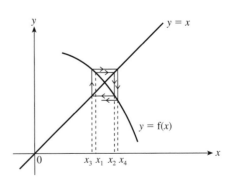

Cobweb diverges

Iteration

- Rearrange in the form $x = f(x)$ and write an iterative formula $x_{n+1} = f(x_n)$.
- Some rearrangements lead to sequences that will converge to a root of the equation.
- x_1 should be chosen somewhere in the region of the root sought.

Numerical integration

Mid-ordinate rule

$$\int_a^b f(x)\,dx \approx h\left(y_{\frac{1}{2}} + y_{\frac{3}{2}} + y_{\frac{5}{2}} + \cdots + y_{n-\frac{1}{2}}\right)$$

Simpson's rule

For even n

$$\int_a^b f(x)\,dx \approx \frac{h}{3}\left(y_0 + y_n + 4(y_1 + y_3 + \cdots + y_{n-1}) + 2(y_2 + y_4 + \cdots + y_{n-2})\right)$$

Proof

Proof is at the heart of mathematics and one of the reasons why the subject is regarded as having such solid foundations. In ancient Greece, over 2500 years ago, the Greek philosophers discovered the power of reason. They were determined to discover truths and, in particular, mathematical truths about the natural world, but more than this they wanted to be certain that the results they found were true – and so they developed the idea of proof.

Throughout the course you have been asked to prove algebraic identities, including many involving trigonometric functions (see Chapter 2).

After working through this chapter you should be able to

■ use the technique of proof by deduction

■ prove identities, including those involving trigonometric functions

■ use algebra in proofs by, for example, expressing an even number in the form $2n$ and an odd number in the form $2n + 1$

■ disprove a false result by finding a counter-example

■ use the technique of proof by contradiction (reductio ad absurdum).

The CD-ROM covers the technique of proof by exhaustion (E8.1.1), necessary and sufficient conditions and use of the implication signs (E8.1.2).

8.1 Introduction

Both mathematicians and scientists try to establish laws and theories with wide or universal application. These can be formulated as a result, for example, of observation or investigation. The theory is initially expressed as a hypothesis, something suspected of being valid. At this point, the treatments of hypotheses in experimental science and in mathematics diverge.

In mathematics, theories must be proved to be accepted as true. Proof is the process of reaching the conclusion suspected of being true by starting with what you know or accept as true and constructing a logical argument to arrive at what you want to prove. The argument should be watertight and convincing.

The theorem known as Pythagoras' theorem states that the square of the hypotenuse of a right-angled triangle is equal to the sum of the squares of the other two sides. Any proof has to be convincing and this enables the result to be used with confidence.

In science, a theory can never be proved. It can be supported (corroborated) by experimental data or disproved if data is obtained that contradicts the theory.

Terminology

In mathematics, a statement may be thought to be true, and attempts may be made to prove it. While the truth of a statement is still under investigation such a statement is called a **hypothesis** or **conjecture** or **proposition** or **premise**.

__Goldbach's conjecture__ states that any even number greater than two can be expressed as the sum of two primes. Although all even numbers up to 4×10^{14} can be expressed as the sum of two primes, no one has been able to prove that every even number can be so expressed. In 2000, a prize of \$1 000 000 was offered for a proof of the conjecture.

Some statements are **definitions**. For example, the sum of the angles at a point is defined as 4 right angles or as 360°.

Many mathematical statements are considered obvious or self-evident or are accepted as true. These are called **axioms**.

For example, if $a = b$ and $b = c$ then $a = c$.

Other propositions are accepted as true but cannot be demonstrated.

Euclid, the renowned Greek geometer, who lived around 300 BC, listed many such propositions on which he built his geometry. One of these states that:
Given a point and a straight line not through the point, there is one and only one line that can be drawn parallel to the given line, passing through the point.

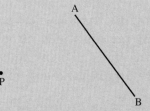

It seems self-evident that there is only one line that can be drawn through P parallel to AB. However certain assumptions have been made. One assumption is that the lines are in a plane (on a flat surface). Geometrical results valid on a plane are not necessarily valid in other geometries. Euclidian geometry is assumed to be in a plane; Einstein's General Theory of Relativity is based on a non-Euclidian geometry, one of curved surfaces.

Using axioms and definitions, and applying logical deduction, many other statements can be proved. These statements are sometimes called **theorems**. Once a result or theorem has been proved it can be used in future proofs.

Example 1   Theorem 1: Vertically opposite angles are equal.

Proof

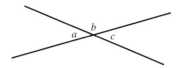

$$a + b = 180°$$ (Angles on a straight line)
$$b + c = 180°$$ (Angles on a straight line)

> By definition a complete revolution is 360° so the sum of the angles on a straight line is 180°.

$$\therefore \quad a + b = b + c$$

> This line follows from the previous two.

$$\therefore \quad\quad a = c$$

> Self-evident from line above.

So vertically opposite angles are equal.

The result that vertically opposite angles are equal may seem so obviously true that a proof is not necessary. A rigorous approach, however, demands that nothing should be accepted without good reason or proof. Some results are not so obvious as the one above, but, after trying many examples, mathematicians may suspect the results are true.

For example, measuring the angles of many triangles, it might be reasonable to conclude that the angle sum is 180°. But even if millions of triangles are drawn and their angles measured, it is not certain that there is *no* possible triangle with a different angle sum.

Attempting to obtain universally valid results from particular instances is a process called **logical induction**. It is a method that may help to formulate intelligent guesses but it *cannot* be used to prove the truth of a statement. To convince mathematicians that, for example, the angle sum of a triangle is always 180°, rigorous argument – in the form of a proof – must be used (see Example 2).

There are several methods that can be used to prove that a hypothesis is true or that it is false. This chapter considers three methods.

- Proof by deduction
- Disproof by counter-example
- Proof by contradiction (*reductio ad absurdum*)

There are other methods of proof, such as proof by exhaustion (see the CD-ROM, E8.1.1) and proof by mathematical induction.

8.2 Proof by deduction

In proof by deduction, each step follows (i.e. is deduced) from the previous one or is justified by quoting an accepted fact or a result previously proved.

Example 2   Theorem 2: The angle sum of a triangle is 180°.

Proof

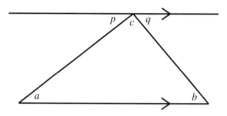

Draw any triangle and let the angles be of size a, b and c. Draw a line parallel to the base as shown. (Euclid proposed that such a line exists and is unique. See page 200.)

$$p = a \qquad \text{(Alternate angles)}$$
$$q = b \qquad \text{(Alternate angles)}$$

> 'Alternate angles are equal' can be quoted. The result can easily be proved using axioms and previously proved results.

$$p + q + c = 180° \qquad \text{(Angles on a straight line)}$$
$$\therefore \quad a + b + c = 180°$$

So the angle sum of any triangle is 180°.

> By definition a complete revolution is 360°, so the sum of the angles on a straight line is 180°.

Using algebra in proofs

In proofs, n (or another letter) is often used to represent 'any number' or 'any integer'. If n represents an integer then even numbers, odd numbers, triangular numbers, consecutive numbers, etc. can be expressed in terms of n.

- An even number can be expressed in the form $2n$, where n is an integer.

- An odd number can be expressed in the form $2n + 1$, where n is an integer.

- The nth triangular number is $\dfrac{n(n+1)}{2}$

- To work with three consecutive integers, use $n - 1$, n, and $n + 1$ or n, $n + 1$, and $n + 2$.

Similarly, any number with digits 'abc' can be written as $100a + 10b + c$. For example

$$574 = 5 \times 100 + 7 \times 10 + 4$$

Some of these expressions are used in Examples 3 and 4.

Example 3   *In this example of proof by deduction, by taking the odd number as $2k + 1$, the general case is considered. The result is therefore true for any specific odd number, e.g. -11, -1, 3, 7, 25 or 123.*

Decide if the following proposition is true or not. If the proposition is true, prove it. If the proposition is not true, find a counter-example.

'The square of an odd number is odd.'

> First, try some examples: $3^2 = 9$, $7^2 = 49$, $25^2 = 625$, $(-11)^2 = 121$. The examples all suggest that the proposition is true, but the proposition must be proved.

Proof   Let the odd number be $2k + 1$.

> An odd number can be expressed in the form $2k + 1$ where k is an integer.

$$(2k + 1)^2 = (2k + 1)(2k + 1)$$
$$= 4k^2 + 4k + 1$$
$$= 2(2k^2 + 2k) + 1$$
$$= 2m + 1$$

where m is an integer.

> The result is put in the form $2m + 1$ to show that it is an odd number. In this case $m = 2k^2 + 2k$.

So the square of an odd number is odd.

Example 4   The digits of a 2-digit number are different. A new number is formed by reversing the digits. Prove that the difference between the squares of the two numbers is a multiple of 99.

Proof   The digits are different, therefore one of the 2-digit numbers is larger than the other 2-digit number. Let the larger number have tens digit a and units digit b. $a \neq b$.

The number with digits 'ab' has value $10a + b$. With the digits reversed, the value is $10b + a$.

The difference of the squares of the two numbers

$$= (10a + b)^2 - (10b + a)^2$$

> $10a + b$ is the larger number.

$$= (11a + 11b)(9a - 9b)$$

> Use difference of squares instead of multiplying out the brackets.

$$= 11(a + b) \times 9(a - b)$$
$$= 99(a^2 - b^2)$$

This number is a multiple of 99 so the difference between the squares of the two numbers is a multiple of 99.

Proving identities

In proving an identity, each line follows from the previous one. To prove an identity, use one of these three methods.

- Start with one side and, by algebraic manipulation, show that it is equal to the other side.
- Show that LHS − RHS = 0.
- Take each side in turn and show that each is equal to the same expression.

Example 5

In this example, the second method is used: showing that LHS − RHS = 0. *The identity could also be proved by taking each side in turn and showing that each is equal to the same expression.*

Prove that $\dfrac{1}{x+1} + \dfrac{2}{x(x-3)} \equiv \dfrac{3}{x-3} - \dfrac{2(x^2+2x-1)}{x(x+1)(x-3)}$.

Proof

LHS − RHS

$$= \frac{1}{x+1} + \frac{2}{x(x-3)} - \frac{3}{x-3} + \frac{2(x^2+2x-1)}{x(x+1)(x-3)}$$

Express as a single fraction.

$$= \frac{x(x-3) + 2(x+1) - 3x(x+1) + 2(x^2+2x-1)}{x(x+1)(x-3)}$$

$$= \frac{x^2 - 3x + 2x + 2 - 3x^2 - 3x + 2x^2 + 4x - 2}{x(x+1)(x-3)}$$

Note: Numerator = 0

$$= 0$$

∴ LHS = RHS

Exercise 8A

1 Prove that the sum of two even numbers is even.

2 Prove that the sum of two odd numbers is even.

3 Prove that the product of two rational numbers is rational.

4 Prove that the sum of the squares of any two consecutive integers is an odd number.

5 Prove that the sum of any two consecutive triangular numbers is a square number.

6 Prove that $xy + x + y = 30$ has no positive integer solutions.

7 Prove that if five consecutive integers are squared, the mean of the squares exceeds the square of their median by 2.

8 Prove that $n^3 - n$ is always a multiple of 6 for $n \geqslant 2$.

9 The digits of a 2-digit number are different. A new number is formed by reversing the digits. Prove that the difference between the two numbers is a multiple of 9.

10 The digits of a 3-digit number are all different. A new number is formed by reversing the digits. Prove that the difference between the two numbers is a multiple of 99.

11 Prove this identity.

$$\frac{2}{5x} + \frac{3x+1}{5(x^2-4)} \equiv \frac{7}{20(x-2)} + \frac{13x+16}{20x(x+2)}$$

12 Given that $\sin x \neq 0$, prove this identity.

$$1 + \cos x \operatorname{cosec} x \cot x \equiv \operatorname{cosec}^2 x$$

8.3 Disproof by counter-example

To investigate the validity of a hypothesis, the first step is usually to test it. If just one example can be found that shows the hypothesis to be false, then the hypothesis is disproved. Such an example, which disproves a hypothesis, is called a counter-example.

Note A counter-example can be used to prove that a hypothesis is false, but not to prove that it is true.

Example 6 Decide if this proposition is true or not.

For $n \geqslant 2$, $2^n - 1$ is a prime number.

If the proposition is true, prove it. If the proposition is not true, find a counter-example.

Solution
For $n = 2$ $\quad 2^2 - 1 = 3$ Try some examples.
For $n = 3$ $\quad 2^3 - 1 = 7$
For $n = 4$ $\quad 2^4 - 1 = 15$

3 and 7 are prime, but 15 is not prime so $n = 4$ provides a counter example. This proves that the proposition is *not* true.

So the hypothesis that 'for $n \geqslant 2$, $2^n - 1$ is a prime number' has been disproved.

Exercise 8B **1** None of these propositions is true. Find a counter-example to disprove each one.

a The product of an odd and an even number is never a perfect square.

b The product of two different irrational numbers is always irrational.

c The product of two different irrational numbers is always rational.

d The sum of two irrational numbers is always irrational.

2 For each of these propositions, decide whether it is true.
If the proposition is true, prove it.
If the proposition is not true, find a counter-example.

   a The sum of two primes is prime.

   b The sum of two primes is never prime.

   c The sum of two rational numbers is rational.

   d $2^{2k} - 1$ is never prime for $k \geqslant 2$.

3 For each of these propositions decide whether it is true.
If the proposition is true, prove it.
If the proposition is not true find a counter-example.

   a $\sqrt{x^2 + y^2} = x + y$

   b If a, b, c and d are positive real numbers and $\dfrac{a}{b} = \dfrac{c}{d}$, then $\dfrac{a}{b} = \dfrac{a+c}{b+d}$.

   c For any real numbers, x and y, $x^2 + y^2 \geqslant 2xy$.

   d $x^2 \geqslant x$ for all values of $x > 0$.

   e If the sum of two numbers, a and b, is 1, then $a^2 + b = b^2 + a$.

   f $x^2 + x + 41$ is prime for all $x \geqslant 0$.

8.4 Proof by contradiction (*reductio ad absurdum*)

In the method of proof by contradiction, a hypothesis is proved to be true by assuming it is *not* true and arriving at a contradiction. This method was used on the CD-ROM accompanying Book 1 (E1.2) to prove that $\sqrt{2}$ is irrational.

Sometimes it is possible to prove a result by more than one method of proof. Examples 7 and 8 could – and would normally – be proved by deduction. They are included here, however, as illustrations of the method of contradiction.

Example 7 Prove, by contradiction, that for all real values of x

$$x^2 + x \geqslant 7x - 9$$

Proof Assume that the statement is not true, i.e. that there exists a real value of x such that

$$x^2 + x < 7x - 9 \quad ①$$

Attempting to solve the inequality gives

$$x^2 + x < 7x - 9$$
$$x^2 - 6x + 9 < 0$$
$$(x - 3)^2 < 0 \quad ②$$

But a perfect square is never negative, so ② has no solution. Therefore no real value of x exists to satisfy ①. Since there is a contradiction, the assumption in ① must be false and therefore the statement $x^2 + x \geqslant 7x - 9$ for all real values of x is true.

Example 8   Prove, by contradiction, that the graph of $y = xe^{x^2}$ has no stationary points.

Proof   Assume that the statement is not true, i.e. that the graph of $y = xe^{x^2}$ has at least one stationary point.

At a stationary point $\dfrac{dy}{dx} = 0$.

$$y = xe^{x^2}$$

> Differentiate the product xe^{x^2}.

$$\frac{dy}{dx} = x \times 2x \times e^{x^2} + e^{x^2}$$

> Factorise the result.

$$= e^{x^2}(2x^2 + 1)$$

But both $e^{x^2} > 0$ and $2x^2 + 1 > 0$ for all real values of x, so $\dfrac{dy}{dx}$ is never zero.

Hence the graph has no stationary point.

Therefore the original assumption must be false and so the statement that 'the graph of $y = xe^{x^2}$ has no stationary points' is true.

Note   Some statements can be proved true or false by deduction or by contradiction. Disproof by counter-example can only be used to prove certain types of statement false (see Exercise 8D, Question 4).

Exercise 8C

1   Prove, by contradiction, that for all real values of x and y

$$x^2 + y^2 \geqslant 2xy$$

2   Prove, by contradiction, that the equation

$$\frac{(x+2)^2}{x} = 1$$

has no real roots.

3   Prove, by contradiction, that for all positive values of x

$$\frac{4x^3 + x}{x^2} \geqslant 4$$

4   Prove, by contradiction, that for all real values of x

$$\frac{2x^2 - 2x + 1}{(1-x)^2} \geqslant 1$$

1 a Prove that, if x and y are connected by the equation

$$x^2 + y^2 - 8x - 4y + 16 = 0$$

then $2 \leqslant x \leqslant 6$ and $0 \leqslant y \leqslant 4$.

b Give a geometrical interpretation for the result in part **a.**

2 Find the errors in these arguments.

a To prove: $\quad\quad\quad\quad -1 = 2$

Given that $\quad\quad 4x + 4 = 1 - 2x$

Then, adding x^2 to both sides, gives

$$x^2 + 4x + 4 = x^2 - 2x + 1$$

So $\quad\quad\quad (x + 2)^2 = (x - 1)^2$

Taking the square root of both sides

$$(x + 2) = (x - 1)$$

Subtracting x $\quad\quad 2 = -1$

b To prove: $\quad\quad 0.0001 > 0.01$

$$4 > 2$$

Hence $\quad\quad 4 \ln 0.1 > 2 \ln 0.1$

And therefore $\quad \ln (0.1)^4 > \ln (0.1)^2$

So $\quad\quad \ln (0.0001) > \ln (0.01)$

Therefore $\quad\quad 0.0001 > 0.01$

3 To calculate 6.5^2 mentally, multiply 6 by 7 (the next consecutive number) and add 0.25.

a Apply the method to calculate 2.5^2, 7.5^2 and 9.5^2.

b Modify the method to calculate 35^2, 85^2 and 105^2.

c Prove that the method works for any number of the form $n + \frac{1}{2}$, where n is a positive integer.

4 Here are three false statements.

Statement 1: If $x < 1$ then $x^2 < 1$ for all real values of x.

Statement 2: There exists a real value of x for which

$$f(x) \equiv x^3 - 3x^2 + 12x + 1$$

is decreasing.

Statement 3: Given that

$$S_n = 1 + 2 + 2^2 + \cdots + 2^n$$

where $n \in \mathbb{Z}, n \geqslant 2$, S_n is prime if n is even and not prime if n is odd.

a Prove that each statement is false.

b Comment on the type of statements that can be disproved by finding a counter-example.

5 For each of these propositions about the months (in English), decide whether it is true. If the proposition is true, prove it. If the proposition is not true find a counter-example.

   a   If the name of a month contains the letter y, then it contains the letter r.

   b   If the name of a month contains the letter j, then it contains the letter u.

   c   If the name of a month contains the letter u, then it contains the letter j.

   d   If the name of a month contains the letter w, then it contains the letter a.

6 Find a counter-example to disprove each of these propositions.

   a   If triangle ABC is isosceles, then AB = AC.

   b   All quadrilaterals with four equal sides are squares.

   c   A shape with rotational symmetry of order 4 has four axes of symmetry.

   d   If a hexagon has six equal angles, it must have six equal sides.

There is an 'Extension exercise' (Ext8) on the CD-ROM.

➤ Key points

Proof

- **Proof by deduction**: Each step follows from the previous one or is justified by quoting an accepted fact or a result previously proved.

- **Disproof by counter-example**: A counter-example can be used to prove that a hypothesis is false, *not* to prove that it is true.

- **Proof by contradiction (*reductio ad absurdum*)**: The hypothesis is proved to be true by assuming it is *not* true and arriving at a contradiction.

There are other methods, such as proof by induction and proof by exhaustion.

Using algebra in proofs

An **even** number can be expressed in the form $2n$, where n is an integer.

An **odd** number can be expressed in the form $2n + 1$, where n is an integer.

To work with 3 **consecutive integers** use, for example, $n - 1$, n, $n + 1$ or n, $n + 1$, $n + 2$.

If the digits of a number are 'abc', the number is equal to $100a + 10b + c$.

Proving identities

Start with one side and by algebraic manipulation show that it is equal to the other side.

Or show that LHS − RHS = 0.

Or take each side in turn and show that each is equal to the same expression.

1 Find $\int_1^2 \left(e^{\frac{x}{2}} + \frac{1}{x} \right) dx$, giving your answer to three significant figures.

AQA Nov 2002

2 The function f is defined for $x > 0$ by

$$f(x) = e^{-2x} + \frac{3}{x} + 3.$$

  a **i** Differentiate f(x) with respect to x to find f$'(x)$.

   ii Hence prove that f is a decreasing function.

  b Find the range of f.

  c Show that the area of the region bounded by the curve $y = e^{-2x} + \frac{3}{x} + 3$,

   the x-axis, and the lines $x = 1$ and $x = 2$ is

   $$\frac{e^2 - 1}{2e^4} + 3(\ln 2 + 1).$$

AQA June 2003

3 A curve has equation $y = e^{3x} - 24x$.

  a Show that the curve cuts the x-axis at the point where $x = \alpha$, with $1 < \alpha < \ln 3$.

  b Determine $\frac{dy}{dx}$ and $\frac{d^2y}{dx^2}$ as functions of x.

  c The curve has a single turning point, P. Find the x-coordinate of P in an exact form, and show that its y-coordinate is $8(1 - 3\ln 2)$. Find the value of $\frac{d^2y}{dx^2}$ at P and hence deduce the nature of the turning point.

  d Find, in terms of e, the area of the region enclosed by the curve, the x-axis, and the lines $x = 2$ and $x = 3$.

AQA Jan 2001

4 The diagram shows a sketch of the curve with equation $y = e^{2x} - 3$ which crosses the y-axis at the point A and the x-axis at the point B.

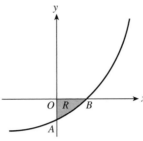

a Find the y-coordinate of A.

b Find the exact value of the x-coordinate of B.

c i Find $\dfrac{\mathrm{d}y}{\mathrm{d}x}$.

   ii Show that the gradient of the curve at B is three times its gradient at A.

d Find the area of the shaded region R bounded by the curve and the coordinate axes. Give your answer in the form $p + q \ln r$, where p, q and r are constants to be determined.

AQA Jan 2002

5 It is given that

$$f(x) = 2x^3 + 3x^2 + 7.$$

a Find the derivative $f'(x)$, factorising your answer.

b Hence, show that $\displaystyle\int_0^2 \frac{x(x+1)}{f(x)}\,\mathrm{d}x = k \ln 5$, stating the value of the constant k.

AQA June 2003

6 a Find $\displaystyle\int (6 \tan \theta - \sec^2 \theta)\,\mathrm{d}\theta$.

b Find the solutions of

$$6 \tan \theta - \sec^2 \theta = 7$$

in the interval $0 \leqslant \theta < 2\pi$, giving each answer in radians to one decimal place.

No credit will be given for simply reading values from a graph.

AQA Jan 2002

7 The diagram shows a sketch of the curve with equation $y = 4 - e^{2x}$ which crosses the y-axis at the point A and the x-axis at the point $B(\ln 2, 0)$.

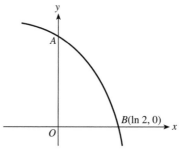

a Find the y-coordinate of A.

b i For the curve $y = 4 - e^{2x}$, find $\dfrac{dy}{dx}$.

   ii Find the value of the gradient of the curve at B.

c

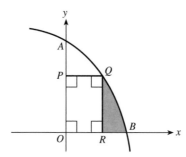

The line $y = 2$ cuts the y-axis at the point P and the curve at the point Q. $OPQR$ is a rectangle.

i Find the exact value of the area of rectangle $OPQR$.

ii Find $\displaystyle\int (4 - e^{2x})\,dx$.

iii Find the area of the shaded region bounded by the curve, the x-axis and the line QR. Give your answer in the form $p \ln 2 + q$, where p and q are constants to be determined.

iv Show that the area of the region above PQ, bounded by the curve, the y-axis and the line PQ is half the area of the shaded region.

AQA Jan 2003

8 The diagram shows a sketch of the curve $y = e^{2x} - 2$.

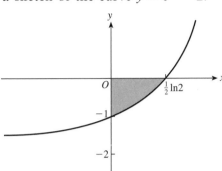

a On separate diagrams, sketch the graphs of the following curves showing the coordinates of the points where the graph intersects the coordinate axes:

i $y = |e^{2x} - 2|$

ii $y = e^{2x} - 5$

b i Find $\int (e^{2x} - 2)\,dx$.

ii Hence show that the area of the shaded region bounded by the curve $y = e^{2x} - 2$ and the coordinate axes is

$$\ln 2 - \tfrac{1}{2}.$$

AQA May 2002

9 The function f is defined by

$$f : x \mapsto \tan 2x - 4x,$$

and has domain $0 \leqslant x \leqslant \dfrac{\pi}{4}$.

a Determine the value of x for which f is stationary.

b Explain why the range of f cannot be $f(x) \geqslant 0$.

c Find $\displaystyle\int_0^{\frac{\pi}{8}} f(x)\,dx$.

10 The curve with equation

$$y = x\sqrt{x^2 + 3}$$

is sketched below. The region R, shaded on the diagram, is bounded by the curve, the x-axis and the line $x = 1$.

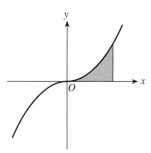

a The region R is rotated through 2π radians about the x-axis. Find the volume of the solid generated.

b Determine an equation of the tangent to the curve at the point where $x = 1$.

c **i** Differentiate $(x^2 + 3)^{\frac{3}{2}}$ with respect to x.

ii Use the result from part **c i** to find the area of region R.

AQA Jan 2002

11 The function f has domain $\dfrac{\pi}{30} \leqslant x \leqslant \dfrac{\pi}{10}$ and is defined by

$$\text{f} : x \mapsto \cot 5x.$$

a Use differentiation to prove that f is a decreasing function.

b Find the range of f.

c The region R is bounded by the curve with equation $y = \text{f}(x)$, the x-axis, and the lines $x = \dfrac{\pi}{30}$ and $x = \dfrac{\pi}{10}$.

i Find the area of R, expressing your answer in the form $k \ln 2$, where the value of k is to be found.

ii Show that the volume of the solid formed by rotating R through 2π radians about the x-axis is

$$\frac{\pi}{15}\left(3\sqrt{3} - \pi\right).$$

AQA June 2001

12 a Find $\displaystyle\int 4x\cos 2x \,\mathrm{d}x$.

b Hence show that $\displaystyle\int_0^{\frac{\pi}{2}} x(1 + 4\cos 2x)\,\mathrm{d}x = \frac{\pi^2}{8} - 2$.

13 The function f is defined by

$$f(x) = \frac{2x}{(1+2x)^3}, \quad x \neq -\tfrac{1}{2}.$$

Using the substitution $u = 1 + 2x$, or otherwise, find

$$\int f(x)\, dx.$$

AQA June 2002

14 a i Draw on the same diagram sketches of the graphs with equations

$$y = 5\,e^{2x} \quad \text{and} \quad y = \frac{4}{x} \quad \text{for } x > 0.$$

ii Explain why this diagram shows that, for $x > 0$, the equation

$$5\,e^{2x} - \frac{4}{x} = 0$$

has just one root, α, and show that $0.3 < \alpha < 0.4$.

b Show, using calculus, that $y = 5\,e^{2x} - \dfrac{4}{x}$ is an increasing function of x for $x > 0$.

c Show that the area of the region enclosed by the curve $y = 5\,e^{2x} - \dfrac{4}{x}$, the x-axis, and the lines $x = \tfrac{1}{2}$ and $x = 2$ can be expressed in the form

$$\frac{5}{2}(e^4 - e) - k \ln 2$$

for some positive integer k whose value is to be determined.

AQA June 2001

15 a By taking logarithms, solve the equation $0.6 = 2^{-x}$, giving your answer to three significant figures.

b A sequence is defined by $x_{n+1} = 2^{-x_n}$, $x_1 = 0.6$.

i Find the values of x_2, x_3, x_4 and x_5, giving your answers to three decimal places.

ii Given that the sequence converges to α, write down an equation in x for which α is a root.

AQA June 2001

16 a Show that the equation $x^3 - 5x^2 - x + 6 = 0$ has a root between 4 and 5.

b Show that the equation $x^3 - 5x^2 - x + 6 = 0$ can be rearranged in the form

$$x = \frac{5x^2 + x - 6}{x^2}.$$

c Use the iterative formula $x_{n+1} = \dfrac{5x_n^2 + x_n - 6}{x_n^2}$ with $x_1 = 4$ to find x_2 and x_3, giving your answers to 3 significant figures.

17 Use Simpson's Rule with five ordinates (four equal strips) to find an approximation to the integral

$$\int_0^2 \ln(x^2 + 1)\,dx$$

giving your answer to 3 decimal places.

AQA June 2003

18 a Use Simpson's Rule with five ordinates (four strips) to find an approximation to the integral

$$I = \int_0^1 x \cos x\,dx,$$

giving your answer to five decimal places.

b Use integration by parts to find the value of I, giving your answer to five decimal places.

AQA Jan 2002

19 Prove that, for A, B, C and D positive real numbers,

a if $\dfrac{A}{B} = \dfrac{C}{D}$

then $\dfrac{A}{B} = \dfrac{A+C}{B+D}$

b if $\dfrac{A}{B} = \dfrac{A+C}{B+D}$

then $\dfrac{A}{B} = \dfrac{C}{D}$

20 a Prove that a sufficient condition for x^2 to be rational is that x is rational.

b Explain, giving an example, why the condition in part **a** is not a necessary condition.

c Prove that a necessary condition for $x + y$ to be irrational is that either x or y or both are irrational.

d Explain, giving an example, why the condition in part **c** is not a sufficient condition.

- **Time $1\frac{1}{2}$ hours.**

- **You are advised to show all your working.**

- **Calculators may be used.**

---

1. (a) (i) Differentiate $1 + \cos x$ with respect to x. *(1 mark)*

   (ii) Hence find $\displaystyle\int \frac{2\sin x}{1 + \cos x}\, \mathrm{d}x$. *(2 marks)*

   (b) Given that $y = \dfrac{2\sin x}{1 + \cos x}$, show that $\dfrac{\mathrm{d}y}{\mathrm{d}x} = \dfrac{2}{1 + \cos x}$. *(4 marks)*

---

2. It is given that $f(x) = \sin^{-1} x + 1 - 9x^2$.

   (a) Show that there is a root of $f(x) = 0$ in the interval $0.3 \leqslant x \leqslant 0.4$. *(2 marks)*
   (b) This root is to be estimated using the iterative formula

   $$x_{n+1} = \sqrt{\frac{\sin^{-1} x_n + 1}{9}}, \quad x_0 = 0.4.$$

   Showing your values of $x_1, x_2, x_3, \ldots$, obtain the value, to three decimal places, of the root. *(4 marks)*

---

3. A function f is defined for all real values of x by $f(x) = e^{2x+3} - 1$.

   (a) (i) Find the range of f. *(1 mark)*
   (ii) Sketch the curve with equation $y = f(x)$, showing the coordinates of any points at which the curve meets the coordinate axes. *(4 marks)*

   (b) The curve with equation $y = f(x)$ has a gradient of 8 at the point P. Find the x-coordinate of P, giving your answer in the form $\ln a + b$, where a is an integer and b is a constant. *(6 marks)*

---

4. It is given that $\cot x + 3 - \operatorname{cosec}^2 x = 0$.

   (a) Show that this equation can be written in the form $\cot^2 x - \cot x - 2 = 0$. *(2 marks)*
   (b) Hence solve the equation $\cot x + 3 - \operatorname{cosec}^2 x = 0$ giving all values of x, where appropriate, to one decimal place in the interval $0° \leqslant x < 360°$. *(5 marks)*

---

5. (a) Use integration by parts to find $\int \ln x \, \mathrm{d}x$. *(4 marks)*

(b) Describe a sequence of geometrical transformations that maps the graph of $y = \ln x$ onto the graph of $y = 3\ln(x+2)$. *(4 marks)*

(c) The region R is bounded by the curve $y = 3\ln(x+2)$, the x-axis from the point $(-1, 0)$ to the point $(3, 0)$ and the line $x = 3$.
Use the substitution $u = x + 2$ to show that the area of R is $3(5\ln 5 - 4)$. *(3 marks)*

(d) (i) Use Simpson's Rule with five ordinates (four strips) to find an approximate value for $\displaystyle\int_{-1}^{3} \left[\ln(x+2)\right]^2 \mathrm{d}x$, giving your answer to four significant figures. *(4 marks)*

(ii) Hence find an approximation of the volume of the solid formed when R is rotated through 2π radians about the x-axis, giving your answer to the nearest integer. *(1 mark)*

---

6. The diagram shows part of the curve with equation $y = 3x\sin^2 x$. The point O is the origin.

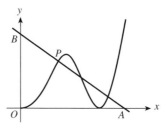

(a) Find $\dfrac{\mathrm{d}y}{\mathrm{d}x}$. *(4 marks)*

(b) The point P on the curve has x-coordinate $\dfrac{\pi}{2}$. Find an equation for the normal to the curve at P, giving your answer in the form $ax + by + c = 0$, where a and b are integers. *(5 marks)*

(c) The normal to the curve at the point P cuts the x-axis at the point A and the y-axis at the point B. Find the exact value of the area of triangle OAB. *(3 marks)*

---

7. The functions f and g are defined with their respective domains by

$$f(x) = \frac{a}{x}, \qquad \text{for real values of } x, \, x \neq 0,$$

$$g(x) = ax + k, \quad \text{for real values of } x,$$

where a and k are positive constants.

(a) Sketch the curve with equation $y = |f(x)|$.   *(2 marks)*

(b) The line $y = g(x)$ is a tangent to the curve $y = |f(x)|$ at the point P,
     and cuts the curve $y = |f(x)|$ at the point Q.
     (i) Show that $k = 2a$.   *(7 marks)*
     (ii) Given that $a = 3$, solve the equation $fg(x) = g^{-1}(x)$, giving your
          answer in the form $a + b\sqrt{c}$ where a, b and c are integers.   *(7 marks)*

END OF QUESTIONS

Algebra

Earlier in the course you saw how algebraic fractions can be added, subtracted, multiplied and divided just like numerical fractions. You also saw how simple algebraic division can be performed and learnt about the factor and remainder theorems.

There have been many occasions in the mathematics you have learnt so far where it has been useful to be able to apply a technique and its reverse process. For example, you have seen how to expand brackets and collect like terms and also how to factorise expressions; both these techniques are used in solving equations.

In this chapter you will apply the techniques you covered earlier in the course to more complicated examples. You will also look at the process of splitting a single fraction up into the sum of two or more simpler, partial fractions. This method will be used both in calculus and in expanding some power series.

After working through this chapter you should be able to

■ *add, subtract, multiply and divide more complicated algebraic fractions*

■ *simplify rational expressions*

■ *divide one polynomial by another linear or quadratic polynomial and hence identify the quotient and remainder*

■ *use the factor theorem and the remainder theorem*

■ *decompose a single fraction into partial fractions.*

9.1 Adding and subtracting algebraic fractions

Remember Fractions can be added or subtracted only if they are expressed with equal denominators.

Fractions can be added by creating a new denominator that is the product of both denominators.

➤
$$\frac{a}{b} + \frac{c}{d} = \frac{ad + bc}{bd}$$

However, this is not always the most efficient way of adding fractions. Instead, the lowest (least) common denominator (LCM) should be used.

Example 1 Calculate $\frac{1}{6} + \frac{2}{15}$.

$$\frac{1}{6} + \frac{2}{15} = \frac{5}{30} + \frac{4}{30}$$
$$= \frac{9}{30}$$
$$= \frac{3}{10}$$

> Instead of using $6 \times 15 = 90$, use 30, the LCM of 6 and 15, to express both fractions with a common denominator.

> Cancel the fraction to its lowest terms.

The procedure is the same for algebraic fractions as for numerical ones.

Example 2   Express $1 + \dfrac{3}{2x - 1} + \dfrac{x + 1}{6x^2 - 3x}$ as a single fraction.

$1 + \dfrac{3}{2x - 1} + \dfrac{x + 1}{6x^2 - 3x}$

$= 1 + \dfrac{3}{2x - 1} + \dfrac{x + 1}{3x(2x - 1)}$

> Factorising $6x^2 - 3x$ gives $3x(2x - 1)$.
> So the denominators are $(2x - 1)$ and $3x(2x - 1)$.
> The LCM of the denominators is $3x(2x - 1)$.

$= \dfrac{3x(2x - 1)}{3x(2x - 1)} + \dfrac{3 \times 3x}{3x(2x - 1)} + \dfrac{x + 1}{3x(2x - 1)}$

> Express 1 as $\dfrac{3x(2x - 1)}{3x(2x - 1)}$.

$= \dfrac{6x^2 - 3x + 9x + x + 1}{3x(2x - 1)}$

$= \dfrac{6x^2 + 7x + 1}{3x(2x - 1)}$

$= \dfrac{(6x + 1)(x + 1)}{3x(2x - 1)}$

> The fraction cannot be cancelled.

> The numerator and denominator are left in factors.

Example 3   **a**   Simplify $\dfrac{6}{x + 2} - \dfrac{3}{x^2 + 2x}$.   **b**   Hence solve $\dfrac{6}{x + 2} - \dfrac{3}{x^2 + 2x} = 1$.

Solution   **a** $\dfrac{6}{x + 2} - \dfrac{3}{x^2 + 2x} = \dfrac{6}{x + 2} - \dfrac{3}{x(x + 2)}$

> Factorise where possible. The LCM of $x + 2$ and $x(x + 2)$ is $x(x + 2)$.

$= \dfrac{6x}{x(x + 2)} - \dfrac{3}{x(x + 2)}$

$= \dfrac{6x - 3}{x(x + 2)}$

> Factorise the numerator.

$= \dfrac{3(2x - 1)}{x(x + 2)}$

b $\dfrac{6}{x + 2} - \dfrac{3}{x^2 + 2x} = 1$

> Rewrite the LHS using the result obtained in part **a**.

$\dfrac{3(2x - 1)}{x(x + 2)} = 1$

> Multiply both sides by $x(x + 2)$.

$3(2x - 1) = x(x + 2)$

> Expand the brackets and collect all the terms on one side of the equation.

$6x - 3 = x^2 + 2x$

$x^2 - 4x + 3 = 0$

$(x - 1)(x - 3) = 0$

$x = 1 \text{ or } x = 3$

In Section 9.5, on partial fractions, the process of adding fractions is reversed; fractions are split into two or more simpler fractions.

9.2 Multiplying and dividing algebraic fractions

Whereas fractions can be added or subtracted only if they are expressed with equal denominators, *any* fractions can be multiplied and divided.

➤
$$\frac{a}{b} \times \frac{c}{d} = \frac{ac}{bd} \quad \text{and} \quad \frac{a}{b} \div \frac{c}{d} = \frac{a}{b} \times \frac{d}{c} = \frac{ad}{bc}$$

Before multiplying, any factor which appears in both the numerator and denominator may be cancelled, and this may make the working easier.

Note Never cancel part of a bracket; only cancel the whole bracket if it is a factor of both the numerator and denominator.

Remember
- Dividing by $\frac{a}{b}$ is the same as multiplying by its reciprocal $\frac{b}{a}$.
- Multiplying or dividing both the numerator and denominator of a fraction by the *same* quantity does not change the value of the fraction.

Examples 4–6 show how fractions can be simplified.

Example 4
$$\frac{x^2 - 4}{3y^2 + y} \times \frac{3y^2 - 2y - 1}{x^2 + x - 6}$$

Factorise each expression where possible.

$$= \frac{(x+2)(x-2)}{y(3y+1)} \times \frac{(3y+1)(y-1)}{(x+3)(x-2)}$$

Cancel any factor which appears in the numerator *and* the denominator. Only cancel whole brackets, never part of a bracket.

$$= \frac{(x+2)(y-1)}{y(x+3)}$$

Example 5
$$\frac{3x + \frac{1}{xy}}{x - \frac{1}{9x^2}}$$

Multiply numerator and denominator by $9x^2y$, the LCM of $9x^2$ and xy.

$$= \frac{9x^2y\left(3x + \frac{1}{xy}\right)}{9x^2y\left(x - \frac{1}{9x^2}\right)}$$

$$= \frac{27x^3y + 9x}{9x^3y - y}$$

$$= \frac{9x(3x^2y + 1)}{y(9x^3 - 1)}$$

Numerator and denominator are given in factors.

Example 6 $\dfrac{(1-x)^2}{x^2+3x-4} \div \dfrac{x^2-8x+7}{x^2+4x}$

Replace $\div$ by $\times$ and invert the second fraction.

$= \dfrac{(1-x)^2}{x^2+3x-4} \times \dfrac{x^2+4x}{x^2-8x+7}$

Factorise each expression.

$= \dfrac{(1-x)^2}{(x+4)(x-1)} \times \dfrac{x(x+4)}{(x-7)(x-1)}$

Cancel any factor which appears in the numerator and the denominator.

$= \dfrac{x}{x-7}$

Note:
$(x-1) = -(1-x)$
but
$(1-x)^2 = (x-1)^2$

Exercise 9A

1 Express each of these as a single fraction in its simplest form.

  a $\dfrac{1}{x} - \dfrac{1}{y}$ **b** $\dfrac{x}{y} + \dfrac{y}{x}$

  c $\dfrac{1}{a^2} + \dfrac{1}{a}$ **d** $\dfrac{1}{ab^2} + \dfrac{1}{a^2 b}$

  e $\dfrac{1}{x-h} + \dfrac{1}{x+h}$ **f** $\dfrac{1}{(x+h)^2} - \dfrac{1}{x^2}$

  g $\dfrac{1}{1-x} - \dfrac{2}{2+x}$ **h** $\dfrac{x}{x^2+2} - \dfrac{2}{2+x}$

  i $\dfrac{n}{n+1} + \dfrac{1}{(n+1)(n+2)}$ **j** $\dfrac{1}{(x+1)^2} + \dfrac{1}{x+1} + 1$

2 Express each of these as a single fraction in its simplest form.

  a $1 + \dfrac{3}{x}$ **b** $2 - \dfrac{4}{x^2}$ **c** $\dfrac{3}{4} + \dfrac{5}{2x-3}$

  d $5 - \dfrac{2}{x^2-1}$ **e** $\dfrac{x+1}{3x-4} - \dfrac{5}{2}$ **f** $\dfrac{x^2}{x^2+2x-3} - 1$

  g $2x + 1 + \dfrac{1}{2x+1}$ **h** $3x - \dfrac{3}{2-x}$ **i** $6xy - \dfrac{12xy}{2-3z^2}$

3 Simplify these expressions.

  a $\dfrac{a+b}{4a+4b}$ **b** $\dfrac{ax^2}{a^2 x}$ **c** $\dfrac{a+b}{a^2-b^2}$

  d $\dfrac{a-b}{b-a}$ **e** $\dfrac{x+2}{x^2+5x+6}$

4 Express each of these as a single fraction in its simplest form.

  a $\dfrac{5x^2}{y} \times \dfrac{y^2}{10x}$ **b** $\dfrac{a-b}{c} \times \dfrac{c^2}{a^2-b^2}$ **c** $\dfrac{x^4}{y^4} \times \left(\dfrac{y}{x}\right)^2$

  d $\dfrac{5x}{4y} \div \dfrac{10x^2}{8y^2}$ **e** $\dfrac{1}{3x-y} \div \dfrac{1}{5y-15x}$ **f** $\dfrac{22xyz}{3a^3 b^2} \times \dfrac{12ab^2 x}{11z} \times \dfrac{15b}{4x^2}$

  g $\dfrac{5t^2-125}{t^2+5t} \div \dfrac{10t^2+40t-50}{3t^2}$

5 Simplify these, expressing each one as a single fraction in its simplest form.

a $\dfrac{\frac{1}{a}}{\frac{a}{b}}$

b $\dfrac{\frac{x}{a}}{\frac{a+b}{a}}$

c $\dfrac{x+1}{1-\frac{1}{x^2}}$

d $\dfrac{\frac{x}{y}-\frac{y}{x}}{\frac{1}{y}-\frac{1}{x}}$

e $\dfrac{3+\frac{1}{x}}{3-\frac{1}{x}}$

f $\dfrac{y-\frac{y}{x}}{\frac{1}{x}-1}$

g $\dfrac{\frac{x}{x+1}-x}{x+\frac{x}{x-1}}$

h $\sqrt{\dfrac{1-\frac{2t}{1+t^2}}{1+\frac{2t}{1+t^2}}}$

6 Express each of these as a single fraction in its simplest form.

a $\dfrac{1}{(x+2)^2}-\dfrac{2}{x+2}+\dfrac{1}{3x-1}$

b $\dfrac{4}{2+3x^2}-\dfrac{1}{1-x}$

c $\dfrac{x}{(x+3)(x-2)}+\dfrac{1}{x-2}$

d $\dfrac{2}{x^2+3x+2}-\dfrac{1}{x+2}$

e $\dfrac{3}{x^2+1}-\dfrac{1}{x-1}+\dfrac{2}{(x-1)^2}$

f $\dfrac{3x}{x^2+2x-3}-\dfrac{x+1}{x^2+5x+6}$

g $\dfrac{2x+1}{x^2+4}+\dfrac{6}{x-1}$

h $\dfrac{1-x}{2x^2+1}-\dfrac{4}{1-2x}+3$

i $x+\dfrac{1}{x+4}-\dfrac{2}{x-1}$

j $2x-3+\dfrac{3}{x-1}-\dfrac{2}{1+x}$

k $\dfrac{2}{x-2}-\dfrac{3x+1}{x^2-7x+10}-\dfrac{1}{x-5}$

l $3-\dfrac{2x}{x^2-1}+\dfrac{7}{x-1}$

m $\dfrac{a}{\sqrt{a+b}}+\dfrac{b}{\sqrt{a+b}}$

n $\dfrac{1}{\sqrt{1+x^2}}-\dfrac{x^2}{(1+x^2)\sqrt{(1+x^2)}}$

7 a Express as a single fraction: $\dfrac{x+10}{x-5}-\dfrac{10}{x}$

b Hence solve for x: $\dfrac{x+10}{x-5}-\dfrac{10}{x}=\dfrac{11}{6}$

c Check the answer by substitution.

8 a Express as a single fraction: $\dfrac{2x}{x-1}+\dfrac{3x-1}{x+2}$

b Hence solve for x: $\dfrac{2x}{x-1}+\dfrac{3x-1}{x+2}=\dfrac{5x-11}{x-2}$

c Check the answer by substitution.

9.3 Division of polynomials

Earlier in the course you learnt how to carry out simple algebraic division by $(x + a)$ or $(x - a)$. This section extends the techniques to cover division by more complicated linear, and quadratic, expressions.

Remember
- Arrange the divisor and dividend in descending powers of the variable, lining up like terms, and leaving gaps in the dividend for 'missing' terms.

- Divide the term on the left of the dividend by the term on the left of the divisor. The result is a term of the quotient.

- Multiply the whole divisor by this term of the quotient and subtract the product from the dividend.

- Bring down as many terms as necessary to form a new dividend.

- Repeat the instructions until all terms of the dividend have been used.

Example 7 Divide $8x^3 - 6x^2 + x + 5$ by $2x + 1$.

$$
\begin{array}{r}
4x^2 - 5x + 3 \\
2x + 1 \overline{)\,8x^3 - 6x^2 + x + 5} \\
\underline{8x^3 + 4x^2} \\
-10x^2 + x \\
\underline{-10x^2 - 5x} \\
6x + 5 \\
\underline{6x + 3} \\
2
\end{array}
$$

So $8x^3 - 6x^2 + x + 5 \equiv (2x + 1)(4x^2 - 5x + 3) + 2$

or $\dfrac{8x^3 - 6x^2 + x + 5}{2x + 1} \equiv 4x^2 - 5x + 3 + \dfrac{2}{2x + 1}$

Remember The sign $\equiv$ means 'is identically equal to'.

Example 8 Find the quotient and remainder when $9x^3 + 5x + 1$ is divided by $3x - 2$.

$$
\begin{array}{r}
3x^2 + 2x + 3 \\
3x - 2 \overline{)\,9x^3 + 5x + 1} \\
\underline{9x^3 - 6x^2} \\
6x^2 + 5x \\
\underline{6x^2 - 4x} \\
9x + 1 \\
\underline{9x - 6} \\
7
\end{array}
$$

> There is no term in x^2 in the dividend, so a gap must be left.

So the quotient is $3x^2 + 2x + 3$ and the remainder is 7.

Example 9 Divide $2x^3 + 3x^2 - 4x + 1$ by $x^2 + 2$.

$$
\begin{array}{r}
2x + 3 \\
x^2 + 2 \overline{)\,2x^3 + 3x^2 - 4x + 1} \\
\underline{2x^3 + 4x } \\
3x^2 - 8x + 1 \\
\underline{3x^2 + 6} \\
-8x - 5
\end{array}
$$

So $2x^3 + 3x^2 - 4x + 1 \equiv (x^2 + 2)(2x + 3) - 8x - 5$

or $\dfrac{2x^3 + 3x^2 - 4x + 1}{x^2 + 2} \equiv 2x + 3 - \dfrac{8x + 5}{x^2 + 2}$

> The minus sign in front of the fraction means the whole fraction is subtracted. $-8x - 5 = -(8x + 5)$

Example 10 Divide $2x^3 + 9x^2 + 5x - 6$ by $x^2 + 3x - 2$.

$$
\begin{array}{r}
2x + 3 \\
x^2 + 3x - 2 \overline{)\,2x^3 + 9x^2 + 5x - 6} \\
\underline{2x^3 + 6x^2 - 4x } \\
3x^2 + 9x - 6 \\
\underline{3x^2 + 9x - 6} \\
0
\end{array}
$$

So $2x^3 + 9x^2 + 5x - 6 \equiv (2x + 3)(x^2 + 3x - 2)$

A simpler method of division

When the degree of the polynomials in the numerator and denominator are the same, the division can be carried out more simply.

Example 11 Divide $x + 2$ by $x + 1$.

> Both numerator and denominator are of degree 1.

$\dfrac{x + 2}{x + 1} = \dfrac{x + 1 + 1}{x + 1}$

> Rewrite the numerator so that it starts with a multiple of the denominator.

$\phantom{\dfrac{x + 2}{x + 1}} = \dfrac{x + 1}{x + 1} + \dfrac{1}{x + 1} = 1 + \dfrac{1}{x + 1}$

Example 12 Express $\dfrac{3x - 5}{x + 2}$ in the form $A + \dfrac{B}{x + 2}$.

> Rewrite the numerator so that it starts with a multiple of the denominator, $x + 2$.

$\dfrac{3x - 5}{x + 2} = \dfrac{3(x + 2) - 11}{x + 2} = 3 - \dfrac{11}{x + 2}$

Example 13 Express $\dfrac{4x^2 - 3x}{x^2 + 5}$ in the form $A + \dfrac{Bx + C}{x^2 + 5}$ where A, B and C are constants.

$$\frac{4x^2 - 3x}{x^2 + 5} = \frac{4(x^2 + 5) - 3x - 20}{x^2 + 5}$$

$$= 4 - \frac{3x + 20}{x^2 + 5}$$

> The minus sign in front of the fraction means the whole fraction is subtracted; $-3x - 20 = -(3x + 20)$.

A similar method can also be used in other cases as the following example shows.

Example 14 Show that $\dfrac{x^3 + 1}{x^2 - 1} = x + \dfrac{1}{x - 1}$.

$$\frac{x^3 + 1}{x^2 - 1} = \frac{\cancel{(x + 1)}(x^2 - x + 1)}{\cancel{(x + 1)}(x - 1)}$$

> Factorise both the numerator and denominator and cancel the common factor.

$$= \frac{x^2 - x + 1}{x - 1}$$

> Rewrite the numerator so that it starts with a multiple of the denominator, $x - 1$.

$$= \frac{x(x - 1) + 1}{x - 1}$$

$$= \frac{x(x - 1)}{x - 1} + \frac{1}{x - 1} = x + \frac{1}{x - 1}$$

Identities

The idea of an identity was introduced in Chapter 3 of Book 1, with some extension material on the CD-ROM for Chapter 7.

The following two examples show the techniques you will need to use in this chapter. In these examples the identities contain unknowns, A, B, etc. which have to be found. One method of finding unknowns is by **substitution**. Since the expressions are equal for *all* values of x, *any* value can be substituted. Choose values, where possible, that eliminate one of the unknowns.

Example 15 Find A and B given that $x + 14 \equiv A(x + 4) + B(x - 1)$.

> Use substitution.

$$x + 14 \equiv A(x + 4) + B(x - 1)$$

> Since this is an identity and thus true for all values of x, any value can be substituted.

Putting $x = -4$
$$-4 + 14 = A \times 0 - 5B$$
$$10 = -5B$$
$$B = -2$$

> Putting $x = -4$ eliminates one of the terms on the RHS since it makes the bracket $(x + 4)$ zero.

Putting $x = 1$
$$1 + 14 = 5A + B \times 0$$
$$15 = 5A$$
$$A = 3$$

> Putting $x = 1$ eliminates the other term on the RHS since it makes the bracket $(x - 1)$ zero.

So $A = 3$ and $B = -2$, and $x + 14 \equiv 3(x + 4) - 2(x - 1)$.

> This can easily be checked by multiplying out the RHS.

Example 16   Find A and B given that $5x^2 + 3x + 1 \equiv A(x^2 + 2) + B(2x^2 + x + 1)$.

Method 1

$5x^2 + 3x + 1 \equiv A(x^2 + 2) + B(2x^2 + x + 1)$

Putting $x = 0$

$1 = A \times 2 + B \times 1$

$1 = 2A + B$ ①

> There are no real values of x which make $(x^2 + 2)$ or $(2x^2 + x + 1)$ zero. However, since this is an identity, any value of x can be substituted. Equations connecting A and B can be found and then solved.

Putting $x = 1$

$5 + 3 + 1 = A(1 + 2) + B(2 + 1 + 1)$

$9 = 3A + 4B$ ②

① × 4 $4 = 8A + 4B$ ③

③ − ② $-5 = 5A$

$A = -1$

Substituting in ①

$1 = -2 + B$

$B = 3$

So $A = -1$ and $B = 3$

> Remember to check the solution.

Method 2

Another technique for solving identities is by **equating coefficients**.

Given that, for example, $ax^2 + bx + c \equiv Ax^2 + Bx + C$, it can be shown that $a = A$, $b = B$ and $c = C$, i.e. the coefficients of the x^2 and x terms must be equal. Similarly the constant terms must be equal.

$5x^2 + 3x + 1 \equiv A(x^2 + 2) + B(2x^2 + x + 1)$

$\equiv Ax^2 + 2A + 2Bx^2 + Bx + B$

$\equiv (A + 2B)x^2 + Bx + (2A + B)$

> Multiplying out brackets and collecting like terms make it easier to compare coefficients. With practice, the coefficients can be worked out mentally.

Equating coefficients of x terms

$B = 3$

> The coefficient of x on the LHS is 3 and on the RHS is B. These must be equal.

Equating constant terms

$1 = 2A + B$ ①

But $B = 3$

Substituting $B = 3$ into ①

> The value of B has been found, so it can be substituted here to find A.

$1 = 2A + 3$

$-2 = 2A$

$A = -1$

So $A = -1$ and $B = 3$

Many identities can be solved by using a combination of both methods: substitution and equating coefficients.

The following example shows how algebraic division can be performed using a given identity.

Example 17 Given that $\dfrac{2x^2}{(x+2)(x-1)} \equiv A + \dfrac{Bx+C}{(x+2)(x-1)}$, find the values of A, B and C.

$$\frac{2x^2}{(x+2)(x-1)} \equiv A + \frac{Bx+C}{(x+2)(x-1)}$$

> Multiply both sides by $(x+2)(x-1)$.

$$2x^2 \equiv A(x+2)(x-1) + Bx + C$$

Putting $x = 1$ in the identity

$$2 = B + C \qquad \text{①}$$

> Substitute values of x which make the factor $(x+2)(x-1)$ zero.

Putting $x = -2$ in the identity

$$8 = -2B + C \qquad \text{②}$$

$2 \times \text{①} \qquad 4 = 2B + 2C \qquad \text{③}$

$\text{②} + \text{③} \qquad 12 = 3C$

$$C = 4$$

Substituting in ①

$$2 = B + 4$$

$$B = -2$$

Putting $x = 0$ in the identity

$$0 = -2A + C$$

> The substitution $x = 0$ makes the term Bx zero.

But $C = 4$ and so

$$0 = -2A + 4$$

$$A = 2$$

The values of A, B and C are 2, -2 and 4 respectively.

9.4 Remainder and factor theorems

In Book 1 you learnt about the remainder and factor theorems and how to use them to factorise polynomials. In this section you will see how this can be useful in simplifying quotients of two polynomials.

Remember

> **Remainder theorem**
>
> If f(x) is divided by $(ax - b)$ the remainder is $\text{f}\left(\dfrac{b}{a}\right)$.
>
> **Factor theorem**
>
> $\text{f}\left(\dfrac{b}{a}\right) = 0 \Leftrightarrow (ax - b)$ is a factor of f(x).

Example 18 The polynomials $f(x) = 3x^3 + 5x^2 - 2x$ and $g(x) = 3x^2 - 7x + 2$ have a common linear factor.

a By considering $f(\frac{1}{3})$ and $g(\frac{1}{3})$, find this factor.

b Hence write $\dfrac{f(x)}{g(x)}$ as a simplified algebraic fraction.

Solution **a** $f(\frac{1}{3}) = 3 \times (\frac{1}{3})^3 + 5 \times (\frac{1}{3})^2 - 2 \times (\frac{1}{3})$

$\quad\quad\quad = \frac{1}{9} + \frac{5}{9} - \frac{2}{3}$

$\quad\quad\quad = 0$

$\quad\therefore\quad (3x - 1)$ is a factor of $f(x)$.   *Using the factor theorem.*

$\quad g(\frac{1}{3}) = 3 \times (\frac{1}{3})^2 - 7 \times (\frac{1}{3}) + 2$

$\quad\quad\quad = \frac{1}{3} - \frac{7}{3} + 2$

$\quad\quad\quad = 0$

$\quad\therefore\quad (3x - 1)$ is a factor of $g(x)$.   *Using the factor theorem.*

Hence $f(x)$ and $g(x)$ have a common factor $(3x - 1)$.

b $\dfrac{f(x)}{g(x)} = \dfrac{3x^3 + 5x^2 - 2x}{3x^2 - 7x + 2}$ *Take out a factor x in the numerator.*

$\quad\quad = \dfrac{x(3x^2 + 5x - 2)}{3x^2 - 7x + 2}$ *Factorise each expression, noting the common factor $(3x - 1)$.*

$\quad\quad = \dfrac{x\cancel{(3x - 1)}(x + 2)}{\cancel{(3x - 1)}(x - 2)}$ *Cancel the common factor.*

$\quad\quad = \dfrac{x(x + 2)}{x - 2}$

Exercise 9B

1 Find the quotient and the remainder when these polynomials are divided by the expression in brackets.

a $2x^3 - 5x^2 + 5x + 9 \quad (2x + 1)$ **b** $15x^3 + 16x^2 - 16x - 4 \quad (3x - 1)$

c $3x^3 + 2x^2 - 21x - 11 \quad (3x + 2)$ **d** $2x^4 - x^3 - 14x^2 + 15x - 5 \quad (2x - 1)$

e $4x^3 - 3x + 4 \quad (2x - 1)$ **f** $3x^3 - 11x^2 - 1 \quad (3x - 2)$

2 Express these as a constant plus a proper fraction as in Examples 11–13.

a $\dfrac{x + 7}{x + 3}$ **b** $\dfrac{x - 3}{x + 5}$ **c** $\dfrac{2x + 1}{x + 3}$ **d** $\dfrac{x^2 - 2}{x^2 + 4}$

e $\dfrac{x^2 + 4}{x^2 + 2x}$ **f** $\dfrac{x^2 - x}{x^2 + 3x}$ **g** $\dfrac{3 - x}{x + 7}$ **h** $\dfrac{1 - 2x}{x - 2}$

3 Find the quotient when these polynomials are divided by the expression in brackets.

a $x^4 - 2x^3 - 4x^2 + 7x - 2 \quad\quad (x^2 + x - 2)$

b $x^4 + 3x^3 + 3x - 1 \quad\quad (x^2 + 1)$

c $3x^4 - 2x^3 + 7x^2 - 4x + 2 \quad\quad (x^2 + 2)$

d $2a^3 - 7a^2 - a + 2 \quad\quad (a^2 - 3a - 2)$

4 Copy and complete these identities.

a $x^3 + x^2 - 5x + 3 \equiv (x^2 - 2x + 1)(\dots)$

b $x^3 + x^2 - 10x + 8 \equiv (x^2 + 3x - 4)(\dots)$

c $x^4 - 2x^3 + 4x^2 - 2x + 3 \equiv (x^2 + 1)(\dots)$

d $x^4 + 3x^3 - 7x^2 - 9x + 12 \equiv (x^2 - 3)(\dots)$

e $2x^4 - 3x^3 + 7x^2 - 12x - 4 \equiv (x^2 + 4)(\dots)$

f $2x^4 + 8x^3 - 9x^2 + 4x - 5 \equiv (2x^2 + 1)(\dots)$

g $12x^4 + 6x^3 - 23x^2 - 4x + 10 \equiv (3x^2 - 2)(\dots)$

h $2x^4 + 11x^3 + 2x^2 - 25x + 12 \equiv (x^2 + 4x - 3)(\dots)$

i $x^4 + x^3 + 7x - 3 \equiv (x^2 + 2x - 1)(\dots)$

5 Find the quotient and the remainder when these polynomials are divided by the expression in brackets.

a $x^4 + x^2 - 10$ $(x^2 - 2)$

b $x^4 + 4x^3 + 7x^2 + 12x + 4$ $(x^2 + 3)$

c $2x^4 - 10x^3 + 11x^2 - 25x$ $(2x^2 + 5)$

d $6x^4 + 15x^3 - 13x^2 + 10x - 6$ $(3x^2 - 2)$

e $8x^4 - 4x - 16$ $(2x^2 - 4)$

f $x^4 - 17x^2 - 12x + 4$ $(x^2 + 4x + 1)$

g $2x^4 - 2x^3 - 7x^2 + 25x$ $(2x^2 + 4x - 3)$

h $8x^4 + 15x - 30$ $(2x^2 - x + 4)$

i $3x^4 + 8x^3 + 14x^2 - x + 6$ $(3x^2 + 2x + 1)$

6 Find the values of A, B and C in these identities.

a $16x + 7 \equiv A(2x - 1) + B(2x + 4)$

b $11 - 3x \equiv A(x + 3) + B(x - 1)$

c $11x^2 - 4x + 5 \equiv A(x^2 + 3) + B(2x^2 - x - 1)$

d $5 - x \equiv A(x - 1)^2 + B(2x - 1)(x - 1) + C(x + 3)$

e $3x^2 - 8x \equiv A(x - 2)^2 + B(x - 1)^2 + C$

f $3x^3 + 6x^2 - 12 \equiv A(x - 1)(x - 2) + Bx(x^2 + 2) + C(x - 2)(x + 2)$

7 Show that $6x^2 - 7x - 5$ and $4 + 5x - 6x^2$ have a common linear factor $2x + 1$.

Hence write $\dfrac{6x^2 - 7x - 5}{4 + 5x - 6x^2}$ as a simplified algebraic fraction.

8 In each of these questions the polynomials $f(x)$ and $g(x)$ have a common linear factor. Find this factor by using the factor theorem and hence simplify the expression $\dfrac{f(x)}{g(x)}$.

a $f(x) = 4x^2 - 1$, $g(x) = 4x^2 + 12x - 7$

b $f(x) = 2x^2 + x - 15$, $g(x) = 2x^2 - 11x + 15$

c $f(x) = 4x^2 - 1$, $g(x) = 2x^3 - 9x^2 - 5x$

d $f(x) = x^3 - 7x^2 - 18x$, $g(x) = 2x^2 + x - 6$

e $f(x) = 3x^2 + 2x - 1$, $g(x) = 3x^4 - 16x^3 + 5x^2$

f $f(x) = 6x^2 - x - 1$, $g(x) = 2x^3 - x^2 - 18x + 9$

9.5 Partial fractions

Earlier it was shown that an expression consisting of algebraic fractions may be rearranged as a single fraction. For example,

$$\frac{4}{x+1} + \frac{2}{x-3} + \frac{1}{(x+1)^2} = \frac{6x^2 - 3x - 13}{(x+1)^2(x-3)}$$

As with many techniques in mathematics, it is often useful to be able to reverse the process. When the denominator of an algebraic fraction can be factorised, it is possible to rewrite the fraction as a sum of fractions with simpler denominators.

The process of splitting into partial fractions involves starting with an expression such as

$$f(x) = \frac{6x^2 - 3x - 13}{(x+1)^2(x-3)}$$

and finding the three fractions with simpler denominators whose sum is f(x); in this case

$$f(x) = \frac{4}{x+1} + \frac{2}{x-3} + \frac{1}{(x+1)^2}$$

Note   Algebraic fractions which may be split into partial fractions must be *proper*, i.e. the numerator must be of lower degree than the denominator. (Improper fractions are discussed on page 238.) Each of the partial fractions is also proper.

The factors of the denominator of the fraction to be split are possible denominators for the partial fractions. Two different cases are considered separately:

■ Denominator having distinct linear factors

■ Denominator containing a repeated linear factor.

The CD-ROM (E9.5.1) considers denominators containing a quadratic factor.

Finding partial fractions: Denominator with distinct linear factors

Three different methods commonly used for finding partial fractions are illustrated here.

■ Substitution method (Examples 19 and 20)

■ Equating coefficients (Example 21)

■ The 'cover-up' method (Example 22)

Example 19 Express $\dfrac{x+3}{(x+1)(x+2)}$ in partial fractions.

> The partial fractions will be of the form
> $$\frac{A}{x+1} \text{ and } \frac{B}{x+2}$$
> where A and B are constants to be found.

Let $\dfrac{x+3}{(x+1)(x+2)} \equiv \dfrac{A}{x+1} + \dfrac{B}{x+2}$　　①

> Notice the equivalence sign since the identity is true for all values of x.

$\dfrac{x+3}{(x+1)(x+2)} \equiv \dfrac{A(x+2) + B(x+1)}{(x+1)(x+2)}$

> This is how the fractions would have been added.

Then　　$x+3 \equiv A(x+2) + B(x+1)$　　②

> Equate the numerator on both sides.

Putting $x = -1$ in ②

$$-1+3 = A(-1+2) + 0$$

$$2 = A$$

> Identity holds for all values of x so substitute a value of x which makes the factor $(x+1)$ zero.

Putting $x = -2$ in ②

$$-2+3 = 0 + B(-2+1)$$

$$1 = -B$$

$$B = -1$$

> The substitution $x = -2$ makes the term with factor $(x+2)$ equal to zero.

So　　$\dfrac{x+3}{x^2+3x+2} \equiv \dfrac{2}{x+1} - \dfrac{1}{x+2}$

> The original expression has been split into partial fractions.

Check:

> Check the result by rearranging the right-hand side.

$\text{RHS} = \dfrac{2}{x+1} - \dfrac{1}{x+2} = \dfrac{2(x+2) - (x+1)}{(x+1)(x+2)}$

> The minus sign outside the bracket changes the sign when the brackets are removed.

$= \dfrac{2x+4-x-1}{(x+1)(x+2)} = \dfrac{x+3}{(x+1)(x+2)}$

$= \text{LHS}$

A quick check can be made by substituting a value of x (other than those already used) in both sides. For instance when $x = 1$,

$$\text{LHS} = \frac{1+3}{2 \times 3} = \frac{4}{6} = \frac{2}{3} \qquad \text{RHS} = \frac{2}{1+1} - \frac{1}{1+2} = 1 - \frac{1}{3} = \frac{2}{3}$$

LHS = RHS, but this is not a rigorous check and only offers an indication of whether the partial fractions are correct.

Note　The expression obtained by equating the numerators is equivalent to multiplying both sides of ① by the denominator, $(x+1)(x+2)$. This method, of multiplying by the denominator, is used in subsequent examples.

Example 20 Express $\dfrac{x+4}{1-x-2x^2}$ in partial fractions.

$$\dfrac{x+4}{1-x-2x^2} \equiv \dfrac{x+4}{(1-2x)(1+x)}$$

> Factorise the denominator.

Let $\dfrac{x+4}{(1-2x)(1+x)} \equiv \dfrac{A}{1-2x} + \dfrac{B}{1+x}$

> Distinct linear factors in denominator

$\therefore \qquad\qquad x+4 \equiv A(1+x) + B(1-2x)$ ①

> Multiply both sides by $(1-2x)(1+x)$.

Putting $x = -1$ in ①

> The identity holds for all values of x, so substitute a value which makes the factor $(x+1)$ zero.

$$-1 + 4 = 0 + B(1 - 2(-1))$$

$$3 = 3B$$

$$B = 1$$

Putting $x = \frac{1}{2}$ in ①

> $x = \frac{1}{2}$ makes the term with factor $(1-2x)$ zero.

$$\tfrac{1}{2} + 4 = A(1 + \tfrac{1}{2})$$

$$\tfrac{9}{2} = A(\tfrac{3}{2})$$

$$A = 3$$

So $\qquad \dfrac{x+4}{1-x-2x^2} \equiv \dfrac{3}{1-2x} + \dfrac{1}{1+x}.$

> *Check*: It may be possible to 'collect the terms' mentally and thus verify the solution by inspection.

Example 21

In this example, the coefficients of powers of x are equated.

Rewrite $\dfrac{6x+1}{(2x-3)(3x-2)}$ in partial fractions.

Let $\dfrac{6x+1}{(2x-3)(3x-2)} \equiv \dfrac{A}{2x-3} + \dfrac{B}{3x-2}$

> Multiply both sides by $(2x-3)(3x-2)$.

$$6x+1 \equiv A(3x-2) + B(2x-3)$$

> The method of substitution would involve fractional values of x, so equating coefficients is preferable.

Equating coefficients of x terms

$$6 = 3A + 2B \qquad ①$$

Equating constant terms

$$1 = -2A - 3B \qquad ②$$

$① \times 2 \qquad\qquad 12 = 6A + 4B \qquad ③$

$② \times 3 \qquad\qquad 3 = -6A - 9B \qquad ④$

$③ + ④ \qquad\qquad 15 = -5B$

$\therefore \qquad\qquad\qquad B = -3$

Substituting in ①

$$6 = 3A - 6$$

$$\therefore \quad A = 4$$

So $\dfrac{6x + 1}{(2x - 3)(3x - 2)} \equiv \dfrac{4}{2x - 3} - \dfrac{3}{3x - 2}$

Check in ②: RHS $= -2A - 3B$
$= -8 - (-9) = 1$
$= $ LHS

Check: You may find, because there are only two fractions, you can verify the solution by inspection.

Note It is sometimes convenient to use both substitution *and* equating coefficients when finding partial fractions. (See Example 23.)

Example 22 *The 'cover-up method' works only when the denominator of the original fraction has distinct linear factors. It is logically equivalent to substitution.*

Rewrite $\dfrac{x - 7}{(x - 4)(x + 2)}$ in partial fractions.

Let $\dfrac{x - 7}{(x - 4)(x + 2)} \equiv \dfrac{A}{x - 4} + \dfrac{B}{x + 2}$

To find A, the $(x - 4)$ factor on the LHS denominator is covered with a finger. The remaining visible part of the fraction is evaluated using $x = 4$, the value which makes $x - 4$, the denominator for A, equal to 0. This gives

$$\frac{4 - 7}{(\text{cover-up})(4 + 2)}$$

Ignoring the 'cover-up' factor, the value of this term is $\dfrac{-3}{6}$ which gives

$$A = \frac{-3}{6} = -\frac{1}{2}$$

Similarly, to find B, factor $(x + 2)$ is covered and the fraction

$$\frac{x - 7}{(x - 4)(\text{cover-up})}$$

is evaluated with $x = -2$, giving

$$\frac{-2 - 7}{(-2 - 4)(\text{cover-up})}$$

So $\qquad B = \dfrac{-9}{-6} = \dfrac{3}{2}$

The original fraction can now be written in partial fractions as

$$\frac{x - 7}{(x - 4)(x + 2)} \equiv \frac{3}{2(x + 2)} - \frac{1}{2(x - 4)}$$

Note: $\frac{3}{2} \times \frac{1}{x+2} = \frac{3}{2(x+2)}$

This result may be checked by combining the fractions on the RHS.

Example 23   Express $\dfrac{5x^2 + 3x + 4}{(x+1)(2x+1)(x-2)}$ as a sum of partial fractions.

Let $\dfrac{5x^2 + 3x + 4}{(x+1)(2x+1)(x-2)} \equiv \dfrac{A}{(x+1)} + \dfrac{B}{(2x+1)} + \dfrac{C}{(x-2)}$

> Multiply both sides by the LHS denominator.

Then $5x^2 + 3x + 4 \equiv A(2x+1)(x-2) + B(x+1)(x-2) + C(x+1)(2x+1)$ ①

Putting $x = 2$ in ①

$$20 + 6 + 4 = 15C$$
$$30 = 15C$$
$$C = 2$$

> Terms containing factor $(x-2)$ become zero.

Putting $x = -1$ in ①

$$5 - 3 + 4 = 3A$$
$$6 = 3A$$
$$A = 2$$

> Terms containing factor $(x+1)$ become zero.

Equating coefficients of x^2 in ①

$$5 = 2A + B + 2C$$
$$5 = 4 + B + 4$$
$$B = -3$$

> Alternatively, substitute $x = -\frac{1}{2}$ in ① to make the factor $(2x+1)$ zero.

So $\dfrac{5x^2 + 3x + 4}{(x+1)(2x+1)(x-2)} \equiv \dfrac{2}{x+1} - \dfrac{3}{2x+1} + \dfrac{2}{x-2}$

Finding partial fractions: Denominator with a repeated linear factor

The previous examples have shown that when the denominator of a fraction has two factors there are two partial fractions and when the denominator has three factors there are three partial fractions.

Consider the fraction $\dfrac{4x^2 - 3x + 2}{(x+1)(x-2)^2}$.

The denominator contains the 'repeated linear factor' $(x-2)^2$.

From the work above, we would expect this fraction to split into three partial fractions.

Two of these are clearly $\dfrac{A}{x+1}$ and $\dfrac{B}{x-2}$ and it can be shown that the third is $\dfrac{C}{(x-2)^2}$.

Note   Where the fraction to be split contains a repeated linear factor in the denominator, find corresponding partial fractions with constant numerators. This form is useful for more advanced work in binomial expansions and in integration (see Chapter 10 and Section 9.6). So, when splitting a fraction with, for example, $(x - c)^2$ in the denominator, there will be two partial fractions with constant numerators:

$$\frac{A}{x - c} + \frac{B}{(x - c)^2}$$

Example 24   Express as the sum of partial fractions

$$\frac{4x^2 - 3x + 2}{(x + 1)(x - 2)^2}$$

> Fractions with denominators $(x + 1)$, $(x - 2)$, $(x - 2)^2$ and constant numerators are required.

Solution   Let $\dfrac{4x^2 - 3x + 2}{(x + 1)(x - 2)^2} \equiv \dfrac{A}{x + 1} + \dfrac{B}{x - 2} + \dfrac{C}{(x - 2)^2}$

> Multiply both sides by $(x + 1)(x - 2)^2$.

$$\frac{4x^2 - 3x + 2}{(x + 1)(x - 2)^2} \times (x + 1)(x - 2)^2 \equiv \left[\frac{A}{x + 1} + \frac{B}{x - 2} + \frac{C}{(x - 2)^2} \right] \times (x + 1)(x - 2)^2$$

$$4x^2 - 3x + 2 \equiv A(x - 2)^2 + B(x + 1)(x - 2) + C(x + 1) \qquad ①$$

Putting $x = 2$ in ①

$$16 - 6 + 2 = C(2 + 1)$$

> Other terms are zero.

$$12 = 3C$$

∴ $C = 4$

Putting $x = -1$ in ①

$$4 + 3 + 2 = 9A$$

> Terms with $(x + 1)$ become zero.

$$9 = 9A$$

∴ $A = 1$

Equating coefficients of x^2 in ①

$$4 = A + B$$

∴ $B = 3$

Check by looking at the coefficients of x on RHS of ①

> The coefficient of x has not been used for equating.

$$-4A - B + C = -4 - 3 + 4 = -3$$

as required.

So $\dfrac{4x^2 - 3x + 2}{(x + 1)(x - 2)^2} \equiv \dfrac{1}{x + 1} + \dfrac{3}{x - 2} + \dfrac{4}{(x - 2)^2}$

Finding partial fractions for improper fractions

An improper fraction has the degree of the numerator equal to or higher than that of the denominator. Before expressing an improper fraction in partial fractions, it is divided to give a quotient and remainder. Then the remainder, which *is* a proper fraction, is dealt with as in the preceding examples.

Example 25

In this example, the degree of the numerator and denominator are both two. It is an improper fraction.

Express $\dfrac{2x^2 + 5}{(x - 4)(x + 2)}$ in partial fractions.

$$\frac{2x^2 + 5}{(x - 4)(x + 2)} \equiv \frac{2x^2 + 5}{x^2 - 2x - 8}$$

> Multiply out the denominator since division will be required.

$$\equiv \frac{2(x^2 - 2x - 8) + 4x + 21}{x^2 - 2x - 8}$$

> Use long division or the alternative method shown here.

$$\equiv 2 + \frac{4x + 21}{x^2 - 2x - 8}$$

> Write as quotient and remainder; the remainder is a proper fraction.

Let $\dfrac{4x + 21}{x^2 - 2x - 8} \equiv \dfrac{4x + 21}{(x - 4)(x + 2)}$

> Restore denominator of remainder to factor form.

$$\equiv \frac{A}{x - 4} + \frac{B}{x + 2}$$

> Find partial fractions for remainder.

$$4x + 21 \equiv A(x + 2) + B(x - 4) \qquad \text{①}$$

> Multiply both sides by denominator.

> Use substitution method.

Putting $x = 4$ in ①

$$16 + 21 = 6A$$

$$\therefore \qquad A = \frac{37}{6}$$

Putting $x = -2$ in ①

$$-8 + 21 = -6B$$

$$\therefore \qquad B = -\frac{13}{6}$$

So $\dfrac{2x^2 + 5}{(x - 4)(x + 2)} \equiv 2 + \dfrac{37}{6(x - 4)} - \dfrac{13}{6(x + 2)}$

> Remember that the RHS starts with '$2 + \ldots$'.

> A and B are fractional.
> $\frac{37}{6} \times \frac{1}{x - 4} = \frac{37}{6(x - 4)}$
> Note that 6 goes in the denominator.

> As before, remember to check.

Example 26   Express $\dfrac{x^3 - x^2 - 8x - 1}{x^2 - 3x - 4}$ in partial fractions.

$$x^2 - 3x - 4 \overline{\smash{)}\begin{array}{l} x + 2 \\ x^3 - x^2 - 8x - 1 \end{array}}$$

$$\underline{x^3 - 3x^2 - 4x}$$

$$2x^2 - 4x - 1$$

$$\underline{2x^2 - 6x - 8}$$

$$2x + 7$$

> Since the fraction is improper, find quotient and remainder first.

> This time the method of long division is used.

So $\dfrac{x^3 - x^2 - 8x - 1}{x^2 - 3x - 4} \equiv x + 2 + \dfrac{2x + 7}{x^2 - 3x - 4}$

> Rewrite as a quotient and remainder.

$$\equiv x + 2 + \dfrac{2x + 7}{(x - 4)(x + 1)}$$

> Factorise the denominator.

Let $\dfrac{2x + 7}{(x - 4)(x + 1)} \equiv \dfrac{A}{x - 4} + \dfrac{B}{x + 1}$

> Split the remainder into partial fractions.

$$A = \dfrac{8 + 7}{5} = 3$$

> Use cover-up method with $x = 4$.

$$B = \dfrac{-2 + 7}{-5} = -1$$

> Use cover-up method with $x = -1$.

So $\dfrac{x^3 - x^2 - 8x - 1}{x^2 - 3x - 4} \equiv x + 2 + \dfrac{3}{x - 4} - \dfrac{1}{x + 1}$

> As before, remember to check.

The CD-ROM (E9.5.2 and E9.5.3) shows how to find partial fractions for higher degree denominators and how to use partial fractions to find the sums of certain types of series.

Exercise 9C

1 Rewrite these as partial fractions.

a $\dfrac{1 - 2x}{x(x + 1)}$ b $\dfrac{2 - 6x}{x(4x - 1)}$ c $\dfrac{4}{(x - 2)(x + 2)}$ d $\dfrac{2x}{(2 + x)(2 - x)}$

e $\dfrac{2x + 1}{x^2 + x - 2}$ f $\dfrac{x - 1}{3x^2 - 11x + 10}$ g $\dfrac{9 - x}{x^2 - 4x + 3}$ h $\dfrac{3 - 4x}{2 + 3x - 2x^2}$

2 Express these as the sum of partial fractions.

a $\dfrac{3x + 1}{(x + 2)(x + 1)(x - 3)}$ b $\dfrac{2x^2 - 5x - 9}{(x - 1)(x + 1)(x + 2)}$ c $\dfrac{5x + 2}{(x + 1)(x^2 - 4)}$

d $\dfrac{3x^2 + 4x - 1}{2x^3 - x^2 - x}$ e $\dfrac{x^2 - 2x - 6}{(x - 2)(x + 1)(x + 2)}$ f $\dfrac{9 - 5x}{(1 - x)(2 - x)(3 - x)}$

3 Express these in partial fractions.

a $\dfrac{x + 1}{(x + 3)^2}$ b $\dfrac{2x^2 - 5x + 7}{(x - 2)(x - 1)^2}$ c $\dfrac{x^2 - 20x - 11}{(x + 1)^2(x - 4)}$

d $\dfrac{3x^2 + 5x + 8}{(1 - x)(x + 1)^2}$ e $\dfrac{-10x^2 + 17x - 15}{2(2x - 1)^2(x + 4)}$ f $\dfrac{x^2 + 13x + 7}{(x + 1)(x^2 - 3x - 4)}$

4 Express these in terms of a quotient and remainder in the form of partial fractions.

a $\dfrac{x}{x-2}$ **b** $\dfrac{x^2+1}{x^2-1}$ **c** $\dfrac{2x^2+3}{x^2-1}$

d $\dfrac{x^2+3}{x(x+1)}$ **e** $\dfrac{x^2-7}{(x+1)(x-2)}$ **f** $\dfrac{x^3+2x^2-2x+2}{(x-1)(x+3)}$

g $\dfrac{x^3-x^2-4x+1}{x^2-4}$ **h** $\dfrac{x^4+2x^3-4x^2-4x+2}{(x-2)(x+3)}$

5 Express these in partial fractions.

a $\dfrac{3x^2-21x+24}{(x+1)(x-2)(x-3)}$ **b** $\dfrac{4x^2+x+1}{x(x^2-1)}$ **c** $\dfrac{5x^2+2}{(3x+1)(x+1)^2}$

d $\dfrac{x}{25-x^2}$ **e** $\dfrac{x^2+1}{(x-2)(x+1)}$

6 Express these in partial fractions.

a $\dfrac{3x+7}{x(x+2)(x-1)}$ **b** $\dfrac{3}{x^2(x+2)}$ **c** $\dfrac{68+11x}{(3+x)(16-x^2)}$

d $\dfrac{2x^2+39x+12}{(2x+1)^2(x-3)}$ **e** $\dfrac{2x^3+x^2-3x+1}{2x^2-3x-2}$

9.6 Using partial fractions in integration

Integrals leading to logarithmic functions often appear in disguise, hidden in functions that can be rewritten using partial fractions.

Example 27 Find $\displaystyle\int \dfrac{5}{(x-2)(x+3)}\,dx$.

Solution Let

$\dfrac{5}{(x-2)(x+3)} \equiv \dfrac{A}{x-2} + \dfrac{B}{x+3}$

> Split the expression into partial fractions (Section 9.5).

> Add the fractions.

$\equiv \dfrac{A(x+3)+B(x-2)}{(x-2)(x+3)}$

> Equate the numerator on both sides.

then $\quad 5 \equiv A(x+3)+B(x-2) \quad$ ①

Putting $x = 2$ in ①

> Putting $x = 2$ makes $(x-2)$ zero.

$$5 = 5A$$
$$A = 1$$

Putting $x = -3$ in ①

> Putting $x = -3$ makes $(x+3)$ zero.

$$5 = -5B$$
$$B = -1$$

So $\qquad \dfrac{5}{(x-2)(x+3)} \equiv \dfrac{1}{x-2} - \dfrac{1}{x+3}$

$$\int \dfrac{5}{(x-2)(x+3)}\,dx = \int \left(\dfrac{1}{x-2} - \dfrac{1}{x+3} \right) dx$$

> Integrate the partial fractions.

$$= \ln(x-2) - \ln(x+3) + c$$

$$= \ln \left(\dfrac{x-2}{x+3} \right) + c$$

> *Remember* : $\log a - \log b = \log \dfrac{a}{b}$

Example 28   Find $\displaystyle\int \dfrac{2x-1}{(x+1)^2}\,dx.$

Solution   Let $\dfrac{2x-1}{(x+1)^2} \equiv \dfrac{A}{x+1} + \dfrac{B}{(x+1)^2}$

> Note the repeated factor. See page 236.

$$\equiv \dfrac{A(x+1) + B}{(x+1)^2}$$

> Add the fractions.

Then $\ 2x - 1 \equiv A(x+1) + B \qquad ①$

> Equate the numerator on both sides.

Putting $x = -1$ in ①

> Find A and B by substituting values and/or equating coefficients.

$$-3 = B$$

Equating x terms in ①

$$2 = A$$

So $\ \dfrac{2x-1}{(x+1)^2} \equiv \dfrac{2}{x+1} - \dfrac{3}{(x+1)^2}$

$$\int \dfrac{2x-1}{(x+1)^2}\,dx = \int \left(\dfrac{2}{x+1} - \dfrac{3}{(x+1)^2} \right) dx$$

> Integrate the partial fractions.

$$= \int \left(\dfrac{2}{x+1} - 3(x+1)^{-2} \right) dx$$

> The first part of the integral gives a log function. The second part can be recognised as the chain rule in reverse (Section 5.4).

$$= 2\ln(x+1) + 3(x+1)^{-1} + c$$

$$= 2\ln(x+1) + \dfrac{3}{x+1} + c$$

Example 29   Show that $\displaystyle\int_2^3 \dfrac{5-x}{(1-x)(3+x)}\,dx = \ln \dfrac{18}{25}.$

Solution   Let $\ \dfrac{5-x}{(1-x)(3+x)} \equiv \dfrac{A}{1-x} + \dfrac{B}{3+x}$

$$\equiv \dfrac{A(3+x) + B(1-x)}{(1-x)(3+x)}$$

Then $\qquad 5 - x \equiv A(3+x) + B(1-x) \qquad ①$

Putting $x = 1$ in ①

$$4 = 4A \quad \Rightarrow \quad A = 1$$

Putting $x = -3$ in ①

$$8 = 4B \quad \Rightarrow \quad B = 2$$

So $\quad \dfrac{5 - x}{(1 - x)(3 + x)} \equiv \dfrac{1}{1 - x} + \dfrac{2}{3 + x}$

$$\int_2^3 \frac{5 - x}{(1 - x)(3 + x)}\, dx = \int_2^3 \left(\frac{1}{1 - x} + \frac{2}{3 + x} \right) dx$$

$$= \Big[-\ln|1 - x| + 2\ln(3 + x) \Big]_2^3$$

$$= -\ln|-2| + 2\ln 6 - (-\ln|-1| + 2\ln 5)$$

$$= 2\ln 6 - 2\ln 5 - \ln 2$$

$$= \ln 36 - \ln 25 - \ln 2$$

$$= \ln \frac{18}{25}$$

$\ln 1 = 0$

$n \ln a = \ln a^n$

$\ln a - \ln b - \ln c = \ln \dfrac{a}{bc}$

Exercise 9D

1 Use partial fractions to find

$$\int \frac{7}{(2x - 1)(3x + 2)}\, dx$$

2 a If

$$\frac{2x - 1}{(x + 1)^2} = \frac{A}{x + 1} + \frac{B}{(x + 1)^2}$$

find the values of A and B.

b Find

$$\int \frac{2x - 1}{(x + 1)^2}\, dx$$

3 Find

a $\displaystyle\int \frac{1}{x^2 - 9}\, dx$ **b** $\displaystyle\int \frac{1}{4x^2 - 9}\, dx$

4 a Find

$$\int \frac{x}{4 - x^2}\, dx$$

without using partial fractions.

b Find this integral using partial fractions.

5 Find these integrals.

a $\displaystyle\int \frac{4x - 33}{(2x + 1)(x^2 - 9)}\, dx$ **b** $\displaystyle\int \frac{5x + 2}{(x - 2)^2(x + 1)}\, dx$

Exercise 9E (Review)

1 Express each of these as a single fraction in its simplest form.

a $\dfrac{3}{x+3} - \dfrac{2}{x-2}$

b $\dfrac{1}{1-x} + \dfrac{2}{1+x}$

c $\dfrac{2x-1}{x^2+1} - \dfrac{1}{x+1}$

d $\dfrac{3}{(x-1)^2} + \dfrac{1}{x-1} + \dfrac{2}{x+1}$

e $\dfrac{1}{(x+2)^2} - \dfrac{2}{x+2} - \dfrac{1}{3x-1}$

f $\dfrac{1}{4x^2-1} + \dfrac{1}{(2x+1)^2}$

2 Simplify these expressions.

a $\dfrac{a-2b}{3a-6b}$

b $\dfrac{x-y}{x^2-y^2}$

c $\dfrac{2a-b}{b-2a}$

d $\dfrac{x+1}{x^2+5x+4}$

e $\dfrac{x^2-2x}{x^2-3x+2}$

f $\dfrac{x^2-4x-21}{x^2+5x+6}$

3 Write each of these as a single fraction in its simplest form.

a $\dfrac{10ab^2}{c} \times \dfrac{c^2}{5a}$

b $\dfrac{x^2y}{z} \times \left(\dfrac{z}{x}\right)^2$

c $\dfrac{4x}{3y} \div \dfrac{8x}{9y^2}$

d $\dfrac{p-q}{pq} \div \dfrac{p^2-q^2}{p^2q}$

e $\dfrac{1}{4a-b} \div \dfrac{1}{8a-2b}$

f $\dfrac{14ab}{3c^2} \times \dfrac{6abc}{2} \times \dfrac{5c}{7a^2b^2}$

4 Find the quotient and the remainder when these polynomials are divided by the expression in brackets.

a $2x^3 + 5x^2 - 14x - 36$ $(2x+5)$
b $3x^3 - 7x^2 + 5x + 10$ $(3x-1)$

c $x^4 + x^3 + 2x^2 + x + 1$ (x^2+1)
d $x^4 + x^3 + x^2 + x + 1$ (x^2+1)

e $x^4 + 3x^3 + 6x^2 + 6x + 2$ (x^2+x+1)

f $x^4 + 6x^2 - 8x + 21$ (x^2-2x+3)

5 Use the factor theorem to factorise $12x^3 + 5x^2 - 2x$ and $3x^3 + 2x^2 - 12x - 8$.

Hence write $\dfrac{12x^3 + 5x^2 - 2x}{3x^3 + 2x^2 - 12x - 8}$ as a simplified algebraic fraction.

6 Express each of these in partial fractions.

a $\dfrac{2(5-2x)}{(2+x)(4-x)}$

b $\dfrac{9}{2x^2-7x-4}$

c $\dfrac{x-4}{(x-1)(x-6)}$

d $\dfrac{3x^2+5x+4}{(x+1)(x-1)(2x+1)}$

e $\dfrac{3x^2+9x+4}{(x-1)(x+3)^2}$

7 Express each of these in terms of a quotient and remainder in the form of partial fractions.

a $\dfrac{2x^2-4x-1}{x(x-1)}$

b $\dfrac{x^3+3x^2-3x-11}{x^2+3x-4}$

c $\dfrac{4x^3+12x^2-2x-8}{(x+2)(2x+1)}$

8 a Given that $\dfrac{x^2 + 1}{(x - 1)(x + 1)} \equiv A + \dfrac{B}{x - 1} + \dfrac{C}{x + 1}$ find the values of A, B and C.

b Hence find $\displaystyle\int \dfrac{x^2 + 1}{(x - 1)(x + 1)}\ dx$.

9 Show that

$$\int_2^3 \dfrac{x - 4}{(x + 2)(x - 1)}\ dx = \ln\dfrac{25}{32}$$

There is a 'Test yourself' exercise (Ty9) and an 'Extension exercise' (Ext9) on the CD-ROM.

➤ Key points

Algebraic fractions

Fractions can be added or subtracted *only* if they are expressed with equal denominators.

The LCM of the denominators is the most efficient denominator to use.

$$\dfrac{a}{b} + \dfrac{c}{d} = \dfrac{ad + bc}{bd}$$

Any fractions can be multiplied and divided.

$$\dfrac{a}{b} \times \dfrac{c}{d} = \dfrac{ac}{bd} \qquad \text{and} \qquad \dfrac{a}{b} \div \dfrac{c}{d} = \dfrac{a}{b} \times \dfrac{d}{c} = \dfrac{ad}{bc}$$

Before multiplying, any factor which appears in both the numerator and denominator may be cancelled, to make the working easier.

Never cancel part of a bracket, only cancel the whole bracket, if it is a factor of both numerator and denominator.

Factorisation

$a(b + c) = ab + ac$

$(a + b)(c + d) = ac + ad + bc + bd$ $\qquad (x + a)(x + b) = x^2 + (a + b)x + ab$

$(a + b)(a - b) = a^2 - b^2$

$(a + b)^2 = a^2 + 2ab + b^2$ $\qquad\qquad\quad (a + b)^3 = a^3 + 3a^2b + 3ab^2 + b^3$

$(a - b)^2 = a^2 - 2ab + b^2$ $\qquad\qquad\quad (a - b)^3 = a^3 - 3a^2b + 3ab^2 - b^3$

$a^3 - b^3 = (a - b)(a^2 + ab + b^2)$

$a^3 + b^3 = (a + b)(a^2 - ab + b^2)$

$a^2 + b^2$ *cannot* be factorised.

$a^2 - b^2$ is the **difference of squares**.

$\dfrac{a - b}{b - a} = -1 \ (b \neq a)$

Division of polynomials

- Arrange the divisor and dividend in descending powers of the variable, lining up like terms, and leaving gaps in the dividend for 'missing' terms.

- Divide the term on the left of the dividend by the term on the left of the divisor. The result is a term of the quotient.

- Multiply the whole divisor by this term of the quotient and subtract the product from the dividend.

- Bring down as many terms as necessary to form a new dividend.

- Repeat the instructions until all terms of the dividend have been used.

The remainder and factor theorems

Remainder theorem

If f(x) is divided by ($ax - b$) the remainder is $f\left(\dfrac{b}{a}\right)$.

Factor theorem

$f\left(\dfrac{b}{a}\right) = 0 \Leftrightarrow (ax - b)$ is a factor of f(x).

Partial fractions

To express a fraction in partial fractions:

1   Check that the fraction is proper. If not, divide by the denominator and express the remainder in partial fractions.

2   Factorise the denominator as far as possible.

3   If the denominator contains

   - a **single linear factor** ($ax + b$), include a partial fraction $\dfrac{A}{ax + b}$

   - a **repeated linear factor** $(ax + b)^2$, include partial fractions $\dfrac{A}{ax + b} + \dfrac{B}{(ax + b)^2}$

4   Make an identity and multiply both sides by the denominator.

5   Use substitution or equating coefficients to find the constants. The 'cover-up' method can be used if the denominator has distinct linear factors only.

10 Binomial expansion

Earlier in the course you saw how to expand $(a + b)^n$ and $(1 + x)^n$ for all positive integers n. It was discovered in the seventeenth century that the result could be extended to obtain the binomial expansion of $(1 + x)^n$ where n is not a positive integer; the expansion is in the form of an infinite series.

After working through this chapter you should

■ be able to use the binomial expansion to expand $(1 + x)^n$ for any rational value of n

■ know that the expansion of $(1 + x)^n$ is valid when $|x| < 1$

■ be able to find the binomial expansion of more complicated expressions, including making use of partial fractions

■ be able to obtain algebraic and numerical approximations by using a binomial expansion.

10.1 Binomial expansion for any rational index

Consider the binomial expansion in the form

$$(1 + x)^n = 1 + nx + \frac{n(n-1)}{2!}x^2 + \frac{n(n-1)(n-2)}{3!}x^3 + \frac{n(n-1)(n-2)(n-3)}{4!}x^4 + \cdots \quad \text{①}$$

The expression was derived for n, a positive integer ($n \in \mathbb{Z}^+$). With n a positive integer the series is finite, having $(n + 1)$ terms.

> Newton was the first to suggest using the binomial expansion with negative and fractional indices in the 1670s.

It may seem surprising that substituting a negative integer or a positive or negative fraction for n gives a meaningful result. The coefficients of the terms in ① can still be evaluated but the series will be infinite, i.e. it will not terminate, since none of the factors n, $(n - 1)$, $(n - 2)$, $\cdots$ is ever zero. To use the binomial expansion for n, when n is negative or a fraction, the series must converge.

It can be proved, using more advanced mathematics, that the series obtained will converge when $|x| < 1$ and that it converges to $(1 + x)^n$.

Note: Write $|x| < 1$ or $-1 < x < 1$.

Remember The binomial expansion of $(1 + x)^n$, when n is negative or a fraction, is valid for $|x| < 1$.

Note In some circumstances the expansion is also valid for $x = 1$ or $x = -1$; discussion of this is beyond the scope of this course.

Example 1 Use the binomial series to expand, in ascending powers of x, as far as the term in x^3

    a $\dfrac{1}{1+x}$ **b** $\dfrac{1}{1-x}$

Solution   **a** $\dfrac{1}{(1+x)} = (1+x)^{-1}$ Put $n = -1$ in ①.

$$= 1 + (-1)(x) + \frac{(-1)(-2)}{2!}(x)^2 + \frac{(-1)(-2)(-3)}{3!}(x)^3 + \cdots$$

$$= 1 - x + x^2 - x^3 + \cdots \quad \text{providing} \quad |x| < 1$$

> The series will converge for $|x| < 1$.

a $\dfrac{1}{(1-x)} = (1-x)^{-1}$ Put $n = -1$ and replace x by $-x$ in ①.

$$= 1 + (-1)(-x) + \frac{(-1)(-2)}{2!}(-x)^2$$

> The series will converge for $|x| < 1$.

$$+ \frac{(-1)(-2)(-3)}{3!}(-x)^3 + \cdots$$

$$= 1 + x + x^2 + x^3 + \cdots \quad \text{providing} \quad |x| < 1$$

> This result could be obtained by replacing x by $-x$ in the answer to part **a**.

Notice the link between the binomial series for $(1 \pm x)^{-1}$ and infinite geometric series.

$1 + x + x^2 + x^3 + \cdots$ is an infinite geometric series with $a = 1$, $r = x$. The sum to infinity will exist if $|x| < 1$.

$$S_\infty = \frac{a}{1-r} = \frac{1}{1-x}$$

$1 - x + x^2 - x^3 + \cdots$ is an infinite geometric series with $a = 1$, $r = -x$. The sum to infinity will exist if $|x| < 1$.

$$S_\infty = \frac{1}{1+x}$$

Example 2 Expand as far as the term in x^3

    a $(1+x)^{-2}$ **b** $(1-x)^{-3}$

Solution   **a** $(1+x)^{-2} = 1 + (-2)x + \dfrac{(-2)(-3)}{2!}x^2 + \dfrac{(-2)(-3)(-4)}{3!}x^3 + \cdots$

$$= 1 - 2x + 3x^2 - 4x^3 + \cdots$$

> *Note*: Coefficients are $1, -2, 3, -4, \ldots$

b $(1-x)^{-3} = 1 + (-3)(-x) + \dfrac{(-3)(-4)}{2!}(-x)^2 + \dfrac{(-3)(-4)(-5)}{3!}(-x)^3 + \cdots$

$$= 1 + 3x + 6x^2 + 10x^3 + \cdots$$

> *Note*: Coefficients are $1, 3, 6, 10, \ldots$

Note The coefficients of the expansion of $(1 \pm x)^{-n}$ for $n \in \mathbb{Z}^+$ can be obtained from the diagonals of Pascal's triangle (see Book 1, Chapter 15).

The first term in each bracket expanded in this section so far has been 1. In Example 4, the bracket will be rearranged to have 1 as its first term so that the standard expansion below can be used.

> **Binomial expansion of $(1 + x)^n$ (for n negative or a fraction, $|x| < 1$)**
>
> $$(1 + x)^n = 1 + nx + \frac{n(n-1)}{2!}x^2 + \frac{n(n-1)(n-2)}{3!}x^3 + \frac{n(n-1)(n-2)(n-3)}{4!}x^4 + \cdots$$

Note Always arrange to expand a bracket whose first term is 1, i.e. of the form $(1 + \cdots)^n$, when the power is negative or a fraction.

The standard expansion can also be written as

$$(1 + x)^n = \binom{n}{0} + \binom{n}{1}x + \binom{n}{2}x^2 + \binom{n}{3}x^3 + \cdots + \binom{n}{r}x^r + \cdots$$

The notation $\binom{n}{r}$ can be used for all rational values of n, whereas ^nC_r is used only for $n \in \mathbb{Z}^+$.

Note For $n, r \in \mathbb{Z}, r \geqslant 0, n \geqslant r$

$$\binom{n}{r} = \frac{n!}{r!(n-r)!} = \frac{n(n-1)(n-2)\cdots(n-r+1)}{r!} = {}^nC_r$$

For $n \in \mathbb{Q}, r \in \mathbb{Z}, r \geqslant 0$

$$\binom{n}{r} = \frac{n(n-1)(n-2)\cdots(n-r+1)}{r!}$$

Example 3 Expand in ascending powers of x, up to the term in x^3, stating the values of x for which the expansion is valid

   a $(1-x)^{\frac{2}{3}}$ **b** $(1+2x)^{-\frac{1}{2}}$

Solution **a**

Put $n = \frac{2}{3}$ and replace x by $-x$ in the general formula.

$$(1-x)^{\frac{2}{3}} = 1 + \frac{2}{3}(-x) + \frac{\left(\frac{2}{3}\right)\left(-\frac{1}{3}\right)}{2!}(-x)^2 + \frac{\left(\frac{2}{3}\right)\left(-\frac{1}{3}\right)\left(-\frac{4}{3}\right)}{3!}(-x)^3 + \cdots$$

$$= 1 - \frac{2}{3}x - \frac{1}{9}x^2 - \frac{4}{81}x^3 + \cdots$$

The expansion is valid for $|-x| < 1$, i.e. for $|x| < 1$.

b

Put $n = -\frac{1}{2}$ and replace x by $2x$ in the general formula.

$$(1+2x)^{-\frac{1}{2}} = 1 + \left(-\frac{1}{2}\right)(2x) + \frac{\left(-\frac{1}{2}\right)\left(-\frac{3}{2}\right)}{2!}(2x)^2 + \frac{\left(-\frac{1}{2}\right)\left(-\frac{3}{2}\right)\left(-\frac{5}{2}\right)}{3!}(2x)^3 + \cdots$$

$$= 1 - x + \frac{3}{2}x^2 - \frac{5}{2}x^3 + \cdots$$

The expansion is valid for $|2x| < 1$, i.e. for $|x| < \frac{1}{2}$.

Example 4 Give the first four terms of the expansion of $(4 + x)^{\frac{3}{2}}$ in ascending powers of x, stating the values for which the expansion is valid.

> When the power is $-$ve or a fraction, take out a factor if necessary so that the bracket will start with 1.

$$(4 + x)^{\frac{3}{2}} = \left[4 \left(1 + \frac{x}{4} \right) \right]^{\frac{3}{2}}$$

> Remember to raise the factor to the same power.

$$= 4^{\frac{3}{2}} \left(1 + \frac{x}{4} \right)^{\frac{3}{2}}$$

> *Note:* $4^{\frac{3}{2}} = \left(\sqrt{4} \right)^3 = 8$

$$= 8 \left(1 + \left(\tfrac{3}{2} \right) \left(\frac{x}{4} \right) + \frac{\left(\frac{3}{2} \right) \left(\frac{1}{2} \right)}{2!} \left(\frac{x}{4} \right)^2 + \frac{\left(\frac{3}{2} \right) \left(\frac{1}{2} \right) \left(-\frac{1}{2} \right)}{3!} \left(\frac{x}{4} \right)^3 + \cdots \right)$$

To the first four terms

$$(4 + x)^{\frac{3}{2}} = 8 \left(1 + \tfrac{3}{8} x + \tfrac{3}{128} x^2 - \tfrac{1}{1024} x^3 \right)$$

$$= 8 + 3x + \tfrac{3}{16} x^2 - \tfrac{1}{128} x^3$$

> For a rough, but not conclusive, check, evaluate the four terms and $(4 + x)^{\frac{3}{2}}$ with $x = 0.001$ say, and compare the values.

The expansion is valid when $\left| \dfrac{x}{4} \right| < 1$, i.e. when $|x| < 4$.

Example 5 Find the term in x^2 in the expansion of $\dfrac{(1 + 2x)}{(3 + x)^2}$.

$$\frac{(1 + 2x)}{(3 + x)^2} = (1 + 2x)(3 + x)^{-2}$$

> Write $(3 + x)^{-2}$ as $3^{-2} \left(1 + \dfrac{x}{3} \right)^{-2}$.

$$= 3^{-2} (1 + 2x) \left(1 + \frac{x}{3} \right)^{-2}$$

> $3^{-2} = \dfrac{1}{3^2} = \dfrac{1}{9}$

$$= \tfrac{1}{9} (1 + 2x) \left(1 + (-2) \frac{x}{3} + \frac{(-2)(-3)}{2!} \left(\frac{x}{3} \right)^2 + \cdots \right)$$

$$= \tfrac{1}{9} (1 + 2x) \left(1 - \frac{2x}{3} + \frac{x^2}{3} + \cdots \right)$$

Term in $x^2 = \tfrac{1}{9} \left(2 \times -\tfrac{2}{3} + \tfrac{1}{3} \right) x^2$

$$= -\tfrac{1}{9} x^2$$

> The term in x^2 comes from multiplying the constant in the 1st bracket by the x^2 term in the 2nd bracket, and the x term in each bracket:
>
> $(1 + 2x) \left(1 - \dfrac{2x}{3} + \dfrac{x^2}{3} + \cdots \right)$

Example 6 Given that

$$\sqrt{\frac{(1+x)}{(1-3x)}} = A + Bx + Cx^2 + \cdots$$

find A, B and C.

$$\sqrt{\frac{(1+x)}{(1-3x)}} = (1+x)^{\frac{1}{2}}(1-3x)^{-\frac{1}{2}}$$

$$= \left(1 + \tfrac{1}{2}x + \frac{\left(\frac{1}{2}\right)\left(-\frac{1}{2}\right)}{2!}x^2 + \cdots\right)\left(1 + \left(-\tfrac{1}{2}\right)(-3x) + \frac{\left(-\frac{1}{2}\right)\left(-\frac{3}{2}\right)}{2!}(-3x)^2 + \cdots\right)$$

Note: $-3x$ is the term being squared, cubed and so on. Put it in a bracket to avoid making mistakes with signs.

$$= \left(1 + \frac{x}{2} - \frac{x^2}{8} + \cdots\right)\left(1 + \frac{3x}{2} + \frac{27x^2}{8} + \cdots\right)$$

'Check' by substituting a small value of x.

$$= 1 + \frac{x}{2} + \frac{3x}{2} + \frac{3x^2}{4} - \frac{x^2}{8} + \frac{27x^2}{8} + \cdots$$

$$= 1 + 2x + 4x^2 + \cdots$$

Compare $A + Bx + Cx^2 + \dots$ with $1 + 2x + 4x^2 + \dots$

So $A = 1$, $B = 2$ and $C = 4$.

$(1 + x)^{\frac{1}{2}}$ is valid for $|x| < 1$.

Remember to consider when the expansion is valid if the expression is raised to a −ve or fractional power.

$(1 - 3x)^{-\frac{1}{2}}$ is valid when $|3x| < 1$, i.e. when $|x| < \frac{1}{3}$.

∴ the expansion is valid when $|x| < \frac{1}{3}$.

Both inequalities must hold.

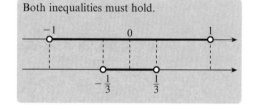

The following example illustrates the use of partial fractions in obtaining a binomial expansion.

Example 7   **a**   Given that $f(x) = \dfrac{7x + 4}{(1 - 2x)(2 + x)}$, express $f(x)$ in partial fractions.

b   Hence expand $f(x)$ in ascending powers of x up to and including the term in x^3.

c   Find the set of values of x for which the series is valid.

Solution   **a**   Let $\dfrac{7x + 4}{(1 - 2x)(2 + x)} \equiv \dfrac{A}{1 - 2x} + \dfrac{B}{2 + x}$

Use one of the methods of Chapter 9 to find $A = 3$ and $B = -2$.

$$\equiv \frac{3}{1 - 2x} - \frac{2}{2 + x}$$

b $f(x) = \dfrac{3}{1-2x} - \dfrac{2}{2+x}$

$= 3(1-2x)^{-1} - 2(2+x)^{-1}$

$= 3(1-2x)^{-1} - 2 \times \dfrac{1}{2}\left(1+\dfrac{x}{2}\right)^{-1}$

> Write $(2+x)^{-1}$ as $2^{-1}(1+\frac{x}{2})^{-1}$ and 2^{-1} as $\frac{1}{2}$.

$= 3(1-2x)^{-1} - \left(1+\dfrac{x}{2}\right)^{-1}$

$= 3\left(1 + 2x + (2x)^2 + (2x)^3 + \dots\right) - \left(1 - \dfrac{x}{2} + \left(\dfrac{x}{2}\right)^2 - \left(\dfrac{x}{2}\right)^3 + \dots\right)$

$= 3 + 6x + 12x^2 + 24x^3 + \dots - \left(1 - \dfrac{x}{2} + \dfrac{x^2}{4} - \dfrac{x^3}{8} + \dots\right)$

> Using the results from Example 1 and ignoring terms in x^4 and higher powers.

$= 2 + \dfrac{13}{2}x + \dfrac{47}{4}x^2 + \dfrac{193}{8}x^3 + \dots$

c   The expansion of $(1-2x)^{-1}$ is valid for $|2x| < 1$, i.e. when $|x| < \frac{1}{2}$.

The expansion of $\left(1 + \dfrac{x}{2}\right)^{-1}$ is valid for $\left|\dfrac{x}{2}\right| < 1$, i.e. when $|x| < 2$.

The expansion of $f(x)$ is valid when both inequalities hold, i.e. when $|x| < \frac{1}{2}$.

10.2   Approximations

One of the main uses of the binomial expansion is for approximating.

For example

$$(1+x)^{-3} = 1 - 3x + 6x^2 - 10x^3 + 15x^4 + \cdots$$

providing $|x| < 1$.

If an approximate value of $(1+x)^{-3}$ is required for small values of x, the terms in x^4, say, and higher powers can be ignored. For a simple approximation, only the first two or three terms of the series are needed.

This is a **linear approximation**: $(1+x)^{-3} \approx 1 - 3x$

This is a **quadratic approximation**: $(1+x)^{-3} \approx 1 - 3x + 6x^2$

Example 8   Find a linear approximation and a quadratic approximation to $(1-x)^{-4}$.

A linear approximation

$$(1-x)^{-4} \approx 1 + (-4)(-x)$$

> The first two terms give a linear approximation.

$$= 1 + 4x$$

> All terms in x^2 and higher powers are ignored.

A quadratic approximation

$$(1-x)^{-4} \approx 1 + (-4)(-x) + \frac{(-4)(-5)}{2!}(-x)^2$$

> The first three terms give a quadratic approximation.

$$= 1 + 4x + 10x^2$$

> All terms in x^3 and higher powers are ignored.

Graphical illustration of approximations

In Example 8, the linear and quadratic approximations to $(1-x)^{-4}$ were $1+4x$ and $1+4x+10x^2$ respectively.

Consider these graphs of $y = (1-x)^{-4}$, $y = 1+4x$ and $y = 1+4x+10x^2$:

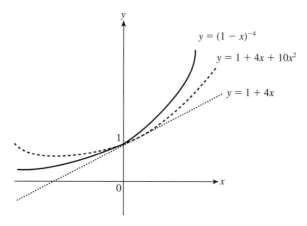

> These graphs, and similar approximations, can be looked at on a graphical calculator or computer.

Since the approximations are only useful for small values of x, the graphs are drawn in the region near $x = 0$.

Note that the linear approximation is a tangent to $y = (1-x)^{-4}$ and that all three curves are very close together in the region of $x = 0$.

Example 9

There may seem little point in being able to calculate, say, $\sqrt{26}$ correct to 3 decimal places when a calculator will very quickly give a much more accurate answer. However, the method used in this example was much more important in the days before calculators, and enables greater accuracy than can be obtained on a normal calculator, as Questions 9 and 11 of Exercise 10A require. It is also still useful to be able to make estimates without a calculator and to be familiar with such series that are of major importance in higher mathematics.

a   Expand $(25 + x)^{\frac{1}{2}}$ as far as the term in x^2.

b   Hence find $\sqrt{26}$ correct to 3 decimal places.

Solution   **a** $(25 + x)^{\frac{1}{2}} = 25^{\frac{1}{2}}\left(1 + \dfrac{x}{25}\right)^{\frac{1}{2}}$

> The power is fractional so take out a factor to make the bracket start with 1. Remember to raise the factor to the power.

$$= 5\left(1 + \frac{1}{2} \times \frac{x}{25} + \frac{\frac{1}{2} \times \left(-\frac{1}{2}\right)}{2!}\left(\frac{x}{25}\right)^2 + \cdots\right)$$

> Remember to consider when the expansion is valid if the expression is raised to a −ve or fractional power.

$$= 5\left(1 + \frac{x}{50} - \frac{x^2}{5000} + \cdots\right)$$

$$= 5 + \frac{x}{10} - \frac{x^2}{1000} + \cdots \quad \text{①}$$

Valid for $-1 < \dfrac{x}{25} < 1$, i.e. $-25 < x < 25$.

> Alternatively write $|x| < 25$.

b   Substituting $x = 1$ in $(25 + x)^{\frac{1}{2}}$ gives $26^{\frac{1}{2}} = \sqrt{26}$.

So an approximation to $\sqrt{26}$ can be found by substituting $x = 1$ in ①.

$\therefore \quad \sqrt{26} = 5 + \frac{1}{10} - \frac{1}{1000} + \cdots$

> A check will confirm that the term in x^3 will not affect the third decimal place.

$\therefore \qquad \approx 5 + 0.1 - 0.001$

$\qquad \qquad = 5.099$ (correct to 3 d.p.)

Example 10   **a**   Express $f(x)$ in partial fractions where $f(x) = \dfrac{1}{x^2 - 3x + 2}$.

b   Hence find an expansion for $f(x)$ in ascending powers of x as far as the term in x^2.

c   Use the expansion in part **b** to find an approximate value of $f(0.01)$ correct to 5 decimal places.

Solution   **a** $\dfrac{1}{x^2 - 3x + 2} = \dfrac{1}{(x - 1)(x - 2)}$

> Factorise the denominator.

Let $\dfrac{1}{(x - 1)(x - 2)} \equiv \dfrac{A}{x - 1} + \dfrac{B}{x - 2}$

> Use one of the methods of Section 9.5 to find $A = -1$ and $B = 1$.

$$\equiv \frac{-1}{x - 1} + \frac{1}{x - 2}$$

> Rewrite with x at the *end* of each denominator ready for binomial expansion.

$$\equiv \frac{1}{1 - x} - \frac{1}{2 - x}$$

> Remember: reversing the sign of numerator *and* denominator leaves a fraction unchanged.

b $\dfrac{1}{1-x} - \dfrac{1}{2-x} = (1-x)^{-1} - (2-x)^{-1}$

The power is −ve so the brackets must start with 1.

$$= (1-x)^{-1} - 2^{-1}\left(1 - \frac{x}{2}\right)^{-1}$$

$$= (1+x+x^2) - \frac{1}{2}\left(1 + \frac{x}{2} + \left(\frac{x}{2}\right)^2 + \cdots\right)$$

Terms in x^3 and higher powers are ignored.

$$= 1 + x + x^2 - \frac{1}{2} - \frac{x}{4} - \frac{x^2}{8} + \cdots$$

$$= \frac{1}{2} + \frac{3x}{4} + \frac{7x^2}{8} + \cdots \quad \text{①}$$

'Check' by substituting a small value of x.

Valid for $|x| < 1$.

c Substituting $x = 0.01$ in ①

$\mathrm{f}(0.01) \approx 0.5 + 0.75 \times 0.01 + 0.875 \times (0.01)^2$

Note: It is easier to work in decimals.

$= 0.5 + 0.0075 + 0.000\,0875$

A check will confirm that the term in x^3 will not affect the fifth decimal place.

$= 0.507\,5875$

$= 0.507\,59$ (correct to 5 d.p.)

Exercise 10A

Numerical answers should be calculated without *a calculator. A calculator can, of course, be used for checking.*

1 Expand these expressions in ascending powers of x, as far as the term in x^3, and state the values of x for which the expansion is valid.

a $(1+x)^{-2}$ **b** $(1+x)^{\frac{1}{3}}$ **c** $(1+x)^{\frac{3}{2}}$ **d** $(1-2x)^{\frac{1}{2}}$

e $\left(1+\dfrac{x}{2}\right)^{-3}$ **f** $(1-3x)^{-\frac{1}{2}}$ **g** $\dfrac{1}{1+3x}$ **h** $\sqrt{1-x^2}$

i $\sqrt[3]{1-x}$ **j** $\dfrac{1}{\sqrt{1+2x}}$ **k** $\dfrac{1}{\left(1+\frac{x}{2}\right)^2}$ **l** $\sqrt{(1-2x)^3}$

m $(4+x)^{-1}$ **n** $(3-x)^{-2}$ **o** $(4+5x)^{\frac{1}{2}}$ **p** $(8+3x)^{\frac{1}{3}}$

2 Using the results that

$$\frac{1}{1+x} = 1 - x + x^2 - \cdots \quad \text{and} \quad \frac{1}{1-x} = 1 + x + x^2 + \cdots$$

deduce the first three terms in these expansions. State the values of x for which each expansion is valid.

a $\dfrac{1}{1+2x}$
  b $\dfrac{1}{1-3x}$
  c $\dfrac{1}{1+\frac{2x}{3}}$

d $\dfrac{1}{1+x^2}$
  e $\dfrac{1}{2+x}$
  f $\dfrac{12}{4-3x}$

3 Find the first three terms in the expansion of

$$\frac{1}{(1+x)^2}$$

Hence deduce these expansions, stating the values of x for which each expansion is valid.

a $\dfrac{1}{(1+2x)^2}$
  b $\dfrac{1}{(1-3x)^2}$
  c $\dfrac{1}{(1+\frac{2x}{3})^2}$

d $\dfrac{1}{(1-x^3)^2}$
  e $\dfrac{1}{(2+x)^2}$
  f $\dfrac{8}{(4-3x)^2}$

4 a Write down the first four terms of the expansion of $(1+x)^{\frac{1}{2}}$.

  b By substituting $x = 0.001$, find $\sqrt{1.001}$, correct to 6 places of decimals.

5 a Write down the first four terms of the expansion of

$$\frac{1}{(1+2x)^2}$$

  b By substituting $x = 0.01$, find $\dfrac{1}{1.02^2}$, correct to 4 places of decimals.

6 Use suitable binomial expansions to find

  a $\sqrt{0.998}$, correct to 6 places of decimals

  b $\sqrt[3]{1.03}$, correct to 4 places of decimals

  c $\dfrac{1}{\sqrt{0.98}}$, correct to 4 places of decimals

7 Find the first four terms of the expansions of these expressions, in ascending powers of x.

  a $\dfrac{1+x}{1-x}$
  b $\dfrac{x+2}{(1+x)^2}$
  c $\dfrac{1-x}{\sqrt{1+x}}$

8 a Expand $(1+x)^{-3}$ as far as the term in x^3.

  b By substituting $x = 0.01$, find $\dfrac{1}{1.01^3}$, correct to 5 decimal places.

  c By substituting $x = 0.001$, find $\dfrac{1}{1.001^3}$, correct to 8 significant figures, giving the answer in standard form.

9 a Expand $(1 + 2x)^{\frac{1}{2}}$ as far as the term in x^2.

  b By substituting $x = 0.01$, find, *without* using a calculator, $\sqrt{1.02}$, correct to 5 decimal places.
Hence, deduce $\sqrt{102}$ correct to 4 decimal places.
(Check the result on a calculator.)

  c By substituting $x = 0.000\,001$, find $\sqrt{1\,000\,002}$, correct to 10 decimal places.
(This is not easy to check on a calculator.)

10 Find the first four terms of the expansion of $(1 - 8x)^{\frac{1}{2}}$ in ascending powers of x.
By substituting $x = \frac{1}{100}$, obtain the value of $\sqrt{23}$, correct to 5 significant figures.

11 Find the first four terms of the expansion of $(1 + 4x)^{\frac{1}{4}}$ in ascending powers of x.
By substituting $x = \frac{1}{10\,000}$, obtain the value of $\sqrt[4]{10\,004}$, correct to 11 significant figures.

12 The third term of the expansion of $(1 + 2x)^n$, in ascending powers of x, is $24x^2$.
Find the two possible values of n and the first four terms of each expansion.

13 The first three terms of the expansion of $(1 + kx)^n$ are $1 + 6x + 27x^2$.
Find k and n and hence find the term in x^3.

14 In the expansion of $(1 + ax)^n$, in ascending powers of x, the coefficient of x is 12 and the coefficient of x^3 is six times the coefficient of x^2.

  a Find a and n.

  b State the values of x for which the expansion is valid.

15 The binomial expansion of
$$\frac{1 + 3x}{(1 - x)^2}$$
is $1 + ax + bx^2 + cx^3 + \cdots$, where a, b and c are constants.

  a Find a, b and c.

  b State the values of x for which the expansion is valid.

16 a Given that
$$f(x) = \frac{1}{(3 + x)(1 - x)}$$
express $f(x)$ in partial fractions.

  b Show that $f(x) = \dfrac{1}{3} + \dfrac{2x}{9} + \dfrac{7x^2}{27} + \cdots$

17 By using partial fractions, or otherwise, find the first three terms in the expansion of
$$\frac{1}{x^2 - 3x - 4}$$

18 By using partial fractions, or otherwise, find the first three terms in the expansion of
$$\frac{4}{(1 - x)(1 - x^2)}$$

19   By using partial fractions, or otherwise, find the first three terms in the expansion of

$$\frac{1}{(1-x)^2(2+x)}$$

20 a   Express in partial fractions

$$\frac{1}{(1+2x)(3-x)}$$

   b   Hence, find the first three terms of the expansion, for $|x| < \frac{1}{2}$, of

$$\frac{1}{(1+2x)(3-x)}$$

21   The integral $I = \displaystyle\int_0^{0.6} \sqrt{1-x^3}\,dx$ has a value of 0.5833 correct to four decimal places.

   a   Obtain the binomial expansion of $(1-x^3)^{\frac{1}{2}}$ in ascending powers of x up to and including the term in x^6.

   b   By integrating the terms of your expression in part **a** show that the approximation to I obtained is also 0.5833 correct to four decimal places.

   c   Use the mid-ordinate rule with 6 intervals to show that the approximation to I obtained is 0.5835 correct to four decimal places.

There is a 'Test yourself' exercise (Ty10) and an 'Extension exercise' (Ext10) on the CD-ROM.

➤ # Key points

Binomial expansion of $(1+x)^n$ for any rational index

For n negative or a fraction, the binomial expansion gives an infinite series which converges for certain values of the variable.

Arrange to expand a bracket whose first term is 1, i.e. of the form $(1 + \cdots)^n$.
Then use this form of the expansion for $|x| < 1$.

$$(1+x)^n = 1 + nx + \frac{n(n-1)}{2!}x^2 + \frac{n(n-1)(n-2)}{3!}x^3 + \frac{n(n-1)(n-2)(n-3)}{4!}x^4 + \cdots$$

Remember to check the values of x for which the expansion is valid.

Approximations

Binomial expansions can be used to find approximations, provided that most of the terms are small enough to be ignored.

Further differentiation

So far in the course you have only considered functions where one variable is given explicitly in terms of another. However, there are many occasions when this is not the case; for example in the case of a circle, where the equation is of the form $x^2 + y^2 = r^2$, or of many other curves such as ellipses and hyperbolae.

After working through this chapter you should be able to

■ *differentiate functions given implicitly*

■ *find stationary points on curves given implicitly*

■ *find the equations of tangents and normals to curves given implicitly*

■ *use the chain rule to find connected rates of change*

■ *apply your knowledge of differentiation to problems of exponential growth and decay*

■ *find the derivatives of a^x and $\log_a x$.*

The CD-ROM (E11.1.1) shows how logarithms can be used to help simplify the differentiation of some complicated products and quotients.

11.1 Implicit functions

The functions considered so far have been of the form $y = f(x)$, where y is given *explicitly* in terms of x; for example $y = x^2 + \sin x$ is an *explicit* function.

When an equation contains two variables, but neither of them is given explicitly in terms of the other, then each is said to be expressed *implicitly* in terms of the other; for example $x^2 + xy + y^2 = 25$ is an *implicit* function.

Many curves are defined by implicit functions, for example, circles, ellipses and hyperbolae. In some cases, it is possible to rearrange the equation to give one variable explicitly in terms of the other. In other cases, this is not possible; it is, however, still possible to find the gradients of such functions.

Implicit differentiation

Remember The chain rule for differentiating composite functions: $\dfrac{dy}{dx} = \dfrac{dy}{dt} \times \dfrac{dt}{dx}$

Consider, for example, this implicit function:

$$x^2 + y^2 = 25 \qquad ①$$

The gradient of the curve is given by $\dfrac{dy}{dx}$.

Differentiating both sides of ① with respect to x

$$\frac{d}{dx}(x^2) + \frac{d}{dx}(y^2) = \frac{d}{dx}(25) \qquad ②$$

To differentiate y^2 with respect to x, the chain rule must be used:

$$\frac{\mathrm{d}}{\mathrm{d}x}(y^2) = \frac{\mathrm{d}}{\mathrm{d}y}(y^2) \times \frac{\mathrm{d}y}{\mathrm{d}x} = 2y\frac{\mathrm{d}y}{\mathrm{d}x}$$

and so ② becomes:

$$2x + 2y\frac{\mathrm{d}y}{\mathrm{d}x} = 0$$

Divide through by the common factor, 2.

$$x + y\frac{\mathrm{d}y}{\mathrm{d}x} = 0$$

Rearrange to isolate $\frac{\mathrm{d}y}{\mathrm{d}x}$.

$$y\frac{\mathrm{d}y}{\mathrm{d}x} = -x$$

$$\frac{\mathrm{d}y}{\mathrm{d}x} = -\frac{x}{y}$$

This technique can be used to differentiate *any* implicit function.

Example 1 Find $\frac{\mathrm{d}y}{\mathrm{d}x}$ for each of these implicit functions.

a $\quad 2x^3 + 3y^2 = 7$
b $\quad \dfrac{x^2}{x - y^2} = 5$

c $\quad 2x^2 + y^3 = 5xy$
d $\quad x\ln y + y^2 = 10$

Solution a $\quad 2x^3 + 3y^2 = 7$

Differentiating with respect to x

Differentiate both sides term by term.

$$\frac{\mathrm{d}}{\mathrm{d}x}(2x^3) + \frac{\mathrm{d}}{\mathrm{d}x}(3y^2) = \frac{\mathrm{d}}{\mathrm{d}x}(7)$$

Use the chain rule to differentiate $3y^2$ with respect to x.

$$6x^2 + \frac{\mathrm{d}}{\mathrm{d}y}(3y^2)\frac{\mathrm{d}y}{\mathrm{d}x} = 0$$

$$6x^2 + 6y\frac{\mathrm{d}y}{\mathrm{d}x} = 0$$

Divide through by the common factor, 6.

$$x^2 + y\frac{\mathrm{d}y}{\mathrm{d}x} = 0$$

Rearrange to obtain $\frac{\mathrm{d}y}{\mathrm{d}x}$.

$$y\frac{\mathrm{d}y}{\mathrm{d}x} = -x^2$$

$$\frac{\mathrm{d}y}{\mathrm{d}x} = -\frac{x^2}{y}$$

b　$\dfrac{x^2}{x-y^2}=5$

First multiply both sides by $(x-y^2)$.

$$x^2=5x-5y^2$$

Differentiating with respect to x

$$\frac{d}{dx}(x^2)=\frac{d}{dx}(5x)-\frac{d}{dx}(5y^2)$$

Use the chain rule to differentiate $5y^2$ with respect to x.

$$2x=5-10y\frac{dy}{dx}$$

Rearrange to obtain $\dfrac{dy}{dx}$.

$$10y\frac{dy}{dx}=5-2x$$

$$\frac{dy}{dx}=\frac{5-2x}{10y}$$

c　$2x^2+y^3=5xy$

Differentiating with respect to x

$$\frac{d}{dx}(2x^2)+\frac{d}{dx}(y^3)=\frac{d}{dx}(5xy)$$

To differentiate $5xy$ use the product rule:
$$\frac{d}{dx}(5xy)=5x\frac{dy}{dx}+y\frac{d}{dx}(5x)$$

$$4x+\frac{d}{dy}(y^3)\frac{dy}{dx}=5x\frac{dy}{dx}+5y$$

$$4x+3y^2\frac{dy}{dx}=5x\frac{dy}{dx}+5y$$

Collect $\dfrac{dy}{dx}$ terms on the RHS, other terms on the LHS.

$$4x-5y=(5x-3y^2)\frac{dy}{dx}$$

$$\frac{dy}{dx}=\frac{4x-5y}{5x-3y^2}$$

d　$x\ln y+y^2=10$

Differentiating with respect to x gives

$$\frac{d}{dx}(x\ln y)+\frac{d}{dx}(y^2)=\frac{d}{dx}(10)$$

Use the product rule to differentiate $x\ln y$.

$$x\frac{d}{dx}(\ln y)+\ln y+\frac{d}{dy}(y^2)\frac{dy}{dx}=0$$

Use the chain rule to differentiate $\ln y$ with respect to x.

$$x\frac{d}{dy}(\ln y)\frac{dy}{dx}+\ln y+2y\frac{dy}{dx}=0$$

$$x\frac{1}{y}\frac{dy}{dx}+\ln y+2y\frac{dy}{dx}=0$$

Collect $\dfrac{dy}{dx}$ terms on the LHS, other terms on the RHS.

$$\left(\frac{x}{y}+2y\right)\frac{dy}{dx}=-\ln y$$

$$\frac{dy}{dx}=\frac{-\ln y}{\dfrac{x}{y}+2y}$$

Multiply numerator and denominator by y to remove fraction in the denominator.

$$\frac{dy}{dx}=-\frac{y\ln y}{x+2y^2}$$

Once the technique of implicit differentiation has been mastered it is possible to reduce the amount of working shown in a solution. For example, in Example 1d, the second and third lines of working could be omitted.

Example 2 Find y' for the implicit function $x^3y^2 + 5x + 6y^3 = 0$.

$$x^3y^2 + 5x + 6y^3 = 0$$

Differentiating with respect to x

$$\frac{d}{dx}(x^3y^2) + \frac{d}{dx}(5x) + \frac{d}{dx}(6y^3) = \frac{d}{dx}(0)$$

$$x^3 2yy' + 3x^2y^2 + 5 + 18y^2y' = 0$$

$$(2x^3y + 18y^2)y' = -3x^2y^2 - 5$$

$$y' = -\frac{3x^2y^2 + 5}{2x^3y + 18y^2}$$

> *Remember:* $y' = \dfrac{dy}{dx}$

> To differentiate x^3y^2 use the product rule and then differentiate y^2, with respect to x, using the chain rule.

> Collect terms in y'.

> *Note*: $-3x^2y^2 - 5 = -(3x^2y^2 + 5)$

Example 3 (Extension) Find $\dfrac{dy}{dx}$, in terms of x and y, for the curve $\tan(x+y) = y^2$.

$$\tan(x+y) = y^2$$

Differentiating with respect to x

$$\frac{d}{dx}\tan(x+y) = \frac{d}{dx}(y^2)$$

$$\sec^2(x+y)\left[1 + \frac{dy}{dx}\right] = 2y\frac{dy}{dx}$$

$$\sec^2(x+y) + \sec^2(x+y)\frac{dy}{dx} = 2y\frac{dy}{dx}$$

$$\sec^2(x+y) = \left[2y - \sec^2(x+y)\right]\frac{dy}{dx}$$

$$\frac{dy}{dx} = \frac{\sec^2(x+y)}{2y - \sec^2(x+y)}$$

Alternatively

$$\frac{dy}{dx} = \frac{1}{2y\cos^2(x+y) - 1}$$

> To differentiate $\tan(x+y)$ with respect to x, use the chain rule.
> $$\frac{d}{dx}\tan(x+y) = \sec^2(x+y)\left[\frac{d}{dx}(x+y)\right]$$
> $$= \sec^2(x+y)\left[1 + \frac{dy}{dx}\right]$$

> Divide through by $\sec^2(x+y)$ and note that
> $$\frac{1}{\sec^2(x+y)} = \cos^2(x+y)$$

Example 4 Find the gradient of the curve $x^2 - 3xy + y^2 = 31$ at the point $(2, -3)$.

$$x^2 - 3xy + y^2 = 31$$

Differentiating with respect to x

$$\frac{d}{dx}(x^2) - \frac{d}{dx}(3xy) + \frac{d}{dx}(y^2) = \frac{d}{dx}(31)$$

$$2x - \left(3x\frac{dy}{dx} + 3y\right) + 2y\frac{dy}{dx} = 0$$

$$2x - 3x\frac{dy}{dx} - 3y + 2y\frac{dy}{dx} = 0 \qquad \text{①}$$

> Take care over negative sign in front of the product $3xy$.

At the point $(2, -3)$, $x = 2$ and $y = -3$, and

$$4 - 6\frac{dy}{dx} + 9 - 6\frac{dy}{dx} = 0$$

$$12\frac{dy}{dx} = 13$$

$$\frac{dy}{dx} = \frac{13}{12}$$

> Note that there is no need to rearrange ① to obtain $\frac{dy}{dx}$ explicitly. It is easier to substitute the values of x and y and then rearrange.

Tangents and normals

Example 5

This example uses implicit differentiation to find the equation of a tangent and a normal to a curve at a given point.

Find the equations of the tangent and normal to the curve $y^2e^x + x^2 = 9$ at the point $(0, 3)$.

$$y^2e^x + x^2 = 9$$

Differentiating with respect to x

$$\frac{d}{dx}(y^2e^x) + \frac{d}{dx}(x^2) = \frac{d}{dx}(9)$$

$$y^2e^x + 2y\frac{dy}{dx}e^x + 2x = 0 \qquad ①$$

At the point $(0, 3)$

$$9 + 6\frac{dy}{dx} + 0 = 0$$

$$\frac{dy}{dx} = -\frac{9}{6} = -\frac{3}{2}$$

> Substitute values of x and y into equation ① before rearranging. Remember $e^0 = 1$.

So the equation of the tangent to the curve at $(0, 3)$ is

$$y - 3 = -\tfrac{3}{2}(x - 0)$$

$$2y - 6 = -3x$$

$$3x + 2y - 6 = 0$$

> Using $y - y_1 = m(x - x_1)$

The gradient of the normal is $\frac{2}{3}$.

So the equation of the normal is

$$y - 3 = \tfrac{2}{3}(x - 0)$$

$$3y - 9 = 2x$$

$$2x - 3y + 9 = 0$$

> *Remember*: the product of the gradients of perpendicular lines is -1.

> Using $y - y_1 = m(x - x_1)$

Stationary points

Example 6   *This example uses implicit differentiation to find the stationary points on a curve.*

Find the stationary points on the curve $x^2 + 4xy + y^2 = -48$.

$$x^2 + 4xy + y^2 = -48$$

Differentiating with respect to x

$$2x + 4x\frac{dy}{dx} + 4y + 2y\frac{dy}{dx} = 0$$

> Divide through by the common factor, 2.

$$x + 2x\frac{dy}{dx} + 2y + y\frac{dy}{dx} = 0 \qquad ①$$

> Collect like terms.

$$(2x + y)\frac{dy}{dx} = -(x + 2y)$$

$$\frac{dy}{dx} = -\frac{x + 2y}{2x + y}$$

> *Note*: If $2x + y = 0$, the gradient is undefined and the tangent to the curve is parallel to the y-axis.

For stationary points, $\dfrac{dy}{dx} = 0$

$$-\frac{x + 2y}{2x + y} = 0$$

> Alternatively substitute $\dfrac{dy}{dx} = 0$ into ①, obtaining $x + 2y = 0$ directly.

$$x + 2y = 0$$

$$x = -2y$$

Substituting this into the equation of the curve

$$(-2y)^2 + 4(-2y)y + y^2 = -48$$

$$4y^2 - 8y^2 + y^2 = -48$$

$$-3y^2 = -48$$

$$y^2 = 16$$

> Substitute $y = \pm 4$ in $x = -2y$ rather than the original equation of the curve.

$$y = \pm 4$$

When $y = 4$, $x = -8$ and when $y = -4$, $x = 8$, so there are stationary points at $(-8, 4)$ and $(8, -4)$.

The CD-ROM (E11.1.2) shows how the second derivative can be used to determine the nature of the stationary points.

1 Differentiate with respect to x

    a y^2 **b** y^3 **c** $3y^4$ **d** xy

    e x^2y **f** xy^2 **g** $\ln y$ **h** $\ln y^5$

    i $\ln x^2y^3$ **j** $\sin y$ **k** $x\cos y$ **l** x^2e^{2y}

    m e^xy^2 **n** $\dfrac{1}{y}$ **o** $\dfrac{x}{y^3}$ **p** $\sin(x+y)$

2 Find $\dfrac{dy}{dx}$ for each of these implicit functions.

    a $x^2 + y^2 = 8$ **b** $2x^3 + 3y^4 = 10$

    c $x^2 + 3xy = 2y^2 + 4$ **d** $x^3 - 2xy^2 + 7x = 0$

    e $4x^2 + 6y^2 = 3x^2y^2$ **f** $3x^3 + 2x^2y + 5xy^2 + 4y^3 = 8$

    g $\dfrac{1}{x} + \dfrac{1}{y} = 2$ **h** $\dfrac{x^2}{2x + 5y^2} = 2$

3 For each of these implicit functions, find $\dfrac{dy}{dx}$ in terms of x and y.

    a $4e^xy - 3xe^y = 10$ **b** $x\tan y = 10$

    c $x\sin y + y\sin x = 1$ **d** $3x\ln y = 2y^2 + 8$

    e $2\sin 2x\cos 3y = 1$ **f** $e^x\ln y = y$

    g $4xy - x\ln y^3 = 8$ **h** $2x\sin^2 y = 3(x+y)^2$

4 Find the gradient of each of these curves at the points specified.

    a $x^2 + y^2 = 8$ at $(2, -2)$ **b** $2x^2 + 3y^3 = 26$ at $(-1, 2)$

    c $e^yx = 1$ at $(1, 0)$ **d** $x\ln y = e$ at (e, e)

    e $y(x + y) = 5$ at $(4, 1)$ **f** $2x\sin y = 1$ at $\left(1, \frac{\pi}{6}\right)$

5 Find the gradient of each for these curves at the points specified.

    a $xy^2 = 20$ at $(5, 2)$ **b** $x^2 + 3xy + 2y^2 = 15$ at $(1, 2)$

    c $(x - 1)^2 + (y + 2)^2 = 2$ at $(2, -3)$ **d** $\sec y = x + y$ at $(1, 0)$

    e $e^xy + x^2y = 2$ at $(0, 2)$ **f** $\dfrac{\sin x}{\sin y} = 2$ at $\left(\dfrac{\pi}{2}, \dfrac{\pi}{6}\right)$

    g $x\ln y^3 = 6$ at $(2, e)$ **h** $\dfrac{8x^2}{4x^2 - 3y^3} = 3y$ at $(3, 2)$

6 $y = \dfrac{3 - x}{x - 1}$ $x \neq 1$, could be written as $xy - y + x - 3 = 0$.

    a Show by differentiating as a quotient, that

$$y' = -\frac{2}{(x - 1)^2}$$

    b Use implicit differentiation of $y(x - 1) = 3 - x$ to find y'.

7 Find the equation of the tangent to the curve $x^2 + y^2 = 5y$ at the point $(2, 4)$.

8 Find the equations of the tangent and normal to the curve $x^2y + 3xy^2 = 2$ at the point $(2, -1)$.

9 Find the equations of the tangents to the curve $xy^2 + x^2y = 30$ at the points where $x = 3$.

10 Find the equations of the tangent and normal to the curve $x^5 + y^5 = 4x^2y^2$ at the point $(2, 2)$.

11 Find the equation of the tangent to the curve $x \ln y + y \ln x = 1$ at the point $(1, e)$.

12 Find the equations of the tangent and normal to the curve $\sin x \sin y = \dfrac{\sqrt{3}}{4}$ at the point $\left(\dfrac{\pi}{3}, \dfrac{\pi}{6}\right)$.

13 Show that one of the points of intersection of the line $3y = 2x + 1$ and the curve $2x^2 - 3xy + y^2 = 5$ is the point $(4, 3)$. Find the equation of the tangent to the curve at this point. Find the area of the triangle bounded by the tangent and the axes.

14 Find the stationary points on each of these curves:

    a $3x^2 + 2y^2 = 6$ **b** $\dfrac{y^2}{5} - \dfrac{x^2}{4} = 1$ **c** $x^2 - y^2 = 6xy - 90$

11.2 Extension: Connected rates of change

The chain rule can be used to solve problems involving connected rates of change.

Imagine watching a video animation of a cube being enlarged such that an edge of the cube, x mm, is increasing at a rate of 2 mm per second.

As x increases, the volume of the cube, V cm³, increases.

The rate at which V is changing,

$\dfrac{\mathrm{d}V}{\mathrm{d}t}$, is related to $\dfrac{\mathrm{d}x}{\mathrm{d}t}$, the rate

at which x is changing.

Note 'Rate of change' usually refers to the rate of change of the variable *with respect to time*. The rate of change with respect to any other variable would be specified.

The rates of change of V and x can be connected using the chain rule

$$\frac{\mathrm{d}V}{\mathrm{d}t} = \frac{\mathrm{d}V}{\mathrm{d}x} \times \frac{\mathrm{d}x}{\mathrm{d}t}$$

If a variable is *increasing*, the rate of change will be *positive*. If it is *decreasing*, the rate of change will be *negative*.

$\dfrac{\mathrm{d}V}{\mathrm{d}x}$ can be found from the relationship $V = x^3$, so $\dfrac{\mathrm{d}V}{\mathrm{d}x} = 3x^2$

$\dfrac{\mathrm{d}x}{\mathrm{d}t} = 2$, since x is increasing at $2\,\text{mm}$ per second.

$$\therefore \quad \frac{\mathrm{d}V}{\mathrm{d}t} = \frac{\mathrm{d}V}{\mathrm{d}x} \times \frac{\mathrm{d}x}{\mathrm{d}t} = 3x^2 \times 2 = 6x^2$$

Since $6x^2$ is positive for all values of x, $\mathrm{d}V/\mathrm{d}t$ gives the rate of *increase* of the volume at any time when the cube is being enlarged.

For example, when $x = 10\,\text{mm}$, the volume is *increasing* at a rate of $(6 \times 10^2 =)\ 600\,\text{mm}^3$ per second.

> **The chain rule can be applied for connected rates of change involving several variables.**
>
> $$\frac{\mathrm{d}a}{\mathrm{d}x} = \frac{\mathrm{d}a}{\mathrm{d}b} \times \frac{\mathrm{d}b}{\mathrm{d}c} \times \cdots \times \frac{\mathrm{d}w}{\mathrm{d}x}$$

Example 7

The radius of a circular ink blot is increasing at $3\,\text{cm}\,\text{s}^{-1}$.

a Find the rate at which the area is increasing when the radius is $2\,\text{cm}$.

b What is the rate of increase of the circumference when the radius is $3\,\text{cm}$?

Give exact answers.

Solution

Let the circle have radius $r\,\text{cm}$, area $A\,\text{cm}^2$ and circumference $C\,\text{cm}$ at time t seconds.

> Define the variables, including their units.

$$\frac{\mathrm{d}r}{\mathrm{d}t} = 3$$

> The radius is given as increasing at $3\,\text{cm}$ per second.

a $\qquad A = \pi r^2$

$\text{so} \quad \dfrac{\mathrm{d}A}{\mathrm{d}r} = 2\pi r$

> It is required to find $\mathrm{d}A/\mathrm{d}t$ when $r = 2$. First find $\mathrm{d}A/\mathrm{d}r$ by using the formula connecting A and r. Then use the chain rule to connect the three variables A, r and t.

$$\frac{\mathrm{d}A}{\mathrm{d}t} = \frac{\mathrm{d}A}{\mathrm{d}r} \times \frac{\mathrm{d}r}{\mathrm{d}t}$$

> Substitute $\dfrac{\mathrm{d}r}{\mathrm{d}t} = 3$.

$$= 2\pi r \times 3 = 6\pi r$$

When $r = 2$

$$\frac{\mathrm{d}A}{\mathrm{d}t} = 6\pi \times 2 = 12\pi$$

The area is increasing at a rate of $12\pi\,\text{cm}^2\,\text{s}^{-1}$ when the radius is $2\,\text{cm}$.

> Remember to include the units in the answer.

b
$$C = 2\pi r$$

so $\dfrac{dC}{dr} = 2\pi$

$$\frac{dC}{dt} = \frac{dC}{dr} \times \frac{dr}{dt}$$

$$= 2\pi \times 3 = 6\pi$$

> It is required to find dC/dt when $r = 3$.
> First write C in terms of r and find dC/dr.
> Then use the chain rule to connect r, C and t.

The circumference is increasing at $6\pi\,\text{cm s}^{-1}$ when $r = 3$.

> Since 6π is constant, the circumference is increasing at $6\pi\,\text{cm s}^{-1}$ throughout the time that the ink blot is enlarging.

Example 8

a A spherical balloon is inflated by a machine which pumps in air at a rate of $15\,\text{cm}^3$ per second. Find, in cm per second, correct to 2 significant figures, the rate at which the radius is increasing when the radius is $10\,\text{cm}$.

b Once inflated, air is released from the balloon at a rate of $20\,\text{cm}^3$ per second. Find the rate at which the surface area is decreasing when $r = 10\,\text{cm}$. (Surface area of a sphere $= 4\pi r^2$)

Solution Let the balloon have volume $V\,\text{cm}^3$, surface area $A\,\text{cm}^2$ and radius $r\,\text{cm}$ at time $t\,$s.

> Include units. Then dV/dt, ... etc. represent numbers.

a $\dfrac{dV}{dt} = 15$

> The volume is increasing at $15\,\text{cm}^3$ per second.

$$V = \tfrac{4}{3}\pi r^3$$

> Remember: $V = \tfrac{4}{3}\pi r^3$

So $\dfrac{dV}{dr} = 4\pi r^2$ ①

> It is required to find dr/dt so use the chain rule to connect the three variables V, r and t.

$$\frac{dV}{dt} = \frac{dV}{dr} \times \frac{dr}{dt}$$ ②

So $15 = 4\pi r^2 \times \dfrac{dr}{dt}$

> Make $\dfrac{dr}{dt}$ the subject.

$$\frac{dr}{dt} = \frac{15}{4\pi r^2}$$

When $r = 10$

$$\frac{dr}{dt} = \frac{15}{4\pi \times 10^2} = 0.012\,(2\text{ s.f.})$$

> Approximate as instructed.

The radius is increasing at a rate of $0.012\,\text{cm}$ per second when the radius is $10\,\text{cm}$.

b $A = 4\pi r^2$

> It is required to find dA/dt when $r = 10$.

so $\dfrac{dA}{dr} = 8\pi r$ ③

$$\frac{dA}{dt} = \frac{dA}{dr} \times \frac{dr}{dt} = 8\pi r \frac{dr}{dt}$$ ④

> First find dr/dt.

From ①

$$\frac{dV}{dr} = 4\pi r^2$$

Also $\quad \dfrac{dV}{dt} = -20$ (given)

Now $\quad \dfrac{dV}{dt} = \dfrac{dV}{dr} \times \dfrac{dr}{dt}$ ②

So $\quad -20 = 4\pi r^2 \times \dfrac{dr}{dt}$

$$\frac{dr}{dt} = -\frac{20}{4\pi r^2} = -\frac{5}{\pi r^2}$$

$$\therefore \quad \frac{dA}{dt} = 8\pi r \times \left(-\frac{5}{\pi r^2} \right) = -\frac{40}{r} \quad ⑤$$

When $r = 10$

$$\frac{dA}{dt} = -\frac{40}{10} = -4$$

The surface area is decreasing at a rate of $4\,\text{cm}^2$ per second when the radius is $10\,\text{cm}$.

Extension Exercise 11B

1 Complete these.

a $\quad \dfrac{dy}{dx} = \dfrac{dy}{dt} \times \dfrac{\square}{dx}$
b $\quad \dfrac{dV}{dt} = \dfrac{dV}{\square} \times \dfrac{dr}{dt}$

c $\quad \dfrac{dy}{dx} = \dfrac{dy}{dz} \times \dfrac{dz}{\square}$
d $\quad \dfrac{dy}{dt} = \dfrac{dy}{dA} \times \dfrac{\square}{\square} \times \dfrac{dc}{dt}$

2 If $A = 6x^2$ and $\dfrac{dA}{dt} = 2$, find $\dfrac{dx}{dt}$ when $x = 3$.

3 The radius of a circle is increasing at $3\,\text{cm s}^{-1}$.

   a Find the rate at which the circumference is increasing when the radius is $5\,\text{cm}$.

   b Find the rate at which the area is increasing when the circumference is $20\pi\,\text{cm}$.

4 A cone has height $6\,\text{cm}$. The radius of the base of the cone is increasing at $5\,\text{cm s}^{-1}$. Find the rate of change of the volume of the cone when the radius of the base is $4\,\text{cm}$.

5 The side of a cube is decreasing at a rate of $6\,\text{cm s}^{-1}$.
   Find the rate of decrease of the volume when the length of a side is $2\,\text{cm}$.

6 A sphere has radius $r\,\text{cm}$. The surface area of the sphere is $4\pi r^2$.
   Find the rate of change of the area when $r = 2$, given that the radius is increasing at a rate of $1\,\text{cm s}^{-1}$.

7 If $y = (x^2 - 3x)^3$, find $\dfrac{dy}{dt}$ when $x = 2$, given that $\dfrac{dx}{dt} = 2$.

8 The volume of a cube is increasing at a rate of $2\,\text{cm}^3\,\text{s}^{-1}$.
Find the rate of change of the length of an edge when the cube has dimensions $3\,\text{cm}$ by $3\,\text{cm}$ by $3\,\text{cm}$.

9 The area of a circle is decreasing at a rate of $0.5\,\text{cm}^2\,\text{s}^{-1}$.
Find the rate of change of the circumference when the radius is $2\,\text{cm}$.

10 At a given instant, the radii of two concentric circles are $8\,\text{cm}$ and $12\,\text{cm}$.
The radius of the outer circle is increasing at a rate of $1\,\text{cm}\,\text{s}^{-1}$ and the radius of the inner circle is increasing at a rate of $2\,\text{cm}\,\text{s}^{-1}$.
Find the rate of change of the area enclosed by the two circles.

11 If $y = \left(x - \dfrac{1}{x}\right)^2$, find $\dfrac{dx}{dt}$ when $x = 2$, given $\dfrac{dy}{dt} = 1$.

12 A hollow right circular cone has height $18\,\text{cm}$ and base radius $12\,\text{cm}$.
It is held vertex downwards beneath a tap leaking at the rate of $2\,\text{cm}^3\,\text{s}^{-1}$.
Find the rate of rise of water level when the depth is $6\,\text{cm}$.

11.3 Exponential and logarithmic functions

This section considers the use of exponential functions in modelling the growth or decay of certain physical, or other, quantities and also extends earlier work on differentiation to obtain the derivatives of more exponential and logarithmic functions.

Exponential growth and decay

There are numerous cases in, for example, science, social science and economics when quantities grow or decay exponentially. These quantities can be modelled as exponential functions of time.

For example, assume at time $t = 0$ there is a population of six cells and that all cells divide in two every hour. This table shows the number of cells, P, at time t hours:

t	0	1	2	3	$\ldots$	t
P	6	6×2	6×4	6×8	$\ldots$	6×2^t

This is an example of *exponential growth*.

The equation expressing P as a function of t is
$P = 6 \times 2^t$.

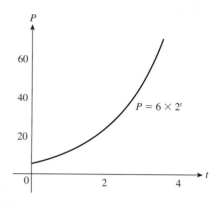

Consider now a population of 10^{12} bacteria which, due to a drug, is halved every day. This table shows the number of bacteria, n, at time t days:

t	0	1	2	3	$\ldots$	t
n	10^{12}	$10^{12} \times \frac{1}{2}$	$10^{12} \times \left(\frac{1}{2}\right)^2$	$10^{12} \times \left(\frac{1}{2}\right)^3$	$\ldots$	$10^{12} \times \left(\frac{1}{2}\right)^t$

This is an example of *exponential decay*.
The equation expressing n as a function of t is
$n = 10^{12} \times \left(\frac{1}{2}\right)^t = 10^{12} \times 2^{-t}$.

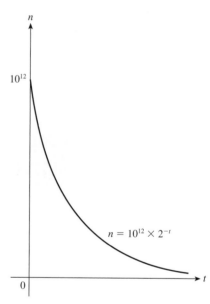

Note Using this model, the population never reaches zero; the curve approaches the horizontal axis asymptotically.

Exponential growth and decay of the form Ae^{kt}

Example 9 *This is an example of exponential growth.*

The population of a town is modelled by the equation $P = 4000e^{0.02t}$, where t is the number of years after 1950.

a Use the equation to predict the population in 2020.

b Estimate when the population will reach 20 000.

c Estimate the rate of growth of the population in 2010.

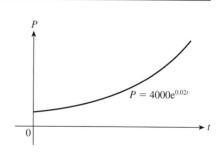

Solution **a** $P = 4000e^{0.02t}$

2020 is 70 years after 1950, so substitute $t = 70$.

When $t = 70$,

$$P = 4000 \times e^{0.02 \times 70}$$

$$= 4000 \times e^{1.4}$$

$$= 16\,200 \, (3 \, \text{s.f.})$$

The population in 2020 will be about 16 200.

b When $P = 20\,000$,

$$20\,000 = 4000 \, e^{0.02t}$$

$$5 = e^{0.02t}$$

Take logs to base e.

$$\ln 5 = 0.02t$$

$$t = 50 \ln 5$$

$$= 80.47\ldots$$

80 years after 1950 is the year 2030.

The population will reach 20 000 in about 2030.

c $P = 4000e^{0.02t}$

$$\frac{dP}{dt} = 4000 \times 0.02 \, e^{0.02t}$$

$$= 80 \, e^{0.02t}$$

When $t = 60$

In 2010, $t = 60$ so substitute into the expression for dP/dt.

$$\frac{dP}{dt} = 80 \times e^{0.02 \times 60}$$

$$= 80e^{1.2}$$

$$= 266 \, (3 \, \text{s.f.})$$

In 2010, the population will be growing at a rate of about 266 per year.

Example 10   *This is an example of exponential decay.*

The mass, m grams, of a radioactive substance, present at time t hours after first being observed, is given by the formula $m = 10e^{1 - 0.01t}$.

a Find the initial mass (denoted by m_0 in the sketch).

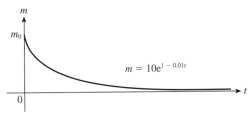

b What mass is present after 30 hours?

c At what time is the mass 10 g?

d At what rate is the mass decreasing when $t = 50$?

Give the answers to three significant figures.

Solution

a $m = 10e^{1-0.01t}$

The initial mass is the mass when $t = 0$.

When $t = 0$

$$m_0 = 10 \times e^1 = 27.182\ldots$$

The initial mass is 27.2 g (3 s.f.).

b When $t = 30$

$$m = 10 \times e^{1-0.01 \times 30}$$

$$= 10e^{0.7}$$

$$= 20.13\ldots$$

The mass after 30 hours is 20.1 g (3 s.f.).

c When $m = 10$

$$10 = 10e^{1-0.01t}$$

$$1 = e^{1-0.01t}$$

$$\ln 1 = 1 - 0.01t$$

$$0 = 1 - 0.01t$$

$$0.01t = 1$$

$$t = 100$$

After 100 hours, the mass will be 10 g.

Take logs to base e.

Use $\ln 1 = 0$

d $m = 10e^{1-0.01t}$

The rate at which the mass is changing is given by $\mathrm{d}m/\mathrm{d}t$.

$$\frac{\mathrm{d}m}{\mathrm{d}t} = 10 \times (-0.01)e^{1-0.01t}$$

$$= -0.1e^{1-0.01t}$$

Use the chain rule:

$$\frac{\mathrm{d}}{\mathrm{d}t}\left(ke^{f(t)}\right) = kf'(t)e^{f(t)}$$

When $t = 50$

$$\frac{\mathrm{d}m}{\mathrm{d}t} = -0.1e^{1-0.01 \times 50}$$

Since m is decreasing, the rate of change is negative.

$$= -0.1e^{0.5}$$

$$= -0.165 \text{ (3 s.f.)}$$

When $t = 50$, the mass is decreasing at 0.165 g/hour.

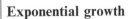

Exponential growth	Exponential decay

$$\frac{dy}{dt} = ky \ \ (k > 0)$$

$$\frac{dy}{dt} = -ky \ \ (k > 0)$$

$$y = Ae^{kt}$$

$$y = Ae^{-kt}$$

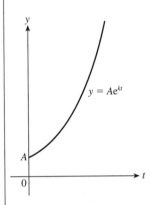

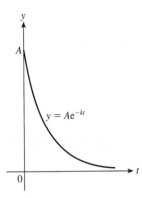

Examples:
- **Rate of growth of bacteria**
- **Rate of increase of population**

Examples:
- **Rate of radioactive decay**
- **Rate of cooling**

One result relating to $y = Ae^{kt}$ is very useful in later work and is investigated further in Section 14.3:

Differentiating with respect to t gives

$$\frac{dy}{dt} = Ake^{kt} = k(Ae^{kt}) = ky$$

$\therefore$ $y = Ae^{kt}$ is a solution of the differential equation $\dfrac{dy}{dt} = ky$.

Extension

The derivative of a^x

Consider $y = a^x$, $a > 0$ and take logarithms to the base e of both sides:

$$\ln y = \ln(a^x) = x \ln a$$

$$x = \frac{1}{\ln a} \ln y$$

> Use $\log(a^b) = b \log a$ and then rearrange to make x the subject.

Differentiating with respect to y

$$\frac{dx}{dy} = \frac{1}{\ln a} \times \frac{1}{y}$$

$$\therefore \quad \frac{dy}{dx} = \ln a \times y$$

$$= \ln a \times a^x$$

> Using $\dfrac{dy}{dx} = \dfrac{1}{\frac{dx}{dy}}$

$$\frac{d}{dx}(a^x) = \ln a \times a^x \qquad (a > 0)$$

Example 11

a $\dfrac{\mathrm{d}}{\mathrm{d}x}(2^x) = \ln 2 \times 2^x$ $\ln 2 = 0.693\,1471\ldots$

b $\dfrac{\mathrm{d}}{\mathrm{d}x}(3^x) = \ln 3 \times 3^x$ $\ln 3 = 1.098\,6123\ldots$

c $\dfrac{\mathrm{d}}{\mathrm{d}x}(4^x) = \ln 4 \times 4^x$ $\ln 4 = 1.386\,2944\ldots$

Special case when $a = \mathrm{e}$:

$$\dfrac{\mathrm{d}}{\mathrm{d}x}(\mathrm{e}^x) = \ln \mathrm{e} \times \mathrm{e}^x = \mathrm{e}^x$$

> Using $\ln \mathrm{e} = 1$.

Example 12   Find the equation of the tangent to the curve $y = 2^{3x+1}$ at $(1, 16)$.

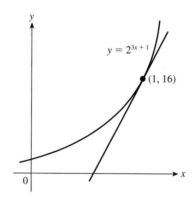

Let $t = 3x + 1$, then $y = 2^t$.

So $\dfrac{\mathrm{d}t}{\mathrm{d}x} = 3$

and $\dfrac{\mathrm{d}y}{\mathrm{d}t} = \ln 2 \times 2^t$

> Use $\dfrac{\mathrm{d}}{\mathrm{d}x}(a^x) = \ln a \times a^x$.

$\qquad\quad = \ln 2 \times 2^{3x+1}$

$\dfrac{\mathrm{d}y}{\mathrm{d}x} = \dfrac{\mathrm{d}y}{\mathrm{d}t} \times \dfrac{\mathrm{d}t}{\mathrm{d}x}$

> Use the chain rule.

$\qquad\quad = \ln 2 \times 2^{3x+1} \times 3$

$\qquad\quad = 3\ln 2 \times 2^{3x+1}$

When $x = 1$

> Find the gradient at $(1, 16)$.

$\qquad \dfrac{\mathrm{d}y}{\mathrm{d}x} = 3\ln 2 \times 2^4 = 48\ln 2$

> $48\ln 2$ is the exact value of the gradient.

Equation of tangent at $(1, 16)$:

$\qquad y - 16 = 48\ln 2(x - 1)$

$\qquad\quad y = (48\ln 2)x + 16 - 48\ln 2$

> To 3 s.f., the equation is $y = 33.3x - 17.3$.

Extension

The derivative of $\log_a x$

Rearrange to make x the subject.

Let $y = \log_a x$

Then $x = a^y$

$$\frac{dx}{dy} = \ln a \times a^y$$

$$= \ln a \times x$$

$$\frac{dy}{dx} = \frac{1}{\frac{dx}{dy}} = \frac{1}{\ln a} \times \frac{1}{x}$$

➤
$$\boxed{\frac{d}{dx}(\log_a x) = \frac{1}{\ln a} \times \frac{1}{x}}$$

Setting $a = e$ provides a special case:

$$\frac{d}{dx}(\log_e x) = \frac{d}{dx}\ln x$$

$\log_e x$ is written as $\ln x$.

$$= \frac{1}{\ln e} \times \frac{1}{x}$$

$\ln e = 1$

$$= \frac{1}{x}$$

➤
$$\boxed{\frac{d}{dx}(\ln x) = \frac{1}{x}}$$

This confirms the result you obtained earlier (see p. 72).

Exercise 11C

1 In a population model, the population, P thousands, is given by $P = 2e^{1+0.1t}$ where t is the time in years.

   a Find the size of the population when $t = 0$.

   b What size is the population when $t = 10$?

   c What is the rate of growth of the population when $t = 5$?

   Give your answers to three significant figures.

2 The mass, m grams, of a substance is decaying exponentially according to the relationship $m = m_0 e^{-2t}$, where t is the time in hours. Initially, the mass is 30 grams. Find m_0 and the rate of decay after 3 hours.

3 Radioactive substances decay at a rate proportional to the amount of radioactive substance present.

   At time t years, the mass of a sample of radium is M grams, where $M = Ae^{-kt}$.

   The half-life of radium is about 1600 years, i.e. it takes about 1600 years for the quantity to reduce by half.

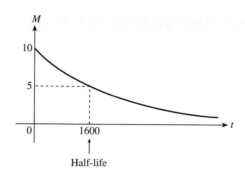

a   Given that $M = 10$ at $t = 0$ and $M = 5$ at $t = 1600$, find A and k.

b   Show that the time taken for the mass of radium to reduce by 90% is 5315 years.

c   Find the rate of decay at $t = 1600$.

d   Explain why the term 'half-life' rather than 'life' is used.

4   Newton's law of cooling leads to the result $\theta = Ae^{-kt}$ where θ is the excess of temperature of the cooling body over its surroundings at time t. For a given liquid, $k = \frac{1}{4}$, θ is measured in °C and t is measured in minutes. The surrounding temperature is 20°C and at $t = 0$, the temperature of the liquid is 60°C.

a   Find A.

b   Find an expression for the temperature, T°C, of the liquid at time t.

c   Calculate the temperature of the liquid when $t = 4$.

d   Find the time taken for the liquid to cool to 30°C.

e   Find the rate of cooling of the liquid when $t = 2$.

5   Differentiate with respect to x
a 5^x b $3(2^x)$ c 2^{3x-4} d $\dfrac{10^x}{2}$ e $\dfrac{1}{6^x}$ f $2^x \times 4^x$

6   Differentiate with respect to x
a $\log_3 x$ b $\log_{10}(x+2)$ c $\log_4(x^2)$ d $\ln 10(\log_{10} x)$

7   An exponential curve has equation $y = 4^x$.
a   Find the gradient of the tangent when $x = 1$.
b   Find the equation of the tangent when $x = 2$.
c   Find the equation of the normal when $x = 0$.

8   Find the equation of the tangent to the curve $y = \log_{10} x$ when $x = 0.1$.

Exercise 11D (Review)

1   Find the gradient of the ellipse $2x^2 + 3y^2 = 14$ at each of the points where $x = 1$.

2   Find the x-coordinates of the stationary points of the curve represented by the equation $x^3 - y^3 - 4x^2 + 3y = 11x + 4$.

3   Find the gradient of the ellipse $x^2 - 3yx + 2y^2 - 2x = 4$ at the point $(1, -1)$.

4   Find the gradient of the tangent at the point $(2, 3)$ to the hyperbola $xy = 6$.

5   Find $\dfrac{dy}{dx}$ in terms of x, when $x^2 + y^2 - 2xy + 3y - 2x = 7$.

6   At what points are the tangents to the circle $x^2 + y^2 - 6y - 8x = 0$ parallel to the y-axis?

7   Find $\dfrac{dy}{dx}$ when

   a $x^2y^3 = 8$ **b** $xy(x - y) = 4$

8   Find $\dfrac{dy}{dx}$ in terms of x and y when $3(x - y)^2 = 2xy + 1$.

9   Find $\dfrac{dy}{dx}$ when $x^2 + 2xy + y^2 = 3$. Explain your answer by factorising the left-hand side of the original equation.

10   Find $\dfrac{dy}{dx}$ when $x^2 - 3xy + y^2 - 2y + 4x = 0$.

11   Find $\dfrac{dy}{dx}$ when $3x^2 - 4xy = 7$.

12   Find the equation of the tangent to the curve $x^2 - y^2 = 9$ at the point $(5, 4)$.

13   Find the equations of the tangent and normal to the curve $x^2 - 3xy + 2y^2 = 3$ at the point $(5, 2)$.

14   Find the equations of the tangents to the ellipse $x^2 + 4y^2 = 4$ which are perpendicular to the line $2x - 3y = 1$.

15   Find the equations of the tangent and normal to the curve $x \tan y + y \sec x = \dfrac{\pi}{4}$ at the point $\left(0, \dfrac{\pi}{4}\right)$.

16   The number of bacteria present in a culture is modelled by $y = y_0 e^{kt}$, where $k > 0$ and t is the number of hours after 12 noon.
   At 1.00 p.m. on the same day, the number of bacteria present has doubled.

   a   According to the formula, how many bacteria are present at 12 noon?

   b   Find the value of k.

   c   At what time will the number of bacteria have increased ten-fold?
       Give the answer to the nearest minute.

   d   The rate of growth when $t = 5$ is cy_0. Find c.

   e   Show that the rate of increase of the number of bacteria is proportional to the number of bacteria present at that time.

17 A circular ink blot on a piece of paper spreads at the rate of $0.5\,\text{cm}^2\,\text{s}^{-1}$.
Find the rate of increase of the radius of the ink blot when the radius is $0.5\,\text{cm}$.

18 The edge of a cube is increasing at a rate of $2\,\text{cm s}^{-1}$.
Find, when the edge is of length $10\,\text{cm}$, the rate of increase of

   a the surface area

   b the volume

   c the sum of the lengths of the edges

   d a diagonal of a face

   e a diagonal of the cube.

There is a 'Test yourself' exercise (Ty11) and an 'Extension exercise' (Ext11) on the CD-ROM.

➤ Key points

Implicit differentiation

The derivative of a function of y, with respect to x, is found by applying the chain rule:

$$\frac{\text{d}}{\text{d}x}\,(\text{f}(y)) = \frac{\text{d}}{\text{d}y}(\text{f}(y)) \times \frac{\text{d}y}{\text{d}x}$$

Exponential growth and decay

Exponential growth

$y = A\text{e}^{kt} \quad (k > 0)$

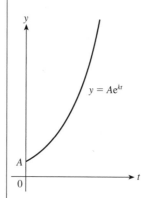

Exponential decay

$y = A\text{e}^{-kt} \quad (k > 0)$

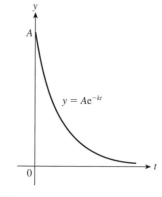

12 Further trigonometry

This chapter completes the work on trigonometry which you need for the course.

After working through this chapter you should

- know the addition, double angle and factor formulae and be able to use them to solve equations and prove identities
- know how to express $a\cos\theta + b\sin\theta$ in the form $R\cos(\theta \pm \alpha)$ or $R\sin(\theta \pm \alpha)$ and make use of this in solving equations
- be able to differentiate more complicated trigonometric functions and simplify your answers by making use of the double angle formulae
- be able to integrate powers of $\sin x$ and $\cos x$ and of related functions.

⬛ The CD-ROM (E12.1) shows how to obtain small angle approximations for $\sin x$, $\tan x$ and $\cos x$ and how to use these approximations to obtain the derivatives of trigonometric functions from first principles.

12.1 The addition formulae

There are many situations where it is necessary to express, for example, $\sin(A + B)$ in terms of $\sin A$, $\cos A$, $\sin B$ and $\cos B$.

Note There is no simple connection. For example,

$$\sin(30° + 60°) = \sin 90° = 1$$

but

$$\sin 30° + \sin 60° = \frac{1}{2} + \frac{\sqrt{3}}{2}$$

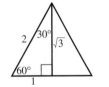

$\sin 30° = \frac{1}{2}$
$\sin 60° = \frac{\sqrt{3}}{2}$

So $\sin(A + B) \neq \sin A + \sin B$

In fact $\sin(A + B) \equiv \sin A \cos B + \cos A \sin B$. This identity is true for *all* values of A and B. As an example, again put $A = 30°$ and $B = 60°$.

$$\sin(30° + 60°) = \sin 30° \cos 60° + \cos 30° \sin 60°$$
$$= \frac{1}{2} \times \frac{1}{2} + \frac{\sqrt{3}}{2} \times \frac{\sqrt{3}}{2} = \frac{1}{4} + \frac{3}{4} = 1$$
$$= \sin 90°$$

Note The fact that $\sin(A + B) \neq \sin A + \sin B$ can be proved using a single counter-example. Proving the identity $\sin(A + B) \equiv \sin A \cos B + \cos A \sin B$ needs more than a single illustration. A proof is required.

To prove $\sin(A + B) = \sin A \cos B + \cos A \sin B$

This proof assumes angles A and B are acute. However, the result holds, in fact, for any angles, and can be proved using vectors (see the CD-ROM).

Consider a triangle OPQ, with $\angle PQO = 90°$, $\angle POQ = A$ and $OP = 1$, placed so that OQ makes an angle B with OL. The feet of the perpendiculars from P and Q to OL are R and S respectively.

T lies on PR such that $\angle PTQ = 90°$.

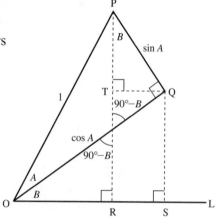

$\angle QPT = B$

In $\triangle OPQ$, $\sin A = \dfrac{PQ}{OP}$, $\cos A = \dfrac{OQ}{OP}$

Note the equal opposite angles $90° - B$ at the intersection of OQ and RP.

and $OP = 1$, so

$\quad PQ = \sin A$

and

$\quad OQ = \cos A$

In $\triangle OQS$

$\quad QS = OQ \sin B = \cos A \sin B$

In $\triangle PTQ$

$\quad PT = PQ \cos B = \sin A \cos B$

In $\triangle OPR$

$\quad PR = OP \sin(A + B) = \sin(A + B)$

Because OP = 1.

But $PR = PT + TR$

$\quad = PT + QS$

TR = QS ∵ they are opposite sides of a rectangle.

$\quad = \sin A \cos B + \cos A \sin B$

So $\sin(A + B) = \sin A \cos B + \cos A \sin B$

Deriving the other addition formulae

Replacing B by $-B$ in

$$\sin(A+B) = \sin A \cos B + \cos A \sin B$$

gives

$$\sin(A-B) = \sin A \cos(-B) + \cos A \sin(-B)$$

But $\quad \cos(-B) = \cos B$ and $\sin(-B) = -\sin B$

So $\quad \sin(A-B) = \sin A \cos B - \cos A \sin B$

Similar results can be derived for $\cos(A+B)$ and $\cos(A-B)$.

$$\cos(A+B) = \cos A \cos B - \sin A \sin B$$
$$\cos(A-B) = \cos A \cos B + \sin A \sin B$$

Note   The formula $\cos(A-B) = \cos A \cos B + \sin A \sin B$ reduces to the identity $\cos^2 A + \sin^2 A = 1$ when $A = B$.
Putting $B = A$ in the LHS gives $\cos 0 = 1$.
Putting $B = A$ in the RHS gives $\cos A \cos A + \sin A \sin A = \cos^2 A + \sin^2 A$.

From the formulae for $\sin(A \pm B)$ and $\cos(A \pm B)$, the results for $\tan(A \pm B)$ can be derived.

$$\tan(A+B) = \frac{\sin(A+B)}{\cos(A+B)}$$

$$= \frac{\sin A \cos B + \cos A \sin B}{\cos A \cos B - \sin A \sin B}$$

Divide every term by $\cos A \cos B$.

$$= \frac{\dfrac{\sin A \cos B}{\cos A \cos B} + \dfrac{\cos A \sin B}{\cos A \cos B}}{\dfrac{\cos A \cos B}{\cos A \cos B} - \dfrac{\sin A \sin B}{\cos A \cos B}}$$

$$= \frac{\tan A + \tan B}{1 - \tan A \tan B}$$

Similarly, it can be shown that

$$\tan(A-B) = \frac{\tan A - \tan B}{1 + \tan A \tan B}$$

➤
$$\sin(A \pm B) = \sin A \cos B \pm \cos A \sin B$$
$$\cos(A \pm B) = \cos A \cos B \mp \sin A \sin B$$
$$\tan(A \pm B) = \frac{\tan A \pm \tan B}{1 \mp \tan A \tan B}$$

The sign $\pm$ on the LHS and RHS is the same for $\sin(A \pm B)$ but the reverse for $\cos(A \pm B)$.

Example 1   Find the value of $\cos 75°$, leaving surds in the answer.

$\cos 75° = \cos(30° + 45°)$

$ = \cos 30° \cos 45° - \sin 30° \sin 45°$

$ = \dfrac{\sqrt{3}}{2} \times \dfrac{1}{\sqrt{2}} - \dfrac{1}{2} \times \dfrac{1}{\sqrt{2}}$

$ = \dfrac{\sqrt{3} - 1}{2\sqrt{2}}$

$ = \dfrac{\sqrt{6} - \sqrt{2}}{4}$

'Leaving surds in the answer' implies special triangles will be used.

The ratios of 30° and 45° can be found from the special triangles.

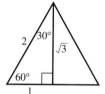

Rationalise the denominator by multiplying numerator and denominator by $\sqrt{2}$.

The result can be checked on a calculator.

Example 2   Rewrite as a single expression and hence evaluate

a $\dfrac{\tan 70° + \tan 65°}{1 - \tan 70° \tan 65°}$

b $\sin 100° \cos 10° - \cos 100° \sin 10°$

Solution   **a** $\dfrac{\tan 70° + \tan 65°}{1 - \tan 70° \tan 65°}$

$ = \tan(70° + 65°)$

$ = \tan 135°$

$ = -1$

Match with the correct formula:

$\tan(A + B) = \dfrac{\tan A + \tan B}{1 - \tan A \tan B}$

$\tan 135° = -\tan 45°$

b $\sin 100° \cos 10° - \cos 100° \sin 10°$

$ = \sin(100° - 10°)$

$ = \sin 90°$

$ = 1$

Match with the correct formula:
$\sin(A - B) = \sin A \cos B - \cos A \sin B$

Example 3   Given that $\sin A = \frac{4}{5}$ and $\cos B = \frac{8}{17}$, where A is obtuse and B is acute, find the value of $\cos(A - B)$.

$\sin A = \frac{4}{5}$ and A is obtuse

$\therefore \quad \cos A = -\frac{3}{5}$

Draw a right-angled triangle with $\sin \alpha = \frac{4}{5}$.

(α is the acute equivalent angle.)

$\cos B = \frac{8}{17}$ and B is acute

$\therefore \quad \sin B = \frac{15}{17}$

$\cos(A - B) = \cos A \cos B + \sin A \sin B$

$\qquad = -\frac{3}{5} \times \frac{8}{17} + \frac{4}{5} \times \frac{15}{17}$

$\qquad = \frac{-24 + 60}{85}$

$\qquad = \frac{36}{85}$

> Draw a right-angled triangle with $\cos B = \frac{8}{17}$.
>
> Use Pythagoras' theorem to find the third side.

Example 4   If $\sin(x + 60°) = \cos(x - 45°)$, find x where $0 < x < 360°$.

$\sin(x + 60°) = \cos(x - 45°)$

$\sin x \cos 60° + \cos x \sin 60° = \cos x \cos 45° + \sin x \sin 45°$

> Use special triangles for sin and cos of 45° and 60°.

$\frac{1}{2}\sin x + \frac{\sqrt{3}}{2}\cos x = \frac{1}{\sqrt{2}}\cos x + \frac{1}{\sqrt{2}}\sin x$

> Multiply through by $2\sqrt{2}$.

$\sqrt{2}\sin x + \sqrt{6}\cos x = 2\cos x + 2\sin x$

> Collect all terms in sin x on one side, cos x on the other side.

$(\sqrt{2} - 2)\sin x = (2 - \sqrt{6})\cos x$

> Use $\frac{\sin x}{\cos x} = \tan x$

$\therefore \qquad\qquad \tan x = \frac{2 - \sqrt{6}}{\sqrt{2} - 2}$

> Multiply numerator and denominator by -1. This is not necessary but makes them both +ve.

$x = \tan^{-1}\left(\frac{\sqrt{6} - 2}{2 - \sqrt{2}}\right)$

$\qquad = 37.5°, \ 180° + 37.5°$

Solution is $x = 37.5°, \ 217.5°$.

Example 5   Given that $\cot\left(x + \frac{\pi}{4}\right) = 4$, find the value of $\cot x$.

> $\cot x = \frac{1}{\tan x}$

$\cot\left(x + \frac{\pi}{4}\right) = \frac{1}{\tan\left(x + \frac{\pi}{4}\right)}$

$\qquad\qquad = \frac{1 - \tan x \tan\frac{\pi}{4}}{\tan x + \tan\frac{\pi}{4}}$

> $\tan\frac{\pi}{4} = 1$

$\qquad\qquad = \frac{1 - \tan x}{\tan x + 1}$

$$\therefore \qquad \frac{1 - \tan x}{\tan x + 1} = 4$$

Multiply both sides by $\tan x + 1$.

$$1 - \tan x = 4 \tan x + 4$$

Collect all terms in $\tan x$ on one side.

$$5 \tan x = -3$$

$$\tan x = -\frac{3}{5}$$

So $\qquad \cot x = -\frac{5}{3}$

Example 6 (Extension)

Without a calculator, find $\sin \theta$ where

$$\theta = \sin^{-1} \frac{5}{13} + \sin^{-1} \frac{4}{5}$$

Let $\sin^{-1} \frac{5}{13} = A$ and $\sin^{-1} \frac{4}{5} = B$.

So $\sin A = \frac{5}{13}$, $\sin B = \frac{4}{5}$, and $\theta = A + B$.

The $\sin^{-1}$ of a +ve quantity will be +ve. Therefore angles A and B are acute.

$$\sin \theta = \sin (A + B)$$

$$= \sin A \cos B + \cos A \sin B$$

To calculate $\cos A$ and $\cos B$ draw triangles and use Pythagoras' theorem to calculate the third side.

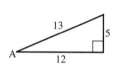

Then $\qquad \cos A = \frac{12}{13}$

and $\qquad \cos B = \frac{3}{5}$

So $\qquad \sin \theta = \sin A \cos B + \cos A \sin B$

$$= \frac{5}{13} \times \frac{3}{5} + \frac{12}{13} \times \frac{4}{5}$$

$$= \frac{15 + 48}{65}$$

$$= \frac{63}{65}$$

**Example 7
(Extension)**

Solve $\cos^{-1} 2x - \cos^{-1} x = \dfrac{\pi}{3}$.

Let $\cos^{-1} 2x = \alpha$ and $\cos^{-1} x = \beta$. Then $\cos\alpha = 2x$, $\cos\beta = x$ and $\alpha - \beta = \dfrac{\pi}{3}$.

$\cos\alpha = 2x \Rightarrow \quad \sin\alpha = \sqrt{1 - (2x)^2}$

$\cos\beta = x \Rightarrow \quad \sin\beta = \sqrt{1 - x^2}$

> Use $\cos\theta = \sqrt{1 - \sin^2\theta}$

$\cos(\alpha - \beta) = \cos\alpha \cos\beta + \sin\alpha \sin\beta$

$\qquad\qquad = 2x \times x + \sqrt{1 - 4x^2}\sqrt{1 - x^2}$

But $\qquad\qquad \alpha - \beta = \dfrac{\pi}{3}$

So $\qquad \cos(\alpha - \beta) = \cos\dfrac{\pi}{3}$

$\qquad\qquad\qquad\qquad = \tfrac{1}{2}$

So $\qquad\qquad \tfrac{1}{2} = 2x^2 + \sqrt{(1 - 4x^2)(1 - x^2)}$

$\sqrt{(1 - 4x^2)(1 - x^2)} = \tfrac{1}{2} - 2x^2$

Squaring both sides gives

$(1 - 4x^2)(1 - x^2) = \tfrac{1}{4} - 2x^2 + 4x^4$

$1 - 5x^2 + 4x^4 = \tfrac{1}{4} - 2x^2 + 4x^4$

$3x^2 = \tfrac{3}{4}$

$x^2 = \tfrac{1}{4}$

$\therefore \qquad\qquad\qquad x = \pm\tfrac{1}{2}$

> Remember, squaring both sides can lead to solutions that do not satisfy the original equation. After squaring both sides, always check whether the solutions satisfy the original equation.

But $x = \tfrac{1}{2}$ does not satisfy the original equation.

Hence, the solution is $x = -\tfrac{1}{2}$.

Exercise 12A

*This exercise gives practice in using the addition formulae. Do **not** use a calculator, except as a check. Give exact answers, using surds where appropriate.*

1 Find the values of

   a $\cos(45° - 30°)$ **b** $\sin(30° + 45°)$ **c** $\cos 105°$

   d $\sin 165°$ **e** $\tan 15°$ **f** $\cot 75°$

2 Find the values of

   a $\sin\left(\dfrac{\pi}{3} + \dfrac{\pi}{4}\right)$ **b** $\cos\left(\dfrac{2\pi}{3} + \dfrac{\pi}{4}\right)$ **c** $\sin\left(\dfrac{\pi}{3} - \dfrac{\pi}{4}\right)$

   d $\cos\dfrac{\pi}{12}$ **e** $\sec\dfrac{\pi}{12}$ **f** $\tan\dfrac{\pi}{12}$

3 If $\sin A = \tfrac{3}{5}$ and $\sin B = \tfrac{5}{13}$, where A and B are acute angles, find the values of

   a $\sin(A + B)$ **b** $\cos(A + B)$ **c** $\cot(A + B)$

4 If $\sin A = \tfrac{4}{5}$ and $\cos B = \tfrac{12}{13}$, where A is obtuse and B is acute, find the values of

   a $\sin(A - B)$ **b** $\tan(A - B)$ **c** $\cot(A + B)$

5 If $\sin A = \dfrac{\sqrt{5}}{5}$ and $\sin B = \dfrac{\sqrt{10}}{10}$, where A and B are acute, find

   a $\cos A$ **b** $\sin(A - B)$ **c** $\tan(A + B)$

6 If, for A and B acute, and a and b positive

$$\sin A = \dfrac{1}{\sqrt{1 + a^2}} \quad \text{and} \quad \sin B = \dfrac{1}{\sqrt{1 + b^2}}$$

find, in terms of a and b

   a $\tan A$ **b** $\cos(A + B)$ **c** $\cot(A - B)$

7 If $\cos A = \frac{3}{5}$ and $\tan B = \frac{12}{5}$, where A and B are both reflex angles, find the values of

   a $\sin(A - B)$ **b** $\tan(A - B)$ **c** $\sec(A + B)$

8 Solve, for values of x between $0°$ and $360°$, these equations.

   a $2 \sin x = \cos(x + 60°)$ **b** $\cos(x + 45°) = \cos x$

   c $\sin(x - 30°) = \frac{1}{2}\cos x$ **d** $3 \sin(x + 10°) = 4 \cos(x - 10°)$

9 If $\tan(A + B) = \frac{1}{7}$ and $\tan A = 3$, find the value of $\tan B$.

10 If A and B are acute, $\tan A = \frac{1}{2}$ and $\tan B = \frac{1}{3}$, find the value of $A + B$.

11 If $\tan A = -\frac{1}{7}$ and $\tan B = \frac{3}{4}$, where A is obtuse and B is acute, find the value of $A - B$.

12 Express as single trigonometrical ratios

   a $\dfrac{1}{2}\cos x - \dfrac{\sqrt{3}}{2}\sin x$ **b** $\dfrac{1}{\sqrt{2}}\sin x + \dfrac{1}{\sqrt{2}}\cos x$

   c $\dfrac{\sqrt{3} + \tan x}{1 - \sqrt{3}\tan x}$ **d** $\cos 16° \sin 42° - \sin 16° \cos 42°$

   e $\dfrac{1}{\cos 24° \cos 15° - \sin 24° \sin 15°}$ **f** $\dfrac{1}{2}\cos 75° + \dfrac{\sqrt{3}}{2}\sin 75°$

13 Find the values of

   a $\cos 75° \cos 15° + \sin 75° \sin 15°$ **b** $\sin 50° \cos 20° - \cos 50° \sin 20°$

   c $\dfrac{\tan 10° + \tan 20°}{1 - \tan 10° \tan 20°}$ **d** $\cos 70° \cos 20° - \sin 70° \sin 20°$

   e $\dfrac{1}{\sqrt{2}}\cos 15° - \dfrac{1}{\sqrt{2}}\sin 15°$ **f** $\dfrac{\sqrt{3}}{2}\cos 15° - \dfrac{1}{2}\sin 15°$

   g $\dfrac{1 - \tan 15°}{1 + \tan 15°}$ **h** $\cos 15° + \sin 15°$

14 Find the value of $\tan A$, when $\tan(A - 45°) = \frac{1}{3}$.

15 Find the value of $\cot B$, when $\cot A = \frac{1}{4}$ and $\cot(A - B) = 8$.

16 For each of these equations, find the value of $\tan x$.

   a $\sin(x + 45°) = 2 \cos(x + 45°)$ **b** $2 \sin(x - 45°) = \cos(x + 45°)$

   c $\tan(x - A) = \frac{3}{2}$, where $\tan A = 2$ **d** $\sin(x + 30°) = \cos(x + 30°)$

17 If $\sin(x + \alpha) = 2\cos(x - \alpha)$, prove that

$$\tan x = \frac{2 - \tan \alpha}{1 - 2\tan \alpha}$$

18 If $\sin(x - \alpha) = \cos(x + \alpha)$, prove that $\tan x = 1$.

19 Putting $A = B$ in the addition formulae, show that

   a $\sin 2A = 2\sin A \cos A$

   b $\cos 2A = \cos^2 A - \sin^2 A$

   c $\tan 2A = \dfrac{2\tan A}{1 - \tan^2 A}$

20 Prove these identities.

   a $\cos(90° - \theta) \equiv \sin \theta$

   b $\sin(180° - \theta) \equiv \sin \theta$

   c $\sin(90° + \theta) \equiv \cos \theta$

   d $\sin(A + B) + \sin(A - B) \equiv 2\sin A \cos B$

   e $\cos(A + B) - \cos(A - B) \equiv -2\sin A \sin B$

   f $\tan A + \tan B \equiv \dfrac{\sin(A + B)}{\cos A \cos B}$

   g $\dfrac{\tan(A + B) - \tan A}{1 + \tan(A + B)\tan A} \equiv \tan B$

12.2 The double angle formulae

Putting $A = B$ in the addition formulae of page 281, as in Question 19 of Exercise 12A, gives the double angle formulae.

$\sin(A + B) = \sin A \cos B + \cos A \sin B$

Putting $A = B$ gives

$$\sin(A + A) = \sin A \cos A + \cos A \sin A$$

▶ $\boxed{\mathbf{\sin 2A = 2\sin A \cos A}}$

$\cos(A + B) = \cos A \cos B - \sin A \sin B$

Putting $A = B$ gives

$$\cos 2A = \cos A \cos A - \sin A \sin A$$
$$= \cos^2 A - \sin^2 A \qquad \text{①}$$

Replacing $\sin^2 A$ by $1 - \cos^2 A$ in ① gives

$$\cos 2A = \cos^2 A - (1 - \cos^2 A)$$
$$= 2\cos^2 A - 1$$

Replacing $\cos^2 A$ by $1 - \sin^2 A$ in ① gives

$$\cos 2A = 1 - \sin^2 A - \sin^2 A$$
$$= 1 - 2\sin^2 A$$

➤
$$\boxed{\begin{aligned}\cos 2A &= \cos^2 A - \sin^2 A \\ &= 2\cos^2 A - 1 \\ &= 1 - 2\sin^2 A\end{aligned}}$$

Note: There are three versions for $\cos 2A$.

$$\tan(A + B) = \frac{\tan A + \tan B}{1 - \tan A \tan B}$$

Putting $A = B$ gives

$$\tan 2A = \frac{\tan A + \tan A}{1 - \tan A \tan A}$$
$$= \frac{2\tan A}{1 - \tan^2 A}$$

➤
$$\boxed{\tan 2A = \frac{2\tan A}{1 - \tan^2 A}}$$

The double angle formulae can then be used to find expressions for other multiples e.g. $\sin 3A$ or $\cos 3A$ (see Question 11 in Exercise 12B).

$$\sin 3A = 3\sin A - 4\sin^3 A$$
$$\cos 3A = 4\cos^3 A - 3\cos A$$

Other useful forms of the double angle formulae can be generated by rearrangement:

■ $\cos 2A = 2\cos^2 A - 1$ can be rearranged as $\cos^2 A = \frac{1}{2}(1 + \cos 2A)$.

■ $\cos 2A = 1 - 2\sin^2 A$ can be rearranged as $\sin^2 A = \frac{1}{2}(1 - \cos 2A)$.

➤
$$\boxed{\begin{aligned}\cos^2 A &= \tfrac{1}{2}(1 + \cos 2A) \\ \sin^2 A &= \tfrac{1}{2}(1 - \cos 2A)\end{aligned}}$$

12.3   The half-angle formulae

Replacing A by $\frac{A}{2}$ in the double angle formulae for $\cos 2A$ gives half-angle formulae for $\cos^2 \frac{A}{2}$ and $\sin^2 \frac{A}{2}$. These do not need to be learnt but it is useful to be able to derive them with ease. Similarly A can be replaced by $\frac{A}{2}$ in the formulae for $\sin 2A$ and $\tan 2A$.

Replacing A by $\dfrac{A}{2}$ in $\cos 2A = 2\cos^2 A - 1$

$$\cos A = 2\cos^2 \frac{A}{2} - 1$$

Rearranging

$$\cos^2 \frac{A}{2} = \frac{1}{2}(1 + \cos A)$$

Replacing A by $\frac{A}{2}$ in $\cos 2A = 1 - 2\sin^2 A$

$$\cos A = 1 - 2\sin^2 \frac{A}{2}$$

Rearranging

$$\sin^2 \frac{A}{2} = \frac{1}{2}(1 - \cos A)$$

➤
$$\cos^2 \frac{A}{2} = \frac{1}{2}(1 + \cos A)$$

$$\sin^2 \frac{A}{2} = \frac{1}{2}(1 - \cos A)$$

Example 8   Express more simply

    a $\cos^2 75° - \sin^2 75°$ **b** $\sec 2\theta \, \text{cosec}\, 2\theta$

Solution   **a** $\cos^2 75° - \sin^2 75° = \cos 2 \times 75°$

                                   $= \cos 150°$

                                     $= -\frac{\sqrt{3}}{2}$

     b $\sec 2\theta \, \text{cosec}\, 2\theta = \dfrac{1}{\cos 2\theta} \times \dfrac{1}{\sin 2\theta}$

                                          $= \dfrac{2}{2 \sin 2\theta \cos 2\theta}$

                                          $= \dfrac{2}{\sin 4\theta}$

                                          $= 2\,\text{cosec}\, 4\theta$

Recognise $\cos^2 A - \sin^2 A = \cos 2A$

$\cos 150° = -\cos 30°$

Use the special triangles.

$\sec A = \dfrac{1}{\cos A}$ $\text{cosec}\, A = \dfrac{1}{\sin A}$

Recognise $2\sin A \cos A = \sin 2A$. So $\sin A \cos A = \frac{1}{2}\sin 2A$. Numerator and denominator have been multiplied by 2.

Example 9   Eliminate θ if $x = \cos 2\theta$ and $y = \sin \theta$.

                 $\cos 2\theta = 1 - 2\sin^2 \theta$

     So $x = 1 - 2y^2$

Find an identity linking $\cos 2\theta (= x)$ and $\sin \theta (= y)$.

Substitute x and y.

Example 10   Solve the equation $\sin 2\theta = \cos \theta$ for $0 \leqslant \theta \leqslant 2\pi$.

$$\sin 2\theta = \cos \theta$$

$$2\sin \theta \cos \theta = \cos \theta$$

$$\cos \theta (2\sin \theta - 1) = 0$$

$$\Rightarrow \qquad \cos \theta = 0 \ \text{or} \ \sin \theta = \frac{1}{2}$$

$$\cos \theta = 0 \Rightarrow \quad \theta = \frac{\pi}{2}, \frac{3\pi}{2}$$

$$\sin \theta = \frac{1}{2} \Rightarrow \quad \theta = \frac{\pi}{6}, \frac{5\pi}{6}$$

Solutions are $\theta = \dfrac{\pi}{6}, \dfrac{\pi}{2}, \dfrac{5\pi}{6}, \dfrac{3\pi}{2}$

> Use the identity $\sin 2\theta = 2\sin \theta \cos \theta$

> Do not divide by $\cos \theta$. It could be zero. Instead, take the common factor $\cos \theta$ outside a bracket.

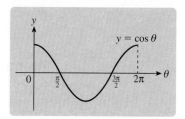

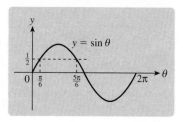

Example 11   Given that $\sin x \neq 0$, simplify

$$\frac{1 - \cos 2x}{\sin 2x}$$

$$\frac{1 - \cos 2x}{\sin 2x} = \frac{2\sin^2 x}{2\sin x \cos x}$$

$$= \frac{\sin x}{\cos x}$$

$$= \tan x$$

> Use $\cos 2x = 1 - 2\sin^2 x$ rearranged as $1 - \cos 2x = 2\sin^2 x$. Also $\sin 2x = 2\sin x \cos x$.

> Cancel $2\sin x$ from both numerator and denominator.

Example 12   Prove that $\cos 2x = \dfrac{1 - \tan^2 x}{1 + \tan^2 x}$.

$$\frac{1 - \tan^2 x}{1 + \tan^2 x} = \frac{1 - \frac{\sin^2 x}{\cos^2 x}}{1 + \frac{\sin^2 x}{\cos^2 x}}$$

$$= \frac{\cos^2 x - \sin^2 x}{\cos^2 x + \sin^2 x}$$

$$= \cos^2 x - \sin^2 x$$

$$= \cos 2x$$

> Start with RHS as it has more to simplify. Express in terms of $\sin x$ and $\cos x$.

> Multiply all terms in numerator and denominator by $\cos^2 x$.

> Denominator $\cos^2 x + \sin^2 x = 1$.

> Recognise $\cos^2 x - \sin^2 x = \cos 2x$.

Exercise 12B   *Question 1 may be done orally.*

1 Express each of these as a single trigonometrical function.

a $2\sin 17° \cos 17°$

b $\dfrac{2\tan 30°}{1 - \tan^2 30°}$

c $2\cos^2 42° - 1$

d $2\sin \frac{1}{2}\theta \cos \frac{1}{2}\theta$

e $1 - 2\sin^2 \dfrac{\pi}{8}$

f $\dfrac{2\tan \frac{1}{2}\theta}{1 - \tan^2 \frac{1}{2}\theta}$

g $\cos^2 \dfrac{\pi}{12} - \sin^2 \dfrac{\pi}{12}$

h $2\sin 2A \cos 2A$

i $2\cos^2 \frac{1}{2}\theta - 1$

j $1 - 2\sin^2 3\theta$

k $\dfrac{\tan 2\theta}{1 - \tan^2 2\theta}$

l $\sin x \cos x$

m $\dfrac{1 - \tan^2 20°}{\tan 20°}$

n $\sec \theta \operatorname{cosec} \theta$

o $1 - 2\sin^2 \frac{1}{2}\theta$

2 Evaluate these *without* a calculator, leaving the answer in surd form.

a $2\sin 15° \cos 15°$

b $\dfrac{2\tan 22\frac{1}{2}°}{1 - \tan^2 22\frac{1}{2}°}$

c $2\cos^2 75° - 1$

d $1 - 2\sin^2 67\frac{1}{2}°$

e $\cos^2 22\frac{1}{2}° - \sin^2 22\frac{1}{2}°$

f $\dfrac{1 - \tan^2 \frac{\pi}{12}}{\tan \frac{\pi}{12}}$

g $\dfrac{1 - 2\cos^2 25°}{1 - 2\sin^2 65°}$

h $\sec \dfrac{\pi}{8} \operatorname{cosec} \dfrac{\pi}{8}$

3 Find the values of $\sin 2\theta$ and $\cos 2\theta$ when

a $\sin \theta = \dfrac{3}{5}$

b $\cos \theta = \dfrac{12}{13}$

c $\sin \theta = -\dfrac{\sqrt{3}}{2}$

4 Find the value of $\tan 2\theta$ when

a $\tan \theta = \dfrac{4}{3}$

b $\tan \theta = \dfrac{8}{15}$

c $\cos \theta = -\dfrac{5}{13}$

5 Find the values of $\cos x$ and $\sin x$ when $\cos 2x$ is

a $\dfrac{1}{8}$

b $\dfrac{7}{25}$

c $-\dfrac{119}{169}$

6 Find the values of $\tan \frac{1}{2}\theta$ when $\tan \theta$ is

a $\dfrac{3}{4}$

b $\dfrac{4}{3}$

c $-\dfrac{12}{5}$

7 If $t = \tan 22\frac{1}{2}°$, use the formula for $\tan 2\theta$ to show that $t^2 + 2t - 1 = 0$.
Deduce the value of $\tan 22\frac{1}{2}°$.

8 Solve these equations for values of θ from $0°$ to $360°$ inclusive.

a $\cos 2\theta + \cos \theta + 1 = 0$

b $\cos \theta = \sin \dfrac{\theta}{2}$

c $\sin 2\theta \cos \theta + \sin^2 \theta = 1$

d $2\sin \theta (5\cos 2\theta + 1) = 3\sin 2\theta$

e $3\cot 2\theta + \cot \theta = 1$

9 Solve these equations for $-\pi \leqslant \theta \leqslant \pi$.

a $\sin 2\theta = \sin \theta$

b $3 \cos 2\theta - \sin \theta + 2 = 0$

c $\sin \dfrac{\theta}{2} = 6 \sin \theta$

d $3 \tan \theta = \tan 2\theta$

e $4 \tan \theta \tan 2\theta = 1$

10 Eliminate θ from each of these equations.

a $x = \cos \theta, \ y = \cos 2\theta$

b $x = 2 \sin \theta, \ y = 3 \cos 2\theta$

c $x = \tan \theta, \ y = \tan 2\theta$

d $x = 2 \sec \theta, \ y = \cos 2\theta$

11 **a** By expanding $\sin(2\alpha + \alpha)$, or otherwise, show that $\sin 3\alpha = 3 \sin \alpha - 4 \sin^3 \alpha$.

b Show that $\cos 3\alpha = 4 \cos^3 \alpha - 3 \cos \alpha$.

12 Prove these identities.

a $\dfrac{\cos 2A}{\cos A + \sin A} = \cos A - \sin A$

b $\dfrac{\sin A}{\sin B} + \dfrac{\cos A}{\cos B} = \dfrac{2 \sin(A + B)}{\sin 2B}$

c $\dfrac{\cos A}{\sin B} + \dfrac{\sin A}{\cos B} = \dfrac{2 \cos(A + B)}{\sin 2B}$

d $\tan \dfrac{A}{2} + \cot \dfrac{A}{2} = 2 \operatorname{cosec} A$

e $\cot A - \tan A = 2 \cot 2A$

f $\dfrac{1}{\cos A + \sin A} + \dfrac{1}{\cos A - \sin A} = \tan 2A \operatorname{cosec} A$

g $\dfrac{\sin A}{1 + \cos A} = \tan \dfrac{A}{2}$

h $\tan 3A = \dfrac{3 \tan A - \tan^3 A}{1 - 3 \tan^2 A}$

i $\operatorname{cosec} 2x - \cot 2x = \tan x$

j $\operatorname{cosec} 2x + \cot 2x = \cot x$

k $\tan x = \sqrt{\dfrac{1 - \cos 2x}{1 + \cos 2x}}$

l $\sin 2x = \dfrac{2 \tan x}{1 + \tan^2 x}$

m $\cos 2x = \dfrac{1 - \tan^2 x}{1 + \tan^2 x}$

n $\cos 2x = \cos^4 x - \sin^4 x$

13 Given that $\sin \theta = \frac{5}{13}$ and that θ is obtuse, find $\sin \dfrac{\theta}{2}$, $\cos \dfrac{\theta}{2}$ and $\tan \dfrac{\theta}{2}$.

14 **a** Show that $\sin \theta = \dfrac{2t}{1 + t^2}$, $\cos \theta = \dfrac{1 - t^2}{1 + t^2}$, $\tan \theta = \dfrac{2t}{1 - t^2}$ where $\tan \dfrac{\theta}{2} = t$.

b Hence solve the equations $2 \cos \theta + 3 \sin \theta - 2 = 0$ for $0° \leqslant \theta \leqslant 360°$.

12.4 Extension: The factor formulae

With the factor formulae, any product of sines and cosines can be converted into a sum or difference of sines and cosines, and vice versa.

> When solving equations, it is usually necessary to work with products. When integrating (Section 12.6), it is usually necessary to work with sums.

To factorise an expression means to express it as a product. Expressions such as $\sin P + \sin Q$ can be expressed as products.

Adding the addition formulae for $\sin(A + B)$ and $\sin(A - B)$ gives

$$\sin(A + B) + \sin(A - B) = \sin A \cos B + \cos A \sin B + \sin A \cos B - \cos A \sin B$$
$$= 2 \sin A \cos B$$

Putting $A + B = P$ and $A - B = Q$

$$\sin P + \sin Q = 2 \sin \frac{P + Q}{2} \cos \frac{P - Q}{2}$$

Similar results for $\sin P - \sin Q$ and for $\cos P \pm \cos Q$ can be obtained.

> $A + B = P$ ①
> $A - B = Q$ ②
> ① + ② $\Rightarrow$
> $2A = P + Q$
> $A = \dfrac{P + Q}{2}$
> Similarly $B = \dfrac{P - Q}{2}$

➤

The factor formulae

$$\sin P + \sin Q = 2 \sin \frac{P + Q}{2} \cos \frac{P - Q}{2}$$

$$\sin P - \sin Q = 2 \cos \frac{P + Q}{2} \sin \frac{P - Q}{2}$$

$$\cos P + \cos Q = 2 \cos \frac{P + Q}{2} \cos \frac{P - Q}{2}$$

$$\cos P - \cos Q = -2 \sin \frac{P + Q}{2} \sin \frac{P - Q}{2}$$

These formulae can be learnt either

- *by recalling how the formulae are derived from the addition formula, i.e. which terms cancel out, or*

- *by observing the pattern and noting that only $\cos P - \cos Q$ has a $-$ve sign in front, or*

- *by remembering, for example, $\sin P + \sin Q$ is 'twice the sine of half the sum multiplied by the cosine of half the difference'.*

The alternative form of the factor formulae (see derivation above) is also useful.

$$\sin(A+B) + \sin(A-B) = 2\sin A \cos B$$
$$\sin(A+B) - \sin(A-B) = 2\cos A \sin B$$
$$\cos(A+B) + \cos(A-B) = 2\cos A \cos B$$
$$\cos(A+B) - \cos(A-B) = -2\sin A \sin B$$

Example 13   By expressing as factors, find the exact value of $\sin 75° + \sin 15°$.

Use $\sin P + \sin Q = 2\sin \dfrac{P+Q}{2} \cos \dfrac{P-Q}{2}$

$$\sin 75° + \sin 15° = 2\sin\frac{75°+15°}{2}\cos\frac{75°-15°}{2}$$
$$= 2\sin 45° \cos 30°$$
$$= 2 \times \frac{1}{\sqrt{2}} \times \frac{\sqrt{3}}{2}$$
$$= \frac{\sqrt{6}}{2}$$

Use the special triangles.

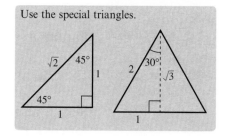

Example 14   Express, as the sum or difference of two sines or two cosines

a $2\cos 4x \cos 2x$ **b** $2\cos 3A \sin A$

Solution   **a** $2\cos 4x \cos 2x$

Use $2\cos A \cos B = \cos(A+B) + \cos(A-B)$

$$= \cos(4x+2x) + \cos(4x-2x)$$
$$= \cos 6x + \cos 2x$$

b $2\cos 3A \sin A$

Use $2\cos A \sin B = \sin(A+B) - \sin(A-B)$

$$= \sin(3A+A) - \sin(3A-A)$$
$$= \sin 4A - \sin 2A$$

Example 15   Solve this equation, for $-180° \leqslant \theta \leqslant 180°$.

$$\sin(\theta + 85°) - \sin(\theta + 25°) = \tfrac{1}{2}$$

Use $\sin P - \sin Q = 2\cos \frac{P+Q}{2} \sin \frac{P-Q}{2}$

$$\sin(\theta + 85°) - \sin(\theta + 25°) = 2\cos\frac{2\theta + 110°}{2}\sin\frac{60°}{2}$$
$$= 2\cos(\theta + 55°)\sin 30°$$

$$\therefore \quad 2\cos(\theta + 55°) \times \tfrac{1}{2} = \tfrac{1}{2}$$
$$\cos(\theta + 55°) = \tfrac{1}{2}$$
$$\theta + 55° = -60°, \ 60°$$
$$\theta = -115°, \ 5°$$

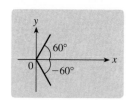

These are the only values for $-180° \leqslant \theta \leqslant 180°$.

Example 16

Prove this identity.

$$\frac{\sin 4A + \sin 2A}{\cos 4A + \cos 2A} = \tan 3A$$

> Start with LHS as it has more to simplify.

$$\frac{\sin 4A + \sin 2A}{\cos 4A + \cos 2A} = \frac{2 \sin \frac{4A+2A}{2} \cos \frac{4A-2A}{2}}{2 \cos \frac{4A+2A}{2} \cos \frac{4A-2A}{2}}$$

> Use the factor formulae to express the numerator and denominator in factors.

$$= \frac{\sin 3A}{\cos 3A}$$

$$= \tan 3A$$

Example 17

Prove that if A, B and C are the angles of a triangle
$\sin 2A + \sin 2B + \sin 2C = 4 \sin A \sin B \sin C$

> First, combine $\sin 2A + \sin 2B$ using the factor formula.

$\sin 2A + \sin 2B + \sin 2C = 2 \sin (A + B) \cos (A - B) + 2 \sin C \cos C$

Since A, B and C are the angles of a triangle

> Now express $\sin C$ and $\cos C$ in terms of A and B.

$$C = 180° - (A + B)$$

so $\sin C = \sin[180° - (A + B)]$

> $\sin x = \sin (180° - x)$

$$= \sin (A + B) \qquad ①$$

and $\cos C = \cos[180° - (A + B)]$

> $\cos x = -\cos (180° - x)$

$$= -\cos (A + B) \qquad ②$$

So $\sin 2A + \sin 2B + \sin 2C$

> Substitute from ① and ②.

$$= 2 \sin C \cos (A - B) - 2 \sin C \cos (A + B)$$

> Take $2 \sin C$ outside the bracket.

$$= 2 \sin C[\cos (A - B) - \cos (A + B)]$$

> $\cos (A - B) - \cos (A + B)$ can be combined using the factor formula.

$$= 2 \sin C[-2 \sin A \sin (-B)]$$

> $\sin B = -\sin(-B)$

$$= 4 \sin A \sin B \sin C$$

Extension Exercise 12C

Questions 1–3 may be done orally.

1 Express each of these as the sum or difference of two sines.

  a $2 \sin x \cos y$ **b** $2 \cos x \sin y$

  c $2 \sin 3\theta \cos \theta$ **d** $2 \sin (S + T) \cos (S - T)$

  e $2 \cos 5x \sin 3x$ **f** $2 \cos (x + y) \sin (x - y)$

2 Express each of these as the sum of two cosines.

  a $2 \cos x \cos y$ **b** $-2 \sin x \sin y$

  c $2 \cos 3\theta \cos \theta$ **d** $-2 \sin (S + T) \sin (S - T)$

  e $2 \sin 5x \sin 3x$ **f** $2 \cos (x + y) \cos (x - y)$

3 Express each of these in factors.

   a $\cos x + \cos y$ **b** $\sin 3x + \sin 5x$ **c** $\sin 2y - \sin 2z$

   d $\cos 5x + \cos 7x$ **e** $\cos 2A - \cos A$ **f** $\sin 4x - \sin 2x$

   g $\cos 3A - \cos 5A$ **h** $\sin 5\theta + \sin 7\theta$

   i $\sin(x + 30°) + \sin(x - 30°)$ **j** $\cos(y + 10°) + \cos(y - 80°)$

   k $\sin 3\theta - \sin 5\theta$ **l** $\cos(x + 30°) - \cos(x - 30°)$

   m $\cos \dfrac{3x}{2} - \cos \dfrac{x}{2}$ **n** $\sin 2(x + 40°) + \sin 2(x - 40°)$

4 Express each of these in factors.

   a $\cos(90° - x) + \cos y$ **b** $\sin A + \cos B$ **c** $\sin 3x + \sin 90°$

   d $1 + \sin 2x$ **e** $\cos A - \sin B$ **f** $\frac{1}{2} + \cos 2\theta$

5 Find the exact value of

   a $2 \sin 75° \cos 15°$ **b** $\cos 165° - \cos 75°$

   c $\sin 255° - \sin 15°$ **d** $\cos 195° \sin 45°$

6 Solve these equations, for values of x from 0 to 2π inclusive.

   a $\cos x + \cos 5x = 0$ **b** $\cos 4x - \cos x = 0$

   c $\sin 3x - \sin x = 0$ **d** $\sin 2x + \sin 3x = 0$

7 Solve these equations, for values of x from $0°$ to $360°$ inclusive.

   a $\sin 4x + \sin 2x = 0$ **b** $\sin(x + 10°) + \sin x = 0$

   c $\cos(2x + 10°) + \cos(2x - 10°) = 0$ **d** $\cos(x + 20°) - \cos(x - 70°) = 0$

8 Prove these identities.

   a $\dfrac{\cos B + \cos C}{\sin B - \sin C} = \cot \dfrac{B - C}{2}$ **b** $\dfrac{\cos B - \cos C}{\sin B + \sin C} = -\tan \dfrac{B - C}{2}$

   c $\dfrac{\sin B + \sin C}{\cos B + \cos C} = \tan \dfrac{B + C}{2}$ **d** $\dfrac{\sin B - \sin C}{\sin B + \sin C} = \cot \dfrac{B + C}{2} \tan \dfrac{B - C}{2}$

9 **a** Express $\sin x + \sin 3x$ as a product.

   b Hence show that

$$\sin x + \sin 2x + \sin 3x \equiv \sin 2x(2 \cos x + 1)$$

   c Use the result in part **b** to solve, for $0 \leqslant x \leqslant \pi$, $\sin x + \sin 2x + \sin 3x = 0$

10 Solve, for $0 \leqslant \theta \leqslant \pi$

   a $\cos \theta + \cos 3\theta + \cos 5\theta = 0$

   b $\sin \theta - \sin 2\theta + \sin 3\theta = 0$

11 Prove these identities.

   a $\cos x + \sin 2x - \cos 3x = \sin 2x(2 \sin x + 1)$

   b $\cos 3\theta + \cos 5\theta + \cos 7\theta = \cos 5\theta(2 \cos 2\theta + 1)$

   c $\cos \theta + 2 \cos 3\theta + \cos 5\theta = 4 \cos^2 \theta \cos 3\theta$

   d $1 + 2 \cos 2\theta + \cos 4\theta = 4 \cos^2 \theta \cos 2\theta$

e $\sin\theta - 2\sin 3\theta + \sin 5\theta = 2\sin\theta(\cos 4\theta - \cos 2\theta)$

f $\cos\theta - 2\cos 3\theta + \cos 5\theta = 2\sin\theta(\sin 2\theta - \sin 4\theta)$

g $\sin x - \sin(x + 60°) + \sin(x + 120°) = 0$

h $\cos x + \cos(x + 120°) + \cos(x + 240°) = 0$

12.5   The form $a\cos\theta + b\sin\theta$

All expressions of the form $a\cos\theta + b\sin\theta$ can be expressed in an alternative form, either $R\cos(\theta \pm \alpha)$ or $R\sin(\theta \pm \alpha)$.

These are transformations: a translation and a stretch, of the sine and cosine curves.

Plotting a few graphs, on a graphical calculator, for example, will illustrate that equations of the form $y = a\cos\theta + b\sin\theta$ are equivalent to transformations of sine (or cosine) curves.

Try plotting $y = 4\cos x + 3\sin x$ and $y = 5\cos(x - \tan^{-1} 0.75)$ on the same axes. The graphs are identical.

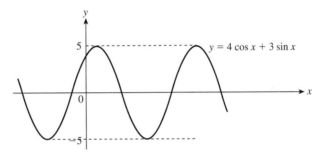

To express $a\cos\theta + b\sin\theta$ in the form $R\cos(\theta - \alpha)$, R and α have to be found.

Let $a\cos\theta + b\sin\theta \equiv R\cos(\theta - \alpha)$

   $\equiv R(\cos\theta\cos\alpha + \sin\theta\sin\alpha)$

> The LHS and RHS are equivalent, so the coefficients of $\cos\theta$ and $\sin\theta$ must be identical.

Equating coefficients of $\cos\theta$ and $\sin\theta$

$$a = R\cos\alpha \qquad ①$$
$$b = R\sin\alpha \qquad ②$$

Squaring and adding

$$a^2 + b^2 = R^2(\cos^2\alpha + \sin^2\alpha)$$
$$= R^2$$

> Squaring and adding eliminates α, because $\cos^2\alpha + \sin^2\alpha = 1$.

So $R = \sqrt{a^2 + b^2}$

> Take +ve R.

Dividing ② by ①

> To eliminate R.

$$\tan\alpha = \frac{b}{a}$$
$$\alpha = \tan^{-1}\frac{b}{a}$$

So $a\cos\theta + b\sin\theta = \sqrt{a^2 + b^2}\cos(\theta - \alpha)$ where $\alpha = \tan^{-1}\frac{b}{a}$.

R is taken as +ve.

Using $R\cos(\theta \pm \alpha)$ is preferable to using $R\sin(\theta \pm \alpha)$ if there is a choice and if there is a subsequent equation to be solved, because $R\cos(\theta \pm \alpha) = c$ is easier to solve than $R\sin(\theta \pm \alpha) = c$.

Matching the sign in $a\cos\theta \pm b\sin\theta$ with the sign in the expansion of $\cos(\theta \pm \alpha)$ or $\sin(\theta \pm \alpha)$ avoids having to deal with negative signs (see Example 20).

The maximum and minimum values of $R\cos(\theta + \alpha)$ are R and $-R$ respectively. These occur when $\cos(\theta + \alpha) = 1$ and $\cos(\theta + \alpha) = -1$ respectively.

Similar results hold for $R\sin(\theta \pm \alpha)$.

Example 18   Find R and α where $0 \leqslant \alpha \leqslant \dfrac{\pi}{2}$ and $R > 0$, given that $4\cos\theta + 3\sin\theta \equiv R\cos(\theta - \alpha)$.

$4\cos\theta + 3\sin\theta \equiv R\cos(\theta - \alpha)$

$\qquad\qquad \equiv R\cos\theta\cos\alpha + R\sin\theta\sin\alpha$

> Expand the RHS and compare with the LHS.

Equating coefficients of $\cos\theta$ and $\sin\theta$

$\qquad 4 = R\cos\alpha \qquad ①$

$\qquad 3 = R\sin\alpha \qquad ②$

Squaring and adding

$\qquad\qquad R^2 = 3^2 + 4^2$

> $R^2(\cos^2\alpha + \sin^2\alpha) = R^2$ since $\cos^2\alpha + \sin^2\alpha = 1$.

$\qquad\qquad\quad = 25$

so $\qquad R = 5$

Dividing ② by ①

$\qquad\qquad \tan\alpha = \tfrac{3}{4}$

so $\qquad \alpha = \tan^{-1}\tfrac{3}{4}$

> α is required in radians.
> With calculator in radian mode find $\alpha = \tan^{-1}\tfrac{3}{4}$.

$\qquad\qquad\quad = 0.644$

So $4\cos\theta + 3\sin\theta = 5\cos(\theta - 0.644)$.

Note $y = 4\cos\theta + 3\sin\theta$ is a translation of $y = \cos\theta$ of 0.644 radians to the right and a stretch, scale factor 5, parallel to the y-axis.

Example 19   **a**   Express $2\sin\theta - 5\cos\theta$ in the form $R\sin(\theta + \alpha)$ where $R > 0$.

b   Hence find the maximum and minimum values of $2\sin\theta - 5\cos\theta$ and the values of θ for which they occur for $0° \leqslant \theta \leqslant 360°$

c   Solve $2\sin\theta - 5\cos\theta = 4$ for $0° \leqslant \theta \leqslant 360°$.

Solution　**a**　$2\sin\theta - 5\cos\theta \equiv R\sin(\theta + \alpha)$

$$\equiv R(\sin\theta\cos\alpha + \cos\theta\sin\alpha)$$

> Expand the RHS and compare with the LHS.

Equating coefficients of $\sin\theta$ and $\cos\theta$

$$2 = R\cos\alpha$$

$$-5 = R\sin\alpha$$

$$R = \sqrt{2^2 + 5^2}$$

$$= \sqrt{29}$$

$$\tan\alpha = -\tfrac{5}{2}$$

So $2\sin\theta - 5\cos\theta = \sqrt{29}\sin(\theta + 291.8°)$.

> $\sin\alpha = -\frac{5}{R}$, ∴ $\sin\alpha$ is −ve
> $\cos\alpha = \frac{2}{R}$, ∴ $\cos\alpha$ is +ve
> $\tan\alpha$ is −ve, ∴$270° \leqslant \alpha \leqslant 360°$

> $\tan^{-1}\left(-\frac{5}{2}\right) = -68.198\ldots°$ and in the 4th quadrant $\tan^{-1}\left(-\frac{5}{2}\right) = 291.8\ldots°$

> To avoid rounding errors, keep an accurate value for $\tan^{-1}\left(-\frac{5}{2}\right)$ on your calculator for later use.

b　The maximum value of $\sin(\theta + 291.8°)$ is 1.

So the maximum value of $\sqrt{29}\sin(\theta + 291.8°)$ is $\sqrt{29}$.

The maximum occurs when $\sin(\theta + 291.8°) = 1$, i.e. when

$$\theta + 291.8° = 90° \text{ or } 450°$$

$$\theta = 450° - 291.8°$$

$$\theta = 158.2°$$

> The sine function has maximum points at $(90°, 1)$, $(450°, 1)$, … and minimum points at $(270°, -1)$, $(630°, -1)$, …

Similarly, the minimum value of $\sqrt{29}\sin(\theta + 291.8°)$ is $-\sqrt{29}$.

The minimum occurs when $\sin(\theta + 291.8°) = -1$, i.e. when

$$\theta + 291.8° = 270° \text{ or } 630°$$

$$\theta = 630° - 291.8°$$

$$\theta = 338.2°$$

c　$$2\sin\theta - 5\cos\theta = 4$$

$$\sqrt{29}\sin(\theta + 291.8°) = 4$$

$$\sin(\theta + 291.8°) = \frac{4}{\sqrt{29}}$$

> $\sin^{-1}\dfrac{4}{\sqrt{29}} \approx 48°$. List enough solutions for $\theta + 291.8°$, remembering that $291.8°$ will be subtracted from them.

$$\theta + 291.8° = 48.0°, 132.0°, 408.0°, 492.0°$$

$$\theta = 116.2°, 200.2°$$

> Give solutions in the range specified.

Note　Always use more accurate values (not the rounded ones) in calculating the answers. The accurate values can be stored on the calculator. In Example 19, the rounded values of $\tan^{-1}\left(-\frac{5}{2}\right)$ and $\sin^{-1}\frac{4}{\sqrt{29}}$ give the correct answer, but this is not always the case.

Choice of $R\cos(\theta \pm \alpha)$ or $R\sin(\theta \pm \alpha)$

When expressing $a\cos\theta + b\sin\theta$ in one of the above forms, the coefficients of $\sin\theta$ and $\cos\theta$ have to be equated. Negative signs can be avoided, thus giving an angle α acute, by choosing the most appropriate form of $R\cos(\theta\pm\alpha)$ and $R\sin(\theta\pm\alpha)$.

Example 20

Express $3\sin\theta - 2\cos\theta$ as a single trigonometrical function.

$$3\sin\theta - 2\cos\theta \equiv R\sin(\theta - \alpha)$$
$$\equiv R\sin\theta\cos\alpha - R\cos\theta\sin\alpha$$

$$
\begin{array}{ccc}
3\sin\theta & & -2\cos\theta \\
\updownarrow & \updownarrow & \updownarrow \\
R\sin\theta\cos\alpha & - & R\cos\theta\sin\alpha
\end{array}
$$

(Then proceed as in previous examples.)

The position of $\sin\theta$ and $\cos\theta$ and the $-$ve sign between the terms all match.

The method of this section can be applied even if equating coefficients does involve negative signs. (See Exercise 12D Questions 5 and 6.)

Exercise 12D

1 Find the value of R and $\tan\alpha$ in these identities.

a $4\cos\theta + 3\sin\theta \equiv R\cos(\theta - \alpha)$ **b** $2\cos\theta - 3\sin\theta \equiv R\cos(\theta + \alpha)$

c $7\cos\theta + \sin\theta \equiv R\sin(\theta + \alpha)$ **d** $3\sin\theta - 4\cos\theta \equiv R\sin(\theta - \alpha)$

2 Solve these equations for $0° \leqslant \theta \leqslant 360°$.

a $\sqrt{3}\cos\theta + \sin\theta = 1$ **b** $5\sin\theta - 12\cos\theta = 6$

c $\sin\theta + \cos\theta = \frac{1}{2}$ **d** $2\cos 2\theta + \sin 2\theta = 1$

3 Solve these equations for $-\pi \leqslant \theta \leqslant \pi$.

a $\sqrt{3}\cos\theta - \sin\theta = 1$ **b** $2\sin\theta + 7\cos\theta = 4$

c $\cos\theta - 7\sin\theta = 2$ **d** $\sqrt{3}\cos 2\theta + \sqrt{6}\sin 2\theta = \sqrt{3}$

4 Prove that
$$\cos\theta - \sin\theta = \sqrt{2}\cos(\theta + 45°) = -\sqrt{2}\sin(\theta - 45°)$$

5 Show that if
$$3\sin x - 4\cos x \equiv R\sin(x + \alpha), \text{ then } R = 5 \text{ and } \tan x = -\tfrac{4}{3}.$$

6 Express $12\sin\theta + 5\cos\theta$ in the form $R\sin(\theta - \alpha)$.

7 a Show that $3\cos\theta + 4\sin\theta$ may be written in the form $5\cos(\theta - \alpha)$ where $\tan\alpha = \frac{4}{3}$.

b Hence find the values of these functions at their local maxima and minima, giving the smallest positive values of θ, in degrees, for which they occur.

i $3\cos\theta + 4\sin\theta$ **ii** $6 + 3\cos\theta + 4\sin\theta$ **iii** $10 - 3\cos\theta - 4\sin\theta$

iv $\dfrac{1}{3\cos\theta + 4\sin\theta}$ **v** $\dfrac{1}{2 + 3\cos\theta + 4\sin\theta}$ **vi** $\dfrac{7}{12 + 3\cos\theta + 4\sin\theta}$

vii $(3\cos\theta + 4\sin\theta)^2$ **viii** $\dfrac{1}{1 + (3\cos\theta + 4\sin\theta)^2}$

8 Find the greatest and least values of these expressions and the values of x, where $0° \leqslant x \leqslant 360°$, for which they occur.

a $8 \cos x - 15 \sin x$

b $4 + \sqrt{2} \cos 2x + \sqrt{2} \sin 2x$

c $\dfrac{1}{4 - \cos x - \sqrt{3} \sin x}$

d $\dfrac{5}{\sqrt{6} \sin x - \sqrt{3} \cos x + 4}$

9 In this diagram, $AC = 2\,m$, $EC = 4\,m$ and $BD = 3.6\,m$.

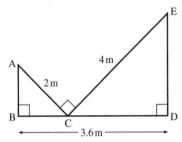

Find angle BAC, correct to the nearest degree.

12.6 Further trigonometric differentiation and integration

In Chapters 5 and 6 you learnt how to differentiate and integrate trigonometric functions. In this section you will see how to use the double angle formulae both to simplify some answers obtained when differentiating and to integrate powers of trigonometric functions.

Example 21 Differentiate with respect to x

a $y = \sin^2 x$

b $y = \ln (\tan 2x)$

Solution

a $y = \sin^2 x$ Use the chain rule.

$\dfrac{dy}{dx} = 2 \sin x \cos x$ Use $\sin 2A = 2 \sin A \cos A$.

$\quad = \sin 2x$

b $y = \ln (\tan 2x)$ Use $\dfrac{d}{dx} (\ln (f(x)) = \dfrac{f'(x)}{f(x)}$.

$\dfrac{dy}{dx} = \dfrac{1}{\tan 2x} \times 2 \sec^2 2x$

$\quad = \dfrac{\cos 2x}{\sin 2x} \times \dfrac{2}{\cos^2 2x}$

$\quad = \dfrac{2}{\sin 2x \cos 2x}$ It is acceptable to leave the answer in this form.

$\quad = \dfrac{4}{2 \sin 2x \cos 2x}$ Use $\sin 2A = 2 \sin A \cos A$.

$\quad = \dfrac{4}{\sin 4x}$

$\quad = 4 \operatorname{cosec} 4x$

Example 22 A particle is moving in a straight line in such a way that its distance, x metres, from a fixed point O, t seconds after the motion begins, is given by $x = \cos t + \cos 2t$.

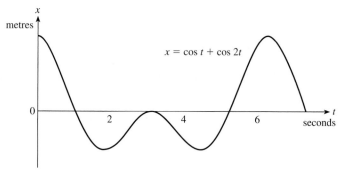

a Find the displacement of the particle from O when $t = 0$.

b At what times during the first 4 seconds of motion is the velocity zero?

c What is the acceleration of the particle $\dfrac{\pi}{2}$ seconds after the motion begins?

> *Note:* At time t, the velocity is given by $\dfrac{\mathrm{d}x}{\mathrm{d}t}$ and the acceleration is given by $\dfrac{\mathrm{d}^2x}{\mathrm{d}t^2}$.

Solution

a $x = \cos t + \cos 2t$

> $\cos 0 = 1$

When $t = 0$

$$x = \cos 0 + \cos 0 = 2$$

$\therefore$ the particle is 2 m from O when $t = 0$.

> From the graph it can be seen that the particle then moves towards O.

b $x = \cos t + \cos 2t$

$$\frac{\mathrm{d}x}{\mathrm{d}t} = -\sin t - 2\sin 2t$$

> Solve $\dfrac{\mathrm{d}x}{\mathrm{d}t} = 0$ to find when the velocity is zero.

$\dfrac{\mathrm{d}x}{\mathrm{d}t} = 0$ when

$$-\sin t - 2\sin 2t = 0$$
$$-(\sin t + 2\sin 2t) = 0$$

> Use $\sin 2A = 2\sin A\cos A$.

$$\sin t + 4\sin t\cos t = 0$$
$$\sin t(1 + 4\cos t) = 0$$

> Do not divide through by $\sin t$; it might equal zero. Instead take it out as a factor.

$\Rightarrow \quad \sin t = 0 \qquad$ or $\quad 1 + 4\cos t = 0$

$\therefore \qquad t = 0,\ \pi,\ 2\pi,\ \dots \quad$ or $\qquad \cos t = -0.25$

> Work in radians.

$$t = 1.823\dots,\ 4.459\dots,\ \dots$$

> $\pi = 3.142\dots$

During the first 4 seconds, the velocity is zero at $t = 0\,\mathrm{s}$, $t = 1.82\,\mathrm{s}$ and $t = 3.14\,\mathrm{s}$ (to 2 d.p.).

> Only the first three values of t are needed.

c
$$\frac{d^2x}{dt^2} = \frac{d}{dt}\left(\frac{dx}{dt}\right)$$

> To find $\frac{d^2x}{dt^2}$, differentiate $\frac{dx}{dt}$ with respect to t.

$$= \frac{d}{dt}(-\sin t - 2\sin 2t)$$

$$= -\cos t - 2\cos 2t \times 2$$

$$= -(\cos t + 4\cos 2t)$$

When $t = \dfrac{\pi}{2}$

> Substitute $t = \dfrac{\pi}{2}$ to find the acceleration at this instant.

$$\frac{d^2x}{dt^2} = -\left(\cos\frac{\pi}{2} + 4\cos\pi\right) = 4$$

> $\cos\dfrac{\pi}{2} = 0,\ \cos\pi = -1$

The acceleration of the particle is $4\,\mathrm{m\,s^{-2}}$.

The derivatives of powers of $\sin x$ and $\cos x$

Example 23 **a** Differentiate $y = 4\cos^3 x$ with respect to x.

b Show that $\dfrac{d}{dx}(5\sin^2 3x) = 15\sin 6x$.

Solution **a** $y = 4\cos^3 x = 4(\cos x)^3$

> Write $\cos^3 x$ as $(\cos x)^3$; it is easier to see the substitution when y is written in this form.

Let $t = \cos x$, then $y = 4t^3$.

So $\dfrac{dt}{dx} = -\sin x$

and $\dfrac{dy}{dt} = 12t^2$

$$= 12(\cos x)^2$$

$$\frac{dy}{dx} = \frac{dy}{dt} \times \frac{dt}{dx}$$

> Without showing the chain rule working:
> $\dfrac{d}{dx}\left(4 \times \left(\Box^3\right)\right) = 4 \times 3\left(\Box^2\right) \times$ derivative of $\Box$

$$= 12(\cos x)^2 \times (-\sin x)$$

$$= -12\sin x\cos^2 x$$

b Let $y = 5\sin^2 3x = 5(\sin 3x)^2$

> Write $\sin^2 3x$ as $(\sin 3x)^2$.

Let $t = \sin 3x$, then $y = 5t^2$.

So $\quad \dfrac{dt}{dx} = 3\cos 3x$

$$\dfrac{d}{dx}\left(\sin\square\right) = \cos\square \times \text{derivative of } \square$$

and $\quad \dfrac{dy}{dt} = 10t$

$$\dfrac{d}{dx}\left(5 \times \left(\square^2\right)\right) = 5 \times 2\left(\square\right) \times \text{derivative of } \square$$

$\qquad\qquad = 10\sin 3x$

$\qquad \dfrac{dy}{dx} = \dfrac{dy}{dt} \times \dfrac{dt}{dx}$

$\qquad\qquad = 10\sin 3x \times 3\cos 3x$

$\qquad\qquad = 30\sin 3x \cos 3x$

$$\sin 2A = 2\sin A \cos A$$

$\qquad\qquad = 15\sin 6x$

So $\quad \dfrac{d}{dx}(5\sin^2 3x) = 15\sin 6x$

$$\boxed{\begin{aligned}\dfrac{d}{dx}(k\sin^n f(x)) &= kn\sin^{n-1} f(x) \times \dfrac{d}{dx}(\sin f(x))\\[2ex]\dfrac{d}{dx}(k\cos^n f(x)) &= kn\cos^{n-1} f(x) \times \dfrac{d}{dx}(\cos f(x))\end{aligned}}$$

Although some may prefer to learn these general results, it is usually better to understand and remember the methods rather than try to quote the formula.

Integration of even powers of $\sin x$ and $\cos x$

To integrate even powers of $\sin x$ and $\cos x$. e.g. $\sin^2 x$, $\cos^2 x$, $\sin^4 x$, $\cos^4 x$, etc., use these trigonometrical identities.

$$\cos 2x = 2\cos^2 x - 1 \Leftrightarrow \cos^2 x = \tfrac{1}{2}(1 + \cos 2x)$$

$$\cos 2x = 1 - 2\sin^2 x \Leftrightarrow \sin^2 x = \tfrac{1}{2}(1 - \cos 2x)$$

Example 24

a $\displaystyle\int \cos^2 x \, dx = \tfrac{1}{2}\int (1 + \cos 2x)\,dx$

$\qquad\qquad\quad = \tfrac{1}{2}\left(x + \tfrac{1}{2}\sin 2x\right) + c$

Use $\cos^2 x = \tfrac{1}{2}(1 + \cos 2x)$

b $\displaystyle\int \sin^2\left(\tfrac{1}{2}x\right) dx = \tfrac{1}{2}\int (1 - \cos x)\,dx$

$\qquad\qquad\qquad = \tfrac{1}{2}(x - \sin x) + c$

$\sin^2 x = \tfrac{1}{2}(1 - \cos 2x)$
$\Rightarrow \sin^2\left(\tfrac{1}{2}x\right) = \tfrac{1}{2}(1 - \cos x)$

c $\displaystyle\int_0^{\frac{\pi}{4}} \sin^4 x \, dx = \int_0^{\frac{\pi}{4}} (\sin^2 x)^2 \, dx$

> Write $\sin^4 x$ as $(\sin^2 x)^2$.

$\displaystyle = \int_0^{\frac{\pi}{4}} \left(\tfrac{1}{2}(1 - \cos 2x)\right)^2 dx$

> Use $\sin^2 x = \tfrac{1}{2}(1 - \cos 2x)$

$\displaystyle = \tfrac{1}{4}\int_0^{\frac{\pi}{4}} (1 - 2\cos 2x + \cos^2 2x) \, dx$

$\displaystyle = \tfrac{1}{4}\int_0^{\frac{\pi}{4}} \left(1 - 2\cos 2x + \tfrac{1}{2}(1 + \cos 4x)\right) dx$

> $\cos^2 x = \tfrac{1}{2}(1 + \cos 2x)$
> $\Rightarrow \cos^2 2x = \tfrac{1}{2}(1 + \cos 4x)$

$\displaystyle = \tfrac{1}{4}\int_0^{\frac{\pi}{4}} \left(\tfrac{3}{2} - 2\cos 2x + \tfrac{1}{2}\cos 4x\right) dx$

$\displaystyle = \tfrac{1}{4}\left[\tfrac{3}{2}x - \sin 2x + \tfrac{1}{8}\sin 4x\right]_0^{\frac{\pi}{4}}$

$\displaystyle = \tfrac{1}{4}\left(\frac{3\pi}{8} - 1 + 0 - 0\right)$

$\displaystyle = \tfrac{3}{32}\pi - \tfrac{1}{4}$

Integrals of the type $\displaystyle\int \cos x \sin^n x \, dx$ and $\displaystyle\int \sin x \cos^n x \, dx$

Using the chain rule

$$\frac{d}{dx}(\sin^n x) = n\sin^{n-1} x \cos x$$

For example

$$\frac{d}{dx}(\sin^3 x) = 3\sin^2 x \cos x$$

Applying the process in reverse gives

$$\int \sin^2 x \cos x \, dx = \tfrac{1}{3}\sin^3 x + c$$

In general

➤ $$\boxed{\int \sin^n x \cos x \, dx = \frac{1}{n+1}\sin^{n+1} x + c}$$

In a similar way, it can be shown that

➤ $$\boxed{\int \cos^n x \sin x \, dx = -\frac{1}{n+1}\cos^{n+1} x + c}$$

> These results are easy to remember and the integration can be done by recognition. Look for the integral of $\cos x$ multiplied by a power of $\sin x$ or the integral of $\sin x$ multiplied by a power of $\cos x$.

Integrals of this type can also be done by substitution and this method may be preferred when the integrand contains trigonometric ratios of multiple angles, as in Example 25.

Example 25   Find $\displaystyle\int \sin 2x \cos^4 2x \, dx$ using the substitution $u = \cos 2x$.

Solution

$\displaystyle\int \sin 2x \cos^4 2x \, dx$

$\displaystyle = \int \sin 2x \cos^4 2x \frac{dx}{du} \, du$

$\displaystyle = \int \cos^4 2x \sin 2x \frac{dx}{du} \, du$

$\displaystyle = \int u^4 \left(-\tfrac{1}{2}\right) du$

$\displaystyle = -\tfrac{1}{2} \int u^4 \, du$

$\displaystyle = -\tfrac{1}{10} u^5 + c$

$\displaystyle = -\tfrac{1}{10} \cos^5 2x + c$

Let $u = \cos 2x$

$\dfrac{du}{dx} = -2 \sin 2x$

$\dfrac{dx}{du} = -\dfrac{1}{2 \sin 2x}$

$\sin 2x \dfrac{dx}{du} = -\dfrac{1}{2}$

Define the substitution.
Find $\dfrac{du}{dx}$ and use $\dfrac{dx}{du} = \dfrac{1}{\dfrac{du}{dx}}$

Substitute for $\sin 2x \dfrac{dx}{du}$ in the integral.

Now substitute $\cos 2x$ for u.

Note   To do this integration by recognition, guess that the integral is a multiple of $\cos^5 2x$ and find the numerical adjustment factor by differentiating $\cos^5 2x$:

$$\frac{d}{dx}(\cos^5 2x) = 5 \cos^4 2x(-2 \sin 2x)$$

$$= -10 \cos^4 x \sin 2x$$

so $\displaystyle\int \sin 2x \cos^4 2x \, dx = -\tfrac{1}{10} \cos^5 2x + c$

Integration of odd powers of $\sin x$ and $\cos x$

The identity $\cos^2 x + \sin^2 x = 1$ is used when integrating odd powers of $\sin x$ and $\cos x$.

$$\cos^2 x + \sin^2 x = 1 \Rightarrow \cos^2 x = 1 - \sin^2 x$$

$$\sin^2 x = 1 - \cos^2 x$$

Example 26   **a** $\displaystyle\int \cos^3 x \, dx = \int \cos x \cos^2 x \, dx$

$\displaystyle = \int \cos x(1 - \sin^2 x) \, dx$

$\displaystyle = \int (\cos x - \cos x \sin^2 x) \, dx$

$\displaystyle = \sin x - \tfrac{1}{3} \sin^3 x + c$

Write $\cos^3 x$ as $\cos x \cos^2 x$ and use the identity $\cos^2 x = 1 - \sin^2 x$.

Recognise type: $\displaystyle\int \cos x \sin^n x \, dx$

b $\displaystyle\int_0^{\frac{\pi}{2}} \sin^5 x \, dx = \int_0^{\frac{\pi}{2}} \sin x (\sin^2 x)^2 x \, dx$

> Write $\sin^5 x$ as $\sin x \sin^4 x = \sin x (\sin^2 x)^2$ and use the identity $\sin^2 x = 1 - \cos^2 x$.

$\displaystyle = \int_0^{\frac{\pi}{2}} \sin x (1 - \cos^2 x)^2 \, dx$

> Multiply out the brackets.

$\displaystyle = \int_0^{\frac{\pi}{2}} \sin x (1 - 2\cos^2 x + \cos^4 x) \, dx$

> Recognise type:
> $\int \sin x \cos^n x \, dx$

$\displaystyle = \int_0^{\frac{\pi}{2}} (\sin x - 2 \sin x \cos^2 x + \sin x \cos^4 x) \, dx$

$\displaystyle = \left[-\cos x + \tfrac{2}{3} \cos^3 x - \tfrac{1}{5} \cos^5 x \right]_0^{\frac{\pi}{2}}$

> Take care with the limits:
> $\cos \dfrac{\pi}{2} = 0, \cos 0 = 1$

$\displaystyle = 0 - (-1 + \tfrac{2}{3} - \tfrac{1}{5})$

$\displaystyle = \tfrac{8}{15}$

Extension

Integrals of the type $\displaystyle\int \sin ax \cos bx \, dx,$

$\displaystyle\int \cos ax \cos bx \, dx, \int \sin ax \sin bx \, dx$

Using the factor formulae (Section 12.4), it may be possible to write the product of two trigonometrical functions as the sum or difference of two functions that can be integrated easily. These are the factor formulae:

$\sin (A + B) + \sin (A - B) = 2 \sin A \cos B \Rightarrow \sin A \cos B = \tfrac{1}{2}(\sin (A + B) + \sin (A - B))$

$\sin (A + B) - \sin (A - B) = 2 \cos A \sin B \Rightarrow \cos A \sin B = \tfrac{1}{2}(\sin (A + B) - \sin (A - B))$

$\cos (A + B) + \cos (A - B) = 2 \cos A \cos B \Rightarrow \cos A \cos B = \tfrac{1}{2}(\cos (A + B) + \cos (A - B))$

$\cos (A + B) - \cos (A - B) = -2 \sin A \sin B \Rightarrow \sin A \sin B = -\tfrac{1}{2}(\cos (A + B) - \cos (A - B))$

Example 27 Find $\displaystyle\int \sin 5x \cos 3x \, dx$.

> Use the factor formula
> $\sin A \cos B = \tfrac{1}{2}(\sin (A + B) + \sin (A - B))$
> with $A = 5x$ and $B = 3x$.

Solution $\sin 5x \cos 3x = \tfrac{1}{2}(\sin (5x + 3x) + \sin (5x - 3x))$

$= \tfrac{1}{2}(\sin 8x + \sin 2x)$

$\therefore \displaystyle\int \sin 5x \cos 3x \, dx = \tfrac{1}{2}\int (\sin 8x + \sin 2x) \, dx$

> The integrand, written as the sum of two sines, can be integrated easily.

$= \tfrac{1}{2}\left(-\tfrac{1}{8}\cos 8x - \tfrac{1}{2}\cos 2x\right) + c$

$= -\tfrac{1}{16}(\cos 8x + 4 \cos 2x) + c$

Integrals of the type $\int \sec^n x \tan x \, dx$ and $\int \tan^n x \sec^2 x \, dx$

$$\frac{d}{dx}(\sec x) = \sec x \tan x$$

so $\quad \dfrac{d}{dx}(\sec^n x) = n \sec^{n-1} x (\sec x \tan x)$ $\qquad$ Use the chain rule.

$$= n \sec^n x \tan x$$

Applying this in reverse:

➤ $\boxed{\displaystyle\int \sec^n x \tan x \, dx = \frac{1}{n} \sec^n x + c}$

$$\frac{d}{dx}(\tan x) = \sec^2 x \qquad \text{Use the chain rule.}$$

so $\quad \dfrac{d}{dx}(\tan^{n+1} x) = (n+1) \tan^n x \sec^2 x$

Applying this in reverse:

➤ $\boxed{\displaystyle\int \tan^n x \sec^2 x \, dx = \frac{1}{n+1} \tan^{n+1} x + c}$

Example 28   **a** $\displaystyle\int \sec^7 x \tan x \, dx = \tfrac{1}{7} \sec^7 x + c$

b $\displaystyle\int \tan^3 x \sec^2 x \, dx = \tfrac{1}{4} \tan^4 x + c$

This technique can also be used to find related integrals, as in Example 29.

Example 29 $\displaystyle\int \tan^3 x \, dx = \int \tan x (\tan^2 x) \, dx$ $\qquad$ Write in this format so that the identity $1 + \tan^2 x = \sec^2 x$ can be used.

$$= \int \tan x (\sec^2 x - 1) \, dx$$

$$= \int (\tan x \sec^2 x - \tan x) \, dx$$

$$= \tfrac{1}{2} \tan^2 x + \ln(\cos x) + c$$

Note

There is an alternative format when finding $\int \tan x \sec^2 x \, dx$:

$$\int (\tan x \sec^2 x) \, dx = \tfrac{1}{2} \sec^2 x + d$$

The answers are equivalent because

$$\tfrac{1}{2} \sec^2 x + d = \tfrac{1}{2}(1 + \tan^2 x) + d$$
$$= \tfrac{1}{2} + \tfrac{1}{2} \tan^2 x + d$$
$$= \tfrac{1}{2} \tan^2 x + c$$

Exercise 12E

1 If $y = (\sin x + \cos x)^2$, show that

$$\frac{d^2 y}{dx^2} + 4 \sin 2x = 0$$

2 Differentiate with respect to x

   a $3 \sin^2 x$ **b** $\tfrac{1}{2} \cos^3 x$ **c** $-2 \cos^4 x$ **d** $4 \sin^2 2x$ **e** $\sqrt{\sin x}$

3 For values of x between 0 and 2π, find the x-coordinates of points on the curve $y = \sin x(2 \cos 2x + 1)$ for which the gradient is zero.

4 Show that

$$\frac{d}{dx} \ln\left(\tan \tfrac{1}{2} x\right) = \frac{1}{\sin x}$$

5 Find these integrals.

   a $\displaystyle\int \sin^2 x \, dx$ **b** $\displaystyle\int \cos^2 3x \, dx$

   c $\displaystyle\int \sin^4 2x \, dx$ **d** $\displaystyle\int \frac{2}{1 + \tan^2 x} \, dx$

6 Evaluate

   a $\displaystyle\int_{-\frac{\pi}{4}}^{\frac{\pi}{4}} \cos^2 2x \, dx$ **b** $\displaystyle\int_{0}^{\frac{\pi}{2}} \cos^4 x \, dx$

7 This diagram shows the part of the curves $y = \cos^2 x$ and $y = \sin^2 x$.

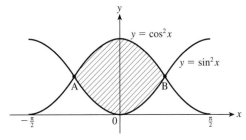

   a Find the coordinates of A and B.

   b Find the area of the shaded region.

8 Find these integrals.

a $\displaystyle\int \cos x \sin^2 x \, dx$ **b** $\displaystyle\int \sin x \cos^5 x \, dx$ **c** $\displaystyle\int \sin 4x \cos^3 4x \, dx$

d $\displaystyle\int \cot x \sin^5 x \, dx$ **e** $\displaystyle\int \cos^3 x \tan x \, dx$ **f** $\displaystyle\int \frac{\sin^3 x}{\sec x} \, dx$

9 a Write down a formula for $\cos x$ in terms of $\cos \dfrac{x}{2}$.

b Show that

$$\int \frac{1}{1+\cos x} \, dx = \tan \frac{x}{2} + c$$

c Find $\displaystyle\int \sqrt{1+\cos x} \, dx$.

10 a Express $\sin 2x$ in terms of $\sin x$ and $\cos x$.

b Find

i $\displaystyle\int \sin 2x \cos^3 x \, dx$ **ii** $\displaystyle\int \sin x \cos \frac{x}{2} \, dx$ **iii** $\displaystyle\int \sin 2x \sin^2 x \, dx$

11 Find $\displaystyle\int x \sin x \cos x \, dx$.

12 Find

a $\displaystyle\int \cos^2 x \, dx$ **b** $\displaystyle\int x \cos^2 x \, dx$

13 The area enclosed by $y = \sin x$ and the x-axis for $0 \leqslant x \leqslant \pi$ is rotated about the x-axis. Find the volume generated.

14 This diagram shows part of the curve $y = 4 \sin x \cos^3 x$.

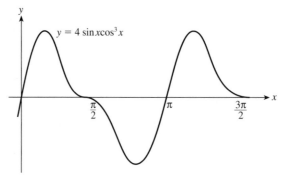

Find the area enclosed by the curve and the x-axis between $x = 0$ and $x = \pi$.

15 Find $\displaystyle\int_0^{\frac{\pi}{2}} \cos x \, (1 + \sin^3 x) \, dx$.

16 **a** Find $\displaystyle\int \cos^4 x \sin^3 x \, dx$.

Hint: write $\sin^3 x$ as $\sin x \,(\sin^2 x)$.

b Find $\displaystyle\int \cos^3 x \sin^2 x \, dx$.

c Find $\displaystyle\int \cos^3 x \sin^3 x \, dx$.

17 Find these integrals.

a $\displaystyle\int \sin^3 x \, dx$ **b** $\displaystyle\int \cos^3 2x \, dx$

18 Evaluate

$$\int_0^{\frac{\pi}{2}} \cos^5 x \, dx$$

19 **a** Factorise $\sin 3x + \sin x$.

b Express $2 \sin 3x \cos 2x$ as the sum of two terms.

c Find

$$\int \sin 3x \cos 2x \, dx$$

20 Find these integrals.

a $\displaystyle\int \cos 3x \sin x \, dx$ **b** $\displaystyle\int 2 \cos \frac{3x}{2} \cos \frac{x}{2} \, dx$ **c** $\displaystyle\int \sin 4x \sin x \, dx$

21 Find these integrals.

a $\displaystyle\int \sec^4 x \tan x \, dx$ **b** $\displaystyle\int \sec^2 x \tan^4 x \, dx$ **c** $\displaystyle\int \sec^2 x \tan^2 x \, dx$

d $\displaystyle\int \sec x \tan^3 x \, dx$ **e** $\displaystyle\int \sec^4 x \, dx$ **f** $\displaystyle\int \tan^4 x \, dx$

Exercise 12F (Review)

1 If $\sin A = \frac{5}{13}$ and $\sin B = \frac{8}{17}$, where A and B are acute, find the values of

   a $\cos (A + B)$ **b** $\sin (A - B)$ **c** $\tan (A + B)$

2 If $\cos A = \frac{15}{17}$ and $\sin B = \frac{20}{29}$, where A is reflex and B is obtuse, find the values of

   a $\sin (A + B)$ **b** $\cos (A - B)$ **c** $\cot (A - B)$

3 Find the values of

   a $\cos 80° \cos 20° + \sin 80° \sin 20°$

   b $\dfrac{\tan 15° + \tan 30°}{1 - \tan 15° \tan 30°}$

   c $\sin 40° \cos 50° + \sin 50° \cos 40°$

4 Find the values of $\sin x$ and $\cos x$ when $\cos 2x$ is

   a $\frac{1}{9}$ **b** $\frac{49}{81}$

5 Find the value of $\tan\theta$ when $\tan 2\theta$ is

   a $-\frac{20}{21}$ **b** $\frac{36}{77}$

6 If $\sin\theta = \frac{35}{37}$, where θ is acute, find the values of

   a $\sin 2\theta$ **b** $\sec 2\theta$

7 Solve these equations, giving values of θ from $0°$ to $360°$ inclusive.

   a $\cot 2\theta = 0.2$ **b** $2\sin 2\theta = 3\sin\theta$

   c $\cos 2\theta + 5\cos\theta = 2$ **d** $2\sin x - \operatorname{cosec} x = 1$

   e $\tan x + 4\cot x = 5$ **f** $2\sec 2x - \cot 2x = \tan 2x$

   g $\sec x + \tan x = 0$ **h** $2\tan\dfrac{\theta}{2} + 3\tan\theta = 0$

8 Solve, for $0 \leqslant x \leqslant 2\pi$

   a $\tan 2\theta + \tan\theta = 0$ **b** $2\operatorname{cosec} 2\theta = 1 + \tan^2\theta$

9 **a** Express $4\cos\theta - 3\sin\theta$ in the form $R\cos(\theta + \alpha)$.

   b Hence solve $4\cos\theta - 3\sin\theta = 1$ for $0° \leqslant \theta \leqslant 360°$.

10 Solve $3\cos\theta + 2\sin\theta = 2.5$ for $-\pi \leqslant \theta \leqslant \pi$.

11 By expressing each of these in the form $R\cos(\theta \pm \alpha)$, find their maximum and minimum values, giving the values of θ between $0°$ and $360°$ for which they occur.

   a $5\cos\theta - 12\sin\theta$ **b** $12\cos\theta - 35\sin\theta$

12 By expressing $\tan 3A$ as $\tan(2A + A)$, prove that

$$\tan 3A = \frac{3\tan A - \tan^3 A}{1 - 3\tan^2 A}$$

13 Given that $B \neq A$, prove this identity.

$$\frac{\sin A + \sin B}{\cos A - \cos B} \equiv \cot\frac{1}{2}(B - A)$$

14 Given that A, B and C are the angles of a triangle, prove this identity.

$$\sin A + \sin(B - C) \equiv 2\sin B\cos C$$

15 Eliminate θ from

   a $x = 2\cos 2\theta,\ y = 3\cos\theta$ **b** $x = 2\tan\theta,\ y = \tan 2\theta$

16 Solve these equations for values of θ from 0 to π inclusive.

   a $\sin 2\theta + \sin 4\theta + \sin 6\theta = 0$ **b** $\cos\frac{1}{2}\theta + 2\cos\frac{3}{2}\theta + \cos\frac{5}{2}\theta = 0$

   c $\sin\theta + \cos 2\theta - \sin 3\theta = 0$

17 In triangle ABC, angle C is $90°$. Show that

   a $\sin 2A = \dfrac{2ab}{c^2}$ **b** $\cos 2A = \dfrac{b^2 - a^2}{c^2}$

   c $\sin\frac{1}{2}A = \sqrt{\dfrac{c - b}{2c}}$ **d** $\cos\frac{1}{2}A = \sqrt{\dfrac{c + b}{2c}}$

18 Prove that $\tan 50° - \tan 40° = 2\tan 10°$.

19 Show that $2\sin 82\frac{1}{2}° \cos 37\frac{1}{2}° = 2\sin 127\frac{1}{2}° \sin 97\frac{1}{2}°$.

20 Prove that $\cos 130° + \cos 110° + \cos 10° = 0$.

21 Find an equation of the tangent to the curve $y = 2\cos^4 x$ at the point with x-coordinate $\frac{\pi}{4}$.

22 Show that

$$\int_0^\pi \cos^2 x \, dx = \frac{\pi}{2}$$

23 Find

a $\int \cos^3 2x \, dx$ **b** $\int \sin^4 2x \, dx$

24 **a** Express $\sin 3x \cos x$ as the sum of two sines.

    b Show that

$$\int_0^{\frac{\pi}{3}} \sin 3x \cos x \, dx = \frac{9}{16}$$

25 **a** Express $\cos 3x \cos 2x$ as the sum of two cosines.

    b Evaluate

$$\int_0^{\frac{\pi}{4}} \cos 3x \cos 2x \, dx$$

26 Evaluate

$$\int_{\frac{\pi}{4}}^{\frac{\pi}{3}} \frac{1}{(1 - \sin x)(1 + \sin x)} \, dx$$

27 This diagram shows part of the curve $y = \cos x + \sin x$.

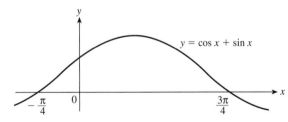

    a Find the area enclosed by the curve and the x-axis between $x = -\frac{\pi}{4}$ and $x = \frac{3\pi}{4}$.

    b Find the volume of the solid generated when this area is rotated through 2π radians about the x-axis.

28 Find the integral $\int \dfrac{\sin 2x}{\cos^2 x} \, dx$.

There is a 'Test yourself' exercise (Ty12) and an 'Extension exercise' (Ext12) on the CD-ROM.

➤ Key points

Addition formulae

$\sin(A \pm B) = \sin A \cos B \pm \cos A \sin B$

$\cos(A \pm B) = \cos A \cos B \mp \sin A \sin B$

$\tan(A \pm B) = \dfrac{\tan A \pm \tan B}{1 \mp \tan A \tan B}$

Double angle formulae

$\sin 2A = 2 \sin A \cos A$

$\begin{aligned} \cos 2A &= \cos^2 A - \sin^2 A \\ &= 2 \cos^2 A - 1 \\ &= 1 - 2 \sin^2 A \end{aligned}$

$\tan 2A = \dfrac{2 \tan A}{1 - \tan^2 A}$

Half-angle formulae

$\cos^2 \dfrac{A}{2} = \dfrac{1}{2}(1 + \cos A)$

$\sin^2 \dfrac{A}{2} = \dfrac{1}{2}(1 - \cos A)$

Extension: Factor formulae

$\sin P + \sin Q = 2 \sin \dfrac{P+Q}{2} \cos \dfrac{P-Q}{2}$ $\qquad \sin(A+B) + \sin(A-B) = 2 \sin A \cos B$

$\sin P - \sin Q = 2 \cos \dfrac{P+Q}{2} \sin \dfrac{P-Q}{2}$ $\qquad \sin(A+B) - \sin(A-B) = 2 \cos A \sin B$

$\cos P + \cos Q = 2 \cos \dfrac{P+Q}{2} \cos \dfrac{P-Q}{2}$ $\qquad \cos(A+B) + \cos(A-B) = 2 \cos A \cos B$

$\cos P - \cos Q = -2 \sin \dfrac{P+Q}{2} \sin \dfrac{P-Q}{2}$ $\qquad \cos(A+B) - \cos(A-B) = -2 \sin A \sin B$

$a\cos\theta + b\sin\theta$

$a\cos\theta + b\sin\theta$ can be expressed in the form $R\cos(\theta\pm\alpha)$ or $R\sin(\theta\pm\alpha)$.

Equating coefficients of $\cos\theta$ and $\sin\theta$ leads to $R = \sqrt{a^2+b^2}$ and $\tan\alpha = \pm\dfrac{b}{a}$ or $\tan\alpha = \pm\dfrac{a}{b}$.

Differentiation of trigonometric functions

$$\frac{d}{dx}(k\sin^n f(x)) = kn\sin^{n-1}f(x) \times \frac{d}{dx}(\sin f(x))$$

$$\frac{d}{dx}(k\cos^n f(x)) = kn\cos^{n-1}f(x) \times \frac{d}{dx}(\cos f(x))$$

Integration of trigonometric functions

Use double angle formulae for even powers of $\sin x$ and $\cos x$

$$\int \sin^2 x\,dx = \frac{1}{2}\int(1-\cos 2x)\,dx \qquad \int \cos^2 x\,dx = \frac{1}{2}\int(1+\cos 2x)\,dx$$

$$\int \sin^n x\cos x\,dx = \frac{1}{n+1}\sin^{n+1}x + c \qquad \int \cos^n x\sin x\,dx = -\frac{1}{n+1}\cos^{n+1}x + c$$

$$\int \sec^n x\tan x\,dx = \frac{1}{n}(\sec^n x) + c \qquad \int \tan^n x\sec^2 x\,dx = \frac{1}{n+1}(\tan^{n+1}x) + c$$

1 Express

$$\frac{14}{(2x+1)(x-3)}$$

in partial fractions.

AQA (AEB) 1997

(Not from the live examinations for the current specification.)

2 Express

$$\frac{2x^2 - 11x - 10}{(3-x)(x+2)^2}$$

in partial fractions.

3 The polynomials $f(x) = 6x^2 + 5x + 1$ and $g(x) = 3x^2 - 5x - 2$ have a common linear factor.

a By considering $f\left(\frac{1}{3}\right)$ and $g\left(\frac{1}{3}\right)$, find this factor.

b Hence write $\dfrac{f(x)}{g(x)}$ as a simplified algebraic fraction.

4 The polynomial $p(x)$ is given by

$$p(x) = (x+3)(x-2)(x-4)$$

a Find the remainder when $p(x)$ is divided by $(x+1)$.

b **i** Express $\dfrac{70}{(x+3)(x-2)(x-4)}$ in the form $\dfrac{A}{x+3} + \dfrac{B}{x-2} + \dfrac{C}{x-4}$.

ii Hence, prove that $\displaystyle\int_5^6 \dfrac{70}{(x+3)(x-2)(x-4)}\,dx = N\ln 3 - M\ln 2$, where N and M are positive integers.

AQA Jan 2004

5 **a** Express $\dfrac{5x-2}{(x-1)(x+2)}$ in partial fractions.

b Hence find the value of $\displaystyle\int_2^3 \dfrac{5x-2}{(x-1)(x+2)}\,dx$, leaving your answer in the form $p\ln 5 + q\ln 2$, where p and q are integers to be found.

AQA June 2001

6 **a** Express $\dfrac{2x^2 - x + 11}{(2x-3)(x+2)}$ in the form $A + \dfrac{B}{2x-3} + \dfrac{C}{x+2}$.

b Hence find $\displaystyle\int_2^6 \dfrac{2x^2 - x + 11}{(2x-3)(x+2)}\,dx$, giving your answer in the form $p + q\ln 2 + r\ln 3$, where p, q and r are integers.

AQA Jan 2002

7 a Given that

$$\frac{25x + 1}{(2x - 1)(x + 1)^2} \equiv \frac{A}{2x - 1} + \frac{B}{x + 1} + \frac{C}{(x + 1)^2}$$

    i show that

$$25x + 1 \equiv A(x + 1)^2 + B(x + 1)(2x - 1) + C(2x - 1)$$

    ii and find the values of A, B and C.

b Hence, find

$$\int_1^2 \frac{25x + 1}{(2x - 1)(x + 1)^2} \, dx$$

leaving your answer in the form $p + q \ln 2$.

AQA Jan 2003

8 a Obtain the binomial expansion of $(1 + 2x)^{-2}$ in ascending powers of x up to and including the term in x^3.

b State the range of values of x for which the full expansion is valid.

AQA June 2003

9 a By considering rectangular strips of width 0.2, use the mid-ordinate rule to obtain an approximation for $\int_0^{0.8} \sqrt{1 + x^3} \, dx$, giving your answer correct to four decimal places.

b Obtain the binomial expansion of $(1 + x^3)^{\frac{1}{2}}$ in ascending powers of x up to and including the term in x^6. By integrating each of the terms in your expansion between the given limits, obtain a second approximation for

$$\int_0^{0.8} \sqrt{1 + x^3} \, dx,$$

giving your answer to four decimal places.

AQA 2003

10 a Determine the binomial expansion of $\left(1 - \frac{x}{10}\right)^{-3}$, in ascending powers of x, up to and including the term in x^3.

b Show that the coefficient of x^n in this expansion is

$$K(n + 1)(n + 2) \times \frac{1}{10^n}$$

for a rational number K whose value is to be determined.

c Determine the value of $\left(\frac{1}{0.99}\right)^3$ correct to fourteen decimal places.

AQA June 2001

11 A curve has implicit equation

$$y^3 + xy = 4x - 2.$$

a Show that the value of $\dfrac{\mathrm{d}y}{\mathrm{d}x}$ at the point (1, 1) is $\frac{3}{4}$.

b Find the equation of the normal to the curve at the point (1, 1).

AQA Jan 2003

12 A curve has implicit equation $x^2 + 2xy - y^3 = 1$.

a Determine an expression for $\dfrac{\mathrm{d}y}{\mathrm{d}x}$ in terms of x and y.

b Hence find the equation of the normal to the curve at the point with coordinates (2, −1).

AQA (AEB) 1997

(Not from the live examinations for the current specification.)

13 a Given that $N = Ae^{kt}$, where A and k are constants, show that $\dfrac{\mathrm{d}N}{\mathrm{d}t} = kN$.

b The number of bacteria, N, in a colony is such that the rate of increase of N is proportional to N. The time, t, is measured in hours from the instant that $N = 2$ million. When $t = 3$, $N = 5$ million. Find the value of t when $N = 8$ million.

AQA Jan 2002

14 a Given that $P = Ae^{-kt}$, where A and k are constants, show that $\dfrac{\mathrm{d}P}{\mathrm{d}t} = -kP$.

b The population P, of a particular species is such that the rate of decrease of P is proportional to P. The time, t, is measured in years from the date when $P = 5000$, and $P = 3500$ when $t = 10$. By using the result in part **a**, or otherwise, find the value of t when $P = 3000$.

AQA June 2002

15 a Simplify $\cos 2\theta \cos \theta - \sin 2\theta \sin \theta$.

b Hence, or otherwise, solve the equation

$$4\cos 2\theta \cos \theta = 4\sin 2\theta \sin \theta + 1,$$

giving all solutions to the nearest degree in the interval $0° < \theta < 180°$.

AQA (AEB) Jan 1998

(Not from the live examinations for the current specification.)

16 When asked to express $\sin \theta - \cos \theta$ in the form $R\cos(\theta - \alpha)$, a student writes

$$\text{``} \sin \theta - \cos \theta \equiv \sqrt{2} \cos\left(\theta - \frac{\pi}{4}\right) \text{''}.$$

Determine, with a reason, whether the student is correct.

17 a Express $5\sin x - 12\cos x$ in the form $R\sin(x - \alpha)$ where R is a positive constant and $0° < \alpha < 360°$.

Give the value of α to the nearest $0.1°$.

b Hence find the solutions in the interval $0° < \theta < 360°$ of the equation

$$5\sin x - 12\cos x = 8.$$

Give each solution to the nearest degree.

18 a Use the identity $\sin(A + B) = \sin A\cos B + \cos A\sin B$ to show that

$$2\sin\left(x + \frac{\pi}{3}\right) = \sin x + \sqrt{3}\cos x.$$

b Show that the equation

$$2\sin\left(x + \frac{\pi}{3}\right) = \cos\left(x + \frac{\pi}{6}\right)$$

can be written in the form

$$3\tan x + \sqrt{3} = 0.$$

c Hence, solve the equation

$$2\sin\left(x + \frac{\pi}{3}\right) = \cos\left(x + \frac{\pi}{6}\right)$$

giving all solutions in terms of π in the interval $0 < x < 2\pi$.

AQA Jan 2003

19 a Express $6\sin^2\theta$ in the form $a + b\cos 2\theta$, where a and b are constants.

b Find the exact value of $\displaystyle\int_0^{\frac{\pi}{12}} 6\sin^2\theta\,d\theta$.

c Solve the equation

$$3 - 3\cos 2\theta = 2\operatorname{cosec}\theta$$

giving all solutions in radians in the interval $0 < \theta < 2\pi$.

No credit will be given for simply reading values from a graph.

AQA June 2002

20 Part of the curve with equation $y = 3\sin 2x + \cos 2x$ is sketched below.

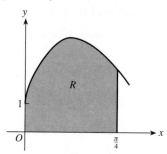

a Find $\dfrac{dy}{dx}$.

b Find the equation of the tangent to the curve at the point where $x = \dfrac{\pi}{4}$.

c **i** Express $(3\sin 2x + \cos 2x)^2$ in the form $A\sin 4x + B\cos 4x + C$, where A, B and C are integers.

ii The region R, shaded above, is bounded by the curve with equation $y = 3\sin 2x + \cos 2x$, the coordinate axes and the line $x = \dfrac{\pi}{4}$.

Find the volume of the solid generated when R is rotated through 2π radians about the x-axis.

AQA Jan 2003

13 Coordinate geometry

So far in the course, all the curves you have met have been defined by Cartesian equations involving two variables, usually x and y. However, there are many occasions when it is more helpful to give each of x and y in terms of a third variable, or parameter. This is particularly useful when describing the path of an object over time, as you are then able to determine not just the curve the object follows, but also where it is at any particular time.

After working through this chapter you should

■ *understand how parametric form can be used to describe the equation of a curve*

■ *be able to convert from the parametric form of the equation of a curve to its corresponding Cartesian form*

■ *differentiate functions given parametrically*

■ *find stationary points on curves given parametrically*

■ *find the equations of tangents and normals to curves given parametrically.*

The CD-ROM (E13.1.1, E13.1.2 and E13.1.3) includes some more advanced work on sketching curves given in parametric and Cartesian form, including sketching rational functions of two quadratics and finding oblique points of inflexion. It also explains how to find the derivative of, and how to integrate, functions defined parametrically.

13.1 Parametric form

Some curves are defined by expressing x and y in terms of a third variable, called a parameter. For example

$$x = t^2$$
$$y = 3t + 1$$

> In this case, t can be eliminated to give the equation $9x = (y-1)^2$, but in other cases this is not possible and, in any case, it is often easier to work with the parametric equations of a curve.

are the **parametric equations** of a curve, and t is called the **parameter**.

Parametric form is often used when parametric equations are of a simpler form than the Cartesian equation of the curve. It is also useful as it allows us to see how a curve is traced out and so determine, for example, where a moving object is at any particular time.

Graphs of curves defined parametrically

Imagine an object, P, moving round the circumference of a circle of radius r. Suppose that the circle has centre the origin, O, as shown in the diagram and consider the situation when the object is at a position such that the angle between OP and the positive x-axis is θ.

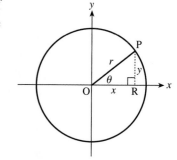

Then, $x = r\cos\theta$ and $y = r\sin\theta$.

These equations are called the **parametric equations** of the circle. Substituting in different values for θ will give different points on the circle:

- $\theta = \dfrac{\pi}{3}$ gives the point $\left(\frac{1}{2}r, \frac{\sqrt{3}}{2}r\right)$

- $\theta = \pi$ gives the point $(-r, 0)$

Note: The whole circle is defined even if θ is restricted to the domain $0 \leqslant \theta < 2\pi$.

θ can be eliminated from the equations by squaring and adding them.

$$x^2 + y^2 = (r\cos\theta)^2 + (r\sin\theta)^2$$
$$= r^2\cos^2\theta + r^2\sin^2\theta$$
$$= r^2(\cos^2\theta + \sin^2\theta)$$
$$= r^2$$

Use $\cos^2\theta + \sin^2\theta = 1$.

So $\quad x^2 + y^2 = r^2$

This is the equation previously obtained for the equation of a circle, centre O and radius r.

Note $\quad (r\cos\theta, r\sin\theta)$ represents a general point on a circle, centre $(0, 0)$ with radius r.

Parametric curves can be plotted using a graphical calculator or computer graph sketching package. The CD-ROM covers more advanced techniques for sketching curves defined parametrically.

The next two examples show how to obtain sketches of simple curves defined parametrically.

Example 1   A curve is defined parametrically by the equations $x = t^2 - 1$, $y = 2t$. Sketch the curve for values of t between -3 and 3.

Draw up a table of values for x and y.

t	-3	-2	-1	0	1	2	3
x	8	3	0	-1	0	3	8
y	-6	-4	-2	0	2	4	6

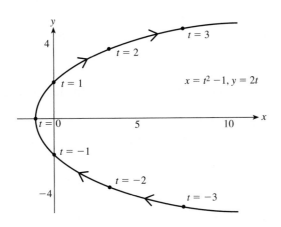

The points can now be plotted and labelled with the values of the parameter, t. The direction in which the curve is traced out for increasing values of t is denoted by the arrows.

This curve looks like a parabola. In the next section you will see how to obtain the Cartesian equation of the curve to confirm that this is the case.

Example 2   A curve is defined parametrically by the equations $x = 4\cos\theta$, $y = 3\sin\theta$. Sketch the curve for values of t between 0 and 2π.

Draw up a table of values for x and y.

Note that most of the table can be completed using the symmetry properties of sine and cosine.

t	0	$\frac{1}{6}\pi$	$\frac{1}{3}\pi$	$\frac{1}{2}\pi$	$\frac{2}{3}\pi$	$\frac{5}{6}\pi$	π	$\frac{7}{6}\pi$	$\frac{4}{3}\pi$	$\frac{3}{2}\pi$	$\frac{5}{3}\pi$	$\frac{11}{6}\pi$	2π
x	4	3.46	2	0	-2	-3.46	-4	-3.46	-2	0	2	3.46	4
y	0	1.5	2.6	3	2.6	1.5	0	-1.5	-2.6	-3	-2.6	-1.5	0

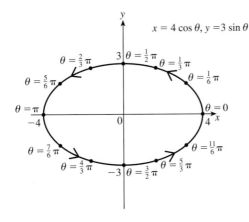

The points can now be plotted and labelled with the values of the parameter, θ. The direction in which the curve is traced out for increasing values of θ is denoted by the arrows.

This curve is an ellipse. In the next section you will see how to eliminate the parameter θ to obtain the Cartesian equation of the curve.

Converting between parametric form and Cartesian form

When the equation of a curve is given in parametric form it is often possible to eliminate the parameter and find a Cartesian equation for the curve.

For example, in Example 1 the curve was defined by the parametric equations $x = t^2 - 1$, $y = 2t$.

Rearranging the second of these equations gives $t = \dfrac{y}{2}$ and substituting this in the first equation gives $x = \left(\dfrac{y}{2}\right)^2 - 1$ or $x = \dfrac{1}{4}y^2 - 1$, which is the Cartesian equation of a parabola as expected.

In Example 2 the parameter can be eliminated by using the fact that $\cos^2\theta + \sin^2\theta = 1$.

In this example $x = 4\cos\theta$, $y = 3\sin\theta$.

Rearranging each equation gives $\dfrac{x}{4} = \cos\theta$, $\dfrac{y}{3} = \sin\theta$ and so $\left(\dfrac{x}{4}\right)^2 + \left(\dfrac{y}{3}\right)^2 = 1$ or $\dfrac{x^2}{16} + \dfrac{y^2}{9} = 1$, which is the Cartesian equation of an ellipse.

It is generally more difficult to find parametric equations for a curve when a Cartesian equation is given.

Example 3   Eliminate the parameter from the parametric equations $x = a + r\cos\theta$, $y = b + r\sin\theta$ and hence find the centre and radius of the circle they represent.

$$x = a + r\cos\theta \qquad y = b + r\sin\theta$$

$$x - a = r\cos\theta \qquad y - b = r\sin\theta$$

> Here, the parameter is θ. Rearrange each equation to isolate the terms in $\cos\theta$ and $\sin\theta$.

$$\therefore \quad (x-a)^2 + (y-b)^2 = r^2\cos^2\theta + r^2\sin^2\theta$$

$$= r^2(\cos^2\theta + \sin^2\theta)$$

> Use $\cos^2\theta + \sin^2\theta = 1$

$$= r^2$$

So the parametric equations represent a circle, centre (a, b) and of radius r.

Example 4   Find the Cartesian equation for each of these curves, by eliminating the parameter from the parametric equations.

a $x = t + 1$	**b** $x = \sin\theta$	**c** $x = \sin\theta$
$y = 3t^2$	$y = \cos 2\theta$	$y = \sin 2\theta$

Solution   **a** $x = t + 1 \qquad y = 3t^2$

> Here, the parameter is t. Rearrange the simpler equation to obtain an expression for t. Substitute this into the other equation.

$t = x - 1$

$\therefore \quad y = 3(x - 1)^2$

b $\quad x = \sin\theta \qquad y = \cos 2\theta$

$$= 1 - 2\sin^2\theta$$

$$= 1 - 2x^2$$

> Here the parameter is θ. $\cos 2\theta$ and $\sin\theta$ need to be linked. Use the double angle formula which expresses $\cos 2\theta$ in terms of $\sin\theta$.

> *Note*: the parametric equations represent only part of the parabola $y = 1 - 2x^2$ as the values of both x and y are restricted to the ranges of $\sin\theta$ and $\cos 2\theta$.

c $\quad x = \sin\theta \qquad y = \sin 2\theta$

$$= 2\sin\theta\cos\theta$$

Therefore

> Use the double angle formula to rewrite $\sin 2\theta$, and then square y to obtain an expression involving $\cos^2\theta$ which can then be rewritten in terms of $\sin^2\theta$.

$$y^2 = 4\sin^2\theta\cos^2\theta$$

$$= 4\sin^2\theta(1 - \sin^2\theta)$$

> Substitute x in place of $\sin\theta$.

$$= 4x^2(1 - x^2)$$

or $\quad y = \pm 2x\sqrt{1 - x^2}$

> **To convert the equation of a curve given in parametric form, to Cartesian form, eliminate (if possible) the parameter.**

Exercise 13A

1 Sketch the curves given by the following parametric equations for the range of values of the parameter given

    a $\quad x = t + 1,\ y = 3t^2 \quad$ for $-3 \leqslant t \leqslant 3$

    b $\quad x = t^2,\ y = \dfrac{1}{t} \quad$ for $-3 \leqslant t \leqslant 3,\ t \neq 0$

    c $\quad x = 5\sin\theta,\ y = 4\cos\theta \quad$ for $0 \leqslant \theta \leqslant 2\pi$

    d $\quad x = t^2,\ y = t^3 \quad$ for $-3 \leqslant t \leqslant 3$

2 Find the Cartesian equations of each of these curves. (a is a constant.)

    a $\quad x = t^2 + 3$ **b** $\quad x = t^3$ **c** $\quad x = 5t$

       $y = t - 2$ $y = t^2$ $y = \dfrac{5}{t}$

    d $\quad x = 3 + t$ **e** $\quad x = at^2$ **f** $\quad x = t^2$

       $y = 4 - 2t$ $y = 2at$ $y = \dfrac{10}{t}$

    g $\quad x = t^2 + 1$ **h** $\quad x = 2\cos\theta$ **i** $\quad x = \sin\theta$

       $y = t^3 + t$ $y = 3\sin\theta$ $y = \cos^3\theta$

    j $\quad x = 3\sec\theta$ **k** $\quad x = a\cos\theta$ **l** $\quad x = 2\cos 2\theta$

       $y = 5\tan\theta$ $y = a\sin 2\theta$ $y = 3\cos\theta$

    m $\quad x = \dfrac{1 - t}{t}$ **n** $\quad x = \dfrac{2 + 3t}{1 + t}$

       $y = \dfrac{1 + t}{t^2}$ $y = \dfrac{3 - 2t}{1 + t}$

3 Find the values of the parameters and the other coordinates of the given points on these curves. (a is a constant.)

a $x = t$, $y = \dfrac{2}{t}$, where $y = 1\frac{1}{2}$

b $x = at^2$, $y = 2at$, where $x = \frac{9}{4}a$

c $x = \dfrac{1+t}{1-t}$, $y = \dfrac{2+3t}{1-t}$, where $y = -\frac{4}{3}$

d $x = a\cos\theta$, $y = b\sin\theta$, where $x = \frac{1}{2}a$

4 Show that the parametric equations

$$\begin{array}{ll} x = 1 + 2t \\ y = 2 + 3t \end{array} \quad \text{and} \quad \begin{array}{l} x = \dfrac{1}{2t-3} \\[2mm] y = \dfrac{t}{2t-3} \end{array}$$

both represent the same straight line, and find its Cartesian equation.

5 Show that the line given parametrically by the equations

$$x = \dfrac{2-t}{1+2t} \qquad y = \dfrac{3+t}{1+2t}$$

passes through the points $(6, 7)$ and $(-2, -1)$. Find the values of t corresponding to these points.

13.2 Parametric differentiation

Consider these parametric equations

$$x = t^2 \qquad y = 3t + 1$$

Each of these can be differentiated with respect to the parameter t, to give

$$\frac{dx}{dt} = 2t \quad \text{and} \quad \frac{dy}{dt} = 3$$

Then, by the chain rule

$$\frac{dy}{dx} = \frac{dy}{dt} \times \frac{dt}{dx} = \frac{dy}{dt} \div \frac{dx}{dt}$$

Use $\dfrac{dt}{dx} = \dfrac{1}{\frac{dx}{dt}}$

Hence $\dfrac{dy}{dx} = \dfrac{3}{2t}$

Example 5 Find $\dfrac{dy}{dx}$, in terms of t, for each of the curves with these parametric equations.

a $x = 5t^2$

$y = 4t^3$

b $x = \dfrac{t}{t-1}$

$y = \dfrac{t^2}{t-1}$

c $x = e^{4t}$

$y = e^{2t} - 1$

Solution **a** $\qquad x = 5t^2 \qquad y = 4t^3$

Differentiate each function with respect to the parameter.

$$\frac{dx}{dt} = 10t \qquad \frac{dy}{dt} = 12t^2$$

$$\frac{dy}{dx} = \frac{dy}{dt} \div \frac{dx}{dt}$$

Or use $\dfrac{dy}{dx} = \dfrac{dy}{dt} \times \dfrac{1}{\frac{dx}{dt}}$

$$= \frac{12t^2}{10t}$$

$$= \frac{6t}{5}$$

b $\qquad x = \dfrac{t}{t-1}$

Use the quotient rule for differentiation.

$$\frac{dx}{dt} = \frac{(t-1) \times 1 - t \times 1}{(t-1)^2}$$

$$= \frac{-1}{(t-1)^2}$$

Use the quotient rule for differentiation.

$$y = \frac{t^2}{t-1}$$

$$\frac{dy}{dt} = \frac{(t-1) \times 2t - t^2 \times 1}{(t-1)^2}$$

Expand brackets and collect like terms to simplify the numerator.

$$= \frac{2t^2 - 2t - t^2}{(t-1)^2}$$

$$= \frac{t^2 - 2t}{(t-1)^2}$$

By the chain rule

$$\frac{dy}{dx} = \frac{dy}{dt} \times \frac{dt}{dx}$$

Use $\dfrac{dt}{dx} = \dfrac{1}{\frac{dx}{dt}}$

$$= \frac{t^2 - 2t}{(t-1)^2} \times \frac{(t-1)^2}{-1}$$

Cancel $(t-1)^2$ terms.

$$= -(t^2 - 2t)$$

$$= 2t - t^2$$

c $\qquad x = e^{4t} \qquad y = e^{2t} - 1$

Differentiate each function with respect to the parameter.

$$\frac{dx}{dt} = 4e^{4t} \qquad \frac{dy}{dt} = 2e^{2t}$$

$$\frac{dy}{dx} = \frac{dy}{dt} \div \frac{dx}{dt} = \frac{2e^{2t}}{4e^{4t}} = \frac{1}{2}e^{-2t}$$

Use $e^a \div e^b = e^{a-b}$

Example 6   Find the gradient of the curve with parametric equations $x = t^2 - t$, $y = t^3 - t$ at the point (2, 0).

$$x = t^2 - t \qquad y = t^3 - t$$

$$\frac{dx}{dt} = 2t - 1 \qquad \frac{dy}{dt} = 3t^2 - 1$$

> Differentiate each function with respect to the parameter.

By the chain rule

$$\frac{dy}{dx} = \frac{dy}{dt} \div \frac{dx}{dt}$$

$$= \frac{3t^2 - 1}{2t - 1}$$

At the point (2, 0)

$$t^2 - t = 2 \quad \text{and} \quad t^3 - t = 0$$

$$t^2 - t - 2 = 0 \qquad t(t^2 - 1) = 0$$

$$t^2 - t - 2 = 0 \qquad t(t^2 - 1) = 0$$

$$(t - 2)(t + 1) = 0 \qquad\qquad t = 0 \text{ or } t = 1 \text{ or } t = -1$$

$$t = 2 \text{ or } t = -1$$

> Substitute in the values of x and y and solve the equations for t.

So, at (2, 0), $t = -1$.

> Compare each equation to find the common solution for t.

When $t = -1$

$$\frac{dy}{dx} = \frac{3 - 1}{-2 - 1}$$

$$= -\frac{2}{3}$$

> Substitute in the value of t to find the gradient at the required point.

Example 7   Find the equation of the tangent to the curve with parametric equations $x = 4 + \cos\theta$, $y = \sin^2\theta$ at the point $\left(4\frac{1}{2}, \frac{3}{4}\right)$.

$$x = 4 + \cos\theta \qquad y = \sin^2\theta$$

$$\frac{dx}{d\theta} = -\sin\theta$$

$$\frac{dy}{d\theta} = 2\sin\theta\cos\theta$$

> To find the gradient of the tangent, first differentiate each function with respect to the parameter, and then use the chain rule to find $\frac{dy}{dx}$.

$$\frac{dy}{dx} = \frac{dy}{d\theta} \div \frac{dx}{d\theta}$$

$$= \frac{2\sin\theta\cos\theta}{-\sin\theta}$$

$$= -2\cos\theta$$

At the point $\left(4\frac{1}{2}, \frac{3}{4}\right)$

> Substitute in the value of x and y and solve the equations for θ.

$$4\frac{1}{2} = 4 + \cos\theta \quad \text{and} \quad \frac{3}{4} = \sin^2\theta$$

$$\cos\theta = \frac{1}{2} \qquad\qquad \sin\theta = \pm\frac{\sqrt{3}}{2}$$

> Take the principal value of θ.

$$\therefore \quad \theta = \frac{\pi}{3}$$

When $\theta = \dfrac{\pi}{3}$

> Substitute in the values of θ to find the gradient at the given point.

$$\frac{dy}{dx} = -2 \times \frac{1}{2} = -1$$

So the equation of the tangent to the curve at $\left(4\frac{1}{2}, \frac{3}{4}\right)$ is

$$y - \frac{3}{4} = -\left(x - 4\frac{1}{2}\right)$$

> Use $y - y_1 = m(x - x_1)$

$$4y - 3 = -4x + 18$$

$$4x + 4y - 21 = 0$$

Example 8   The diagram shows a sketch of the curve with parametric equations $x = t^2 + t$, $y = t^3 - 2t$ along with the tangent at the point with parameter $t = 2$. Find the equation of this tangent and the coordinates of the point where it meets the curve again.

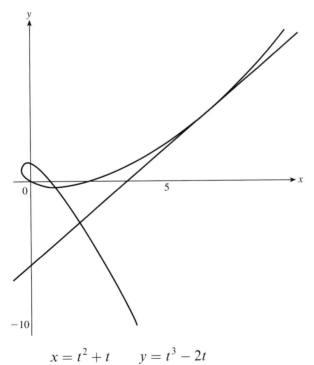

$$x = t^2 + t \qquad y = t^3 - 2t$$

> Differentiate each function with respect to the parameter.

$$\frac{dx}{dt} = 2t + 1 \qquad \frac{dy}{dt} = 3t^2 - 2$$

$$\frac{dy}{dx} = \frac{dy}{dt} \div \frac{dx}{dt}$$

$$= \frac{3t^2 - 2}{2t + 1}$$

When $t = 2$, $x = 6$, $y = 4$ and

(6, 4) is the point where $t = 2$.

$$\frac{dy}{dx} = \frac{3 \times 4 - 2}{4 + 1} = 2$$

So the equation of the tangent at (6, 4) is

$$y - 4 = 2(x - 6)$$

Use $y - y_1 = m(x - x_1)$

$$y - 4 = 2x - 12$$

$$y = 2x - 8$$

This tangent meets the curve when

To find where this tangent meets the curve again solve simultaneously the equations of the curve and the tangent.

$$t^3 - 2t = 2(t^2 + t) - 8$$

$$t^3 - 2t^2 - 4t + 8 = 0$$

Substitute the parametric equations of the curve into the equation of the tangent.

$$(t - 2)(t^2 - 4) = 0$$

$$(t - 2)(t - 2)(t + 2) = 0$$

$$t = \pm 2$$

$(t - 2)$ must be a factor of the LHS because the tangent touches the curve at the point where $t = 2$. Because the tangent *touches* the curve, $t = 2$ is actually a double root.

$t = 2$ represents the point where the tangent touches the curve, so the curve meets the curve again when $t = -2$, i.e. at the point (2, −4).

Example 9   Find, and classify, the stationary points on the curve with parametric equations $x = t^2 + 4$, $y = t^3 - 12t$.

$$x = t^2 + 4, \quad y = t^3 - 12t$$

Differentiate each function with respect to the parameter.

$$\frac{dx}{dt} = 2t, \quad \frac{dy}{dt} = 3t^2 - 12$$

$$\frac{dy}{dx} = \frac{dy}{dt} \div \frac{dx}{dt}$$

$$= \frac{3t^2 - 12}{2t}$$

$$= \frac{3(t^2 - 4)}{2t}$$

At a stationary point $\dfrac{dy}{dx} = 0$

Therefore $\dfrac{3(t^2 - 4)}{2t} = 0$

$$t^2 - 4 = 0$$

For a fraction to be zero the numerator must be zero.

and hence $\quad t = \pm 2$

When $t = -2$, $x = 4 + 4 = 8$, $y = -8 + 24 = 16$

and when $t = 2$, $x = 4 + 4 = 8$, $y = 8 - 24 = -16$

so there are stationary points when $t = -2$ at (8, 16) and when $t = 2$ at (8, −16).

To determine the nature of the stationary points consider the gradient either side of each point.

t	-1	-2	-3	1	2	3
x	5	8 $(y=16)$	13	5	8 $(y=-16)$	13
$\dfrac{dy}{dx} = \dfrac{3(t^2-4)}{2t}$	4.5	0	-2.5	-4.5	0	2.5

Maximum point   Minimum point

The curve has a maximum point at $(8, 16)$ and a minimum point at $(8, -16)$.

The CD-ROM (E13.2.1 and E13.2.2) shows how to find the second derivative of functions defined parametrically, and how to find the equations of tangents and normals at a general point.

Exercise 13B

1 Find $\dfrac{dy}{dx}$, in terms of t, for each of the curves with these parametric equations.

a $x = 3t^2$
  $y = 2t^3$

b $x = t^2 + t$
  $y = 2t - t^2$

c $x = 2\sin t$
  $y = 4\cos t$

d $x = t\cos t$
  $y = t^2 \sin t$

e $x = e^{-t} + 4$
  $y = \ln t$

f $x = (t+1)^2$
  $y = (t+1)^3$

g $x = 4\sqrt{t}$
  $y = 5t^2 + 4$

h $x = t\ln t$
  $y = 2t + t^2$

2 Find the gradient of each of the curves with these parametric equations at the point defined by the value of t given.

a $x = t^2$
  $y = t^4 - 1$
  $[t = 2]$

b $x = \sin t$
  $y = \tan t$
  $[t = \frac{\pi}{4}]$

c $x = \sqrt{t+1}$
  $y = t^2$
  $[t = 3]$

d $x = t(t+1)$
  $y = (t-1)^3$
  $[t = 1]$

e $x = te^t$
  $y = (t+1)^2$
  $[t = 0]$

f $x = \dfrac{1}{t+1}$
  $y = (t-2)^2$
  $[t = 1]$

g $x = \cos t$
  $y = \sin 2t$
  $[t = \frac{\pi}{6}]$

h $x = 4e^t$
  $y = e^{4t} - 8t$
  $[t = 2]$

3 Find $\dfrac{dy}{dx}$, in terms of t, when a, b and c are constants, and

a $x = at^2$
  $y = 2at$

b $x = ct$
  $y = \dfrac{c}{t}$

c $x = a\cos t$
  $y = b\sin t$

d $x = a\sec t$
  $y = b\tan t$

e $x = a\cos^3 t$
  $y = a\sin^3 t$

4 Find $\dfrac{dy}{dx}$, in terms of t, for each of the curves with these parametric equations.

a $x = \dfrac{t}{t-1}$

$y = \dfrac{t^2}{1-t}$

b $x = \dfrac{2t}{t+2}$

$y = \dfrac{3t}{t+3}$

c $x = \dfrac{2t}{1+t^2}$

$y = \dfrac{1-t^2}{1+t^2}$

d $x = \dfrac{2}{\sqrt{1+t^2}}$

$y = \dfrac{t}{\sqrt{1+t^2}}$

e $x = \dfrac{t}{1-t}$

$y = \dfrac{1-2t}{1-t}$

5 **a** If $x = t^2$, $y = t^3$ find $\dfrac{dy}{dx}$ in terms of t.

b If $y = x^{\frac{3}{2}}$ find $\dfrac{dy}{dx}$.

c Is there any connection between the two results?

6 Find the equations of the tangents and normals to the curves given parametrically by these equations at the points given (a, c constant):

a $x = t^2$

$y = t^3$

$(1, -1)$

b $x = t^2$

$y = \dfrac{1}{t}$

$\left(\tfrac{1}{4}, 2\right)$

c $x = at^2$

$y = 2at$

$(a, -2a)$

d $x = ct$

$y = \dfrac{c}{t}$

$(-c, -c)$

e $x = t^2 - 4$

$y = t^3 - 4t$

$(-3, -3)$

f $x = 3\cos\theta$

$y = 2\sin\theta$

$\left(\tfrac{3}{2}, \sqrt{3}\right)$

g $x = \dfrac{3at}{1+t^3}$

$y = \dfrac{3at^2}{1+t^3}$

$\left(\tfrac{2}{3}a, \tfrac{4}{3}a\right)$

h $x = 4 + \ln t$

$y = t^3 + e^t$

$(4, e+1)$

7 The diagram shows a sketch of the curve with parametric equations $x = t^3 - 12t$, $y = 3t^2$ along with the normal at the point with parameter $t = 2$.

Find the equation of this normal and the coordinates of the point where it cuts the curve again.

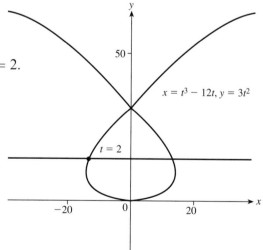

8 The diagram shows a sketch of the curve with parametric equations $x = 3t^2$, $y = 4t^3$ along with the tangent at the point with parameter $t = 1$.

Find the equation of this tangent and the coordinates of the point where it meets the curve again.

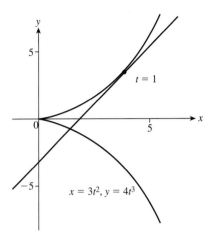

9 Find, and classify, the stationary points on the curves with these parametric equations.

a $x = t^2$, $y = t^3 - 3t$

b $x = 3\cos\theta$, $y = 2\sin\theta$

Exercise 13C (Review)

1 Find the Cartesian equations of each of these curves.

a $x = t^5$
  $y = t^2 - 4$

b $x = 3t$
  $y = \dfrac{3}{t}$

c $x = 4\sin\theta$
  $y = 3\cos^2\theta$

d $x = \cos 2\theta$
  $y = 2\sin\theta$

2 Sketch each of these curves.

a $x = t^3 + 1$
  $y = t^2$

b $x = 2t$
  $y = \dfrac{2}{t}$

c $x = 12t - t^3$
  $y = t^2 - 9$

d $x = 5\cos\theta - 5$
  $y = 3\sin\theta + 2$

3 Find $\dfrac{dy}{dx}$, in terms of t, for each of the curves given parametrically by these equations.

a $x = t^2 - 2t$, $y = t^4 - 4t$

b $x = \cos t$, $y = t(4 - \sin t)$

c $x = e^{2t} - 4$, $y = t^4 - 8t$

4 Find the equation of the tangent to the curve given parametrically by $x = 3\cos\theta$, $y = \sin 4\theta$ at the point with parameter $\theta = \dfrac{\pi}{3}$.

5 Find the equation of the normal to the curve with parametric equations $x = t^2 + 1$, $y = \ln(t + 3)$ at the point $(5, 0)$.

There is a 'Test yourself' exercise (Ty13) and an 'Extension exercise' (Ext13) on the CD-ROM.

➤ Key points

Parametric form

- The parametric equations of a circle, centre the origin and radius r are:

$$x = r \cos \theta, \ y = r \sin \theta$$

- The parametric equations of a circle, centre (a, b) and radius r are:

$$x = a + r \cos \theta, \ y = b + r \sin \theta$$

- To obtain the Cartesian equation of a curve given in parametric form, eliminate (if possible) the parameter from the parametric equations $x = f(t)$ and $y = g(t)$.

Parametric differentiation

To differentiate a function defined parametrically, apply the chain rule.

If $x = f(t)$ and $y = g(t)$ then:

$$\frac{dy}{dx} = \frac{dy}{dt} \times \frac{dt}{dx}$$

$$= g'(t) \times \frac{1}{f'(t)}$$

$$= \frac{g'(t)}{f'(t)}$$

14 Differential equations

In the course so far you have learnt how to differentiate and integrate a number of functions. In this chapter you will see how to apply the techniques you have learnt to solve differential equations, that is equations containing a derivative such as $\dfrac{dy}{dx}$.

These equations are of particular importance in science and you will see how they can be used to help model a number of physical, and other, problems. Differential equations are also commonly used when modelling situations in economics and the social sciences.

After working through this chapter you should be able to

■ *find the general solution of first-order differential equations, where the variables can be separated*

■ *find the particular solution to a differential equation by using the conditions given*

■ *form differential equations, when given appropriate information about rates of change*

■ *solve problems involving exponential growth and decay.*

14.1 First-order differential equations

A differential equation in x and y is an equation containing at least one of the derivatives

$$\frac{dy}{dx}, \frac{d^2y}{dx^2}, \frac{d^3y}{dx^3}, \dots$$

The **order of a differential equation** is the order of the highest derivative occurring in it.

First-order differential equations

$$\frac{dy}{dx} = 5x^2 \qquad \frac{dy}{dx} - xy = 0 \qquad y\frac{dy}{dx} = e^x$$

Second-order differential equations

$$\frac{d^2y}{dx^2} = \cos 2x \qquad \frac{d^2y}{dx^2} = 3\frac{dy}{dx} - 4y$$

This section considers the solution of first-order differential equations only.

Consider the curve $y = f(x)$, such that the gradient at the point (x, y) is x.

Since the gradient is given by $\dfrac{dy}{dx}$, this leads to the differential equation

$$\frac{dy}{dx} = x$$

Integrating with respect to x gives

$$y = \frac{1}{2}x^2 + c$$

This is called the **general solution** and it contains an **arbitrary integration constant**, c.

The differential equation $\dfrac{dy}{dx} = x$ can be illustrated diagrammatically by drawing gradients at several points. For example, a line with gradient 1 is drawn at points with x-coordinate 1, such as $(1, 1)$, $(1, 2)$, $(1, 3)$; a line with gradient -2 at points with x-coordinate -2, such as $(-2, -1)$, $(-2, 0)$, $(-2, 3)$, and so on.

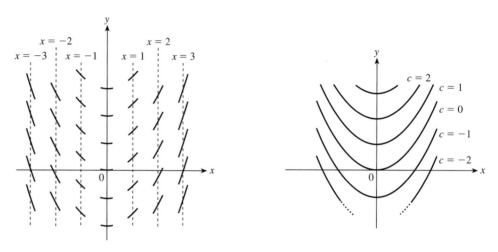

This helps to show that the general solution $y = \frac{1}{2}x^2 + c$ represents a **family of curves**.

Remember that for *every* curve in the family, the gradient is x at the point (x, y),

i.e. $\dfrac{dy}{dx} = x$.

A **particular solution** relates to a specific member of the family of curves and can be found from additional information.

For example, if it is known that the curve passes through $(2, 3)$, then the specific curve is found by substituting $x = 2$ and $y = 3$ into the general equation. This enables the value of c to be calculated:

$$y = \tfrac{1}{2}x^2 + c$$
$$3 = \tfrac{1}{2}(2)^2 + c$$
$$c = 1$$

So, the particular solution is $y = \frac{1}{2}x^2 + 1$.

First-order differential equations of the form $\dfrac{dy}{dx} = g(x)$

Here are some examples of first-order differential equations that can be written in the form $\dfrac{dy}{dx} = g(x)$.

$$\frac{dy}{dx} = 3x^2 + 8x$$

$g(x) = 3x^2 + 8x$

$$\frac{dy}{dx} + 6 = \frac{1}{x}$$

Rewrite as $\dfrac{dy}{dx} = \dfrac{1}{x} - 6$. Then $g(x) = \dfrac{1}{x} - 6$.

$$x^2 \frac{dy}{dx} - 4 = 0$$

Rewrite as $\dfrac{dy}{dx} = \dfrac{4}{x^2}$. Then $g(x) = \dfrac{4}{x^2}$.

This is the simplest form of differential equation; it can be attempted directly by integrating both sides with respect to x.

> If $\dfrac{dy}{dx} = g(x)$, then $y = \displaystyle\int g(x)\,dx$

Example 1 Find the particular solution of the differential equation $\dfrac{dy}{dx} = 3x^2 + 8x$, given that $y = 5$ when $x = -1$.

Solution

$$\frac{dy}{dx} = 3x^2 + 8x$$

Integrate with respect to x.

$$y = x^3 + 4x^2 + c$$

Find the general solution containing an arbitrary constant.

When $x = -1$, $y = 5$

so $\quad 5 = (-1)^3 + 4(-1)^2 + c$

Use the specific information to find the particular solution.

$$5 = -1 + 4 + c$$

$$c = 2$$

The particular solution is $y = x^3 + 4x^2 + 2$.

Example 2 Find the displacement, s cm, from O of a particle at time t s, if its velocity, v cm s^{-1}, is given by the differential equation

$$v = \frac{ds}{dt} = t^2 - t + 4$$

and the displacement is 100 cm at time 6 s. Find also the initial displacement.

Solution

$$\frac{ds}{dt} = t^2 - t + 4$$

Integrate to find the general solution of the differential equation.

$$s = \frac{1}{3}t^3 - \frac{1}{2}t^2 + 4t + c$$

When $t = 6$, $s = 100$

so $\quad 100 = \dfrac{1}{3}(6)^3 - \dfrac{1}{2}(6)^2 + 4(6) + c$

$\qquad 100 = 78 + c$

$\qquad c = 22$

$\therefore \qquad s = \dfrac{1}{3}t^3 - \dfrac{1}{2}t^2 + 4t + 22$

When $t = 0$, $s = 22$.
Initially the particle is 22 cm from O.

> Use the specific information to find the particular solution.

> This is the particular solution.

> The initial displacement is the displacement when $t = 0$.

First-order differential equations of the form $f(y)\dfrac{dy}{dx} = g(x)$

Consider the differential equations

$$y^2\dfrac{dy}{dx} - 2x = 0 \qquad \text{and} \qquad \dfrac{dy}{dx} = ye^x$$

Both can be written in the form

$$f(y)\dfrac{dy}{dx} = g(x)$$

in which $f(y)$ is a function of y only and $g(x)$ is a function of x only:

$$y^2\dfrac{dy}{dx} - 2x = 0 \Rightarrow y^2\dfrac{dy}{dx} = 2x$$

> $f(y) = y^2$, $g(x) = 2x^2$

$$\dfrac{dy}{dx} = ye^x \Rightarrow \dfrac{1}{y}\dfrac{dy}{dx} = e^x$$

> $f(y) = \dfrac{1}{y}$, $g(x) = e^x$

If an equation can be written in the form

$$f(y)\dfrac{dy}{dx} = g(x)$$

a method known as **separating the variables** can be used.

If $\qquad f(y)\dfrac{dy}{dx} = g(x)$

then $\quad \displaystyle\int f(y)\dfrac{dy}{dx}\,dx = \int g(x)\,dx$

$\therefore \qquad \displaystyle\int f(y)\,dy = \int g(x)\,dx$

> Integrate both sides with respect to x, using $\displaystyle\int f(y)\dfrac{dy}{dx}\,dx = \int f(y)\,dy$ (Section 5.5)

> The left-hand side is now the integral of a function of y with respect to y. The right-hand side is the integral of a function of x with respect to x.

Summarising

> If $f(y)\dfrac{dy}{dx} = g(x)$, then $\displaystyle\int f(y)\,dy = \int g(x)\,dx$

In practice, when a differential equation in x and y in the form

$$\frac{dy}{dx} = \ldots$$

can be solved by separating the variables, the terms in y are taken to the left-hand side (with dy to give $\int \ldots dy$) and the terms in x are taken to the right-hand side (with dx to give $\int \ldots dx$).

Example 3   Find the solution of the differential equation $\dfrac{dy}{dx} = 2xy^2$, expressing y in terms of x.

Solution

$$\frac{dy}{dx} = 2xy^2$$

> The general solution is required. Write the differential equation in the form
> $f(y)\dfrac{dy}{dx} = g(x)$ and separate the variables.

$$\int \frac{1}{y^2}\,dy = \int 2x\,dx$$

$$-\frac{1}{y} = x^2 + c$$

> There is no need to put an integration constant each side; put it on one side or the other.

$$y = -\frac{1}{x^2 + c}$$

> Rearrange to make y the subject.

Example 4   Solve $\dfrac{dy}{dx} = \dfrac{e^{2x}}{y}$, given that $y = 3$ when $x = 0$.

Solution

$$\frac{dy}{dx} = \frac{e^{2x}}{y}$$

> A particular solution is required, but first find the general solution by separating the variables.

$$\int y\,dy = \int e^{2x}\,dx$$

$$\frac{y^2}{2} = \frac{1}{2}e^{2x} + c$$

> Multiply throughout by 2 to avoid the fractions. There is no need to write $2c$, just change the constant to a different letter.

$$y^2 = e^{2x} + d$$

When $x = 0$, $y = 3$

> Use the condition to find d.

so $9 = 1 + d$

$$d = 8$$

The particular solution is $y^2 = e^{2x} + 8$.

Example 5

Unless a solution is requested in a specific format, such as y expressed as a function of x, it can be left in a convenient form. Note particularly the commonly occurring formats shown in this example.

Solve these differential equations.

a $\dfrac{dy}{dx} = \dfrac{x}{y}$ **b** $\dfrac{dy}{dx} = \dfrac{y}{x}$ **c** $\dfrac{dy}{dx} - xy = 0$

Solution

a $\dfrac{dy}{dx} = \dfrac{x}{y}$

> Separate the variables and write as $\int \ldots dy = \int \ldots dx$.

$$\int y\,dy = \int x\,dx$$

$$\frac{1}{2}y^2 = \frac{1}{2}x^2 + c$$

> Multiply through by 2.

$$y^2 = x^2 + d$$

b $\dfrac{dy}{dx} = \dfrac{y}{x}$

$$\int \frac{1}{y}\,dy = \int \frac{1}{x}\,dx$$

$$\ln y = \ln x + c$$

> Let $c = \ln k$ and use $\ln a + \ln b = \ln ab$.

$$= \ln x + \ln k$$

$$= \ln kx$$

> This is a common way of writing the integration constant when expressions involve logarithms and leads to a neat form of the solution.

$$\therefore \quad y = kx$$

c $\dfrac{dy}{dx} - xy = 0$

> Rearrange so that it is easier to separate the variables.

$$\frac{dy}{dx} = xy$$

$$\int \frac{1}{y}\,dy = \int x\,dx$$

$$\ln y = \frac{1}{2}x^2 + c$$

> Use $\ln a = b \Rightarrow a = e^b$

$$y = e^{\frac{1}{2}x^2 + c}$$

> Use $e^{a+b} = e^a \times e^b$

$$y = e^{\frac{1}{2}x^2} \times e^c$$

> Write $A = e^c$

$$\therefore \quad y = Ae^{\frac{1}{2}x^2}$$

> This is a common format. The choice of constant is arbitrary, although A is often used here.

It is interesting to look at the families of curves resulting from these differential equations. The curves below have been drawn using integer values of the constant from -6 to 6.

$$\frac{dy}{dx} = \frac{x}{y}$$

$$\Rightarrow y^2 = x^2 + d$$

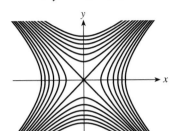

$$\frac{dy}{dx} = \frac{y}{x}$$

$$\Rightarrow y = kx$$

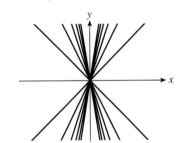

$$\frac{dy}{dx} = xy$$

$$\Rightarrow y = Ae^{\frac{1}{2}x^2}$$

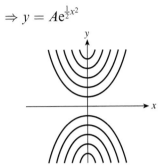

Example 6 Given that

$$\cos^2 t \frac{dx}{dt} = \frac{1}{x}$$

and $x = 4$ when $t = \dfrac{\pi}{4}$, find x when $t = \dfrac{\pi}{3}$, giving the answer correct to 3 significant figures.

Solution $\qquad \cos^2 t \dfrac{dx}{dt} = \dfrac{1}{x}$

> The variables can be separated, so put terms in x on the left (with dx) and terms in t on the right (with dt).

$$\int x \, dx = \int \frac{1}{\cos^2 t} \, dt$$

> $\dfrac{1}{\cos^2 t} = \sec^2 t$

$$\int x \, dx = \int \sec^2 t \, dt$$

$$\frac{1}{2} x^2 = \tan t + c$$

> Multiply by 2 to avoid the fraction.

$$x^2 = 2 \tan t + d$$

> This is the general solution.

$x = 4$ when $t = \dfrac{\pi}{4}$

so $\qquad 16 = 2 \tan \dfrac{\pi}{4} + d$

> $\tan \dfrac{\pi}{4} = 1$

$$d = 14$$

$\therefore \qquad x^2 = 2 \tan t + 14$

> This is the particular solution.

When $t = \dfrac{\pi}{3}$

Substitute into the particular solution.

$$x^2 = 2\tan\dfrac{\pi}{3} + 14$$

$$= 17.46\ldots$$

$$x = \pm 4.18 \ (3 \text{ s.f.})$$

First-order differential equations of the form $f(y)\dfrac{dy}{dx} = 1$

Here are some examples.

$$y^2\dfrac{dy}{dx} = 1$$

$f(y) = y^2$

$$\dfrac{dy}{dx} = y + 2$$

Rewrite as $\dfrac{1}{y+2}\dfrac{dy}{dx} = 1$. Then $f(y) = \dfrac{1}{y+2}$.

$$\dfrac{dy}{dx} - e^y = 0$$

Rewrite as $e^{-y}\dfrac{dy}{dx} = 1$. Then $f(y) = e^{-y}$.

To solve this type of differential equation, separate the variables, considering $f(y)\dfrac{dy}{dx} = g(x)$ with $g(x) = 1$.

➤
> **If** $\quad f(y)\dfrac{dy}{dx} = 1$, **then** $\displaystyle\int f(y)\,dy = \int 1\,dx$
>
> $\qquad\qquad\qquad\Rightarrow \displaystyle\int f(y)\,dy = x + c$

Example 7 Find the general solution of $\dfrac{dy}{dx} = y + 2$.

Solution

$$\dfrac{dy}{dx} = y + 2$$

Separate the variables and write as $\int \ldots dy = \int \ldots dx$.

$$\Rightarrow \quad \int \dfrac{1}{y+2}\,dy = \int 1\,dx$$

$$\ln(y+2) = x + c$$

Use $A = e^c$

$$y + 2 = Ae^x$$

$$y = Ae^x - 2$$

Note An alternative method involves recalling that $\dfrac{\mathrm{d}x}{\mathrm{d}y} = \dfrac{1}{\frac{\mathrm{d}y}{\mathrm{d}x}}$. The equation is rewritten in

the form $\dfrac{\mathrm{d}x}{\mathrm{d}y} = \mathrm{f}(y)$. Then both sides are integrated with respect to y, giving

$x = \displaystyle\int \mathrm{f}(y)\,\mathrm{d}y$.

$\dfrac{\mathrm{d}y}{\mathrm{d}x} = y + 2 \Rightarrow \quad \dfrac{\mathrm{d}x}{\mathrm{d}y} = \dfrac{1}{y+2}$

> Integrate both sides with respect to y.

$$x = \int \frac{1}{y+2}\,\mathrm{d}y$$

$$x = \ln(y+2) + c$$

$$\ln(y+2) = x - c$$

> Use $A = \mathrm{e}^{-c}$

$$y + 2 = A\mathrm{e}^{x}$$

$$y = A\mathrm{e}^{x} - 2$$

> The resulting general solution is the same; the first method is most commonly used.

Example 8 Solve $\dfrac{\mathrm{d}y}{\mathrm{d}x} = 3y(y+2)$, giving the solution in the form $y = \mathrm{f}(x)$.

Solution

$$\frac{\mathrm{d}y}{\mathrm{d}x} = 3y(y+2)$$

> When separating the variables in this example it is better to leave the 3 where it is and not take it to the denominator with the y terms.

$$\int \frac{1}{y(y+2)}\,\mathrm{d}y = \int 3\,\mathrm{d}x$$

> Split $\dfrac{1}{y(y+2)}$ into partial fractions (Sections 9.5 and 9.6).

Now $\quad \dfrac{1}{y(y+2)} \equiv \dfrac{A}{y} + \dfrac{B}{y+2}$

$$1 \equiv A(y+2) + By$$

When $y = 0$ $\qquad 1 = 2A$

so $\qquad A = \dfrac{1}{2}$

When $y = -2$ $\qquad 1 = -2B$

so $\qquad B = -\dfrac{1}{2}$

$\therefore \qquad \dfrac{1}{y(y+2)} = \dfrac{1}{2}\left(\dfrac{1}{y} - \dfrac{1}{y+2}\right)$

The integration becomes

$$\frac{1}{2}\int\left(\frac{1}{y}-\frac{1}{y+2}\right)dy = \int 3\,dx$$

Multiply through by 2 to eliminate the fraction.

$$\int\left(\frac{1}{y}-\frac{1}{y+2}\right)dy = \int 6\,dx$$

$$\ln y - \ln(y+2) = 6x + c$$

$$\ln\left(\frac{y}{y+2}\right) = 6x + c$$

$$\frac{y}{y+2} = e^{6x+c}$$

Write e^c as A.

$$\frac{y}{y+2} = Ae^{6x}$$

As requested, rearrange to make y the subject of the formula.

$$y = (y+2)Ae^{6x}$$

Take all the terms in y to one side and all other terms to the other side.

$$y = yAe^{6x} + 2Ae^{6x}$$

$$y - yAe^{6x} = 2Ae^{6x}$$

Then take out y as a factor.

$$y(1 - Ae^{6x}) = 2Ae^{6x}$$

$$y = \frac{2Ae^{6x}}{1 - Ae^{6x}}$$

Exercise 14A

This exercise contains integration techniques introduced in Chapters 5 and 6.

1 Find the general solution of these differential equations.

a $\dfrac{dy}{dx} = 4x - 2$ b $\dfrac{dy}{dx} = \dfrac{e^{-x}}{y}$ c $x\dfrac{dx}{dt} = t + 1$

d $\dfrac{dy}{dx} = \dfrac{\cos x}{y}$ e $\dfrac{1}{x}\dfrac{dy}{dx} = \cos^2 y$ f $\dfrac{dv}{dt} + v^2 = 0$

g $x^4\dfrac{dy}{dx} = \sqrt{y}$ h $\dfrac{dy}{dx} = e^y(x+2)$ i $\dfrac{dx}{dy} = \dfrac{2y}{2x-1}$

j $\dfrac{dy}{dx} = \dfrac{x^3}{y^2}$ k $e^{2y}\dfrac{dy}{dx} = x$ l $\cos y\dfrac{dy}{dx} = e^x$

2 Find the general solution of these differential equations.

a $\dfrac{dy}{dx} = y$ b $(x+3)\dfrac{dy}{dx} = y$ c $\dfrac{dy}{dx} = \dfrac{1}{xy}$

d $\cos y\dfrac{dy}{dx} = \dfrac{\sin y}{x}$ e $x\dfrac{dy}{dx} = \cot y$ f $x\dfrac{dy}{dx} = y + 2$

g $\dfrac{dy}{dx} = (2x+1)y$ h $t^2\dfrac{d\theta}{dt} = \theta$ i $\dfrac{dA}{dy} = (y+4)(2A+3)$

3 a Express in partial fractions.

$$\frac{2x+1}{(x+2)(x+1)}$$

b If $\dfrac{1}{2x+1}\dfrac{\mathrm{d}y}{\mathrm{d}x} = \dfrac{y}{(x+2)(x+1)}$

express y in terms of x.

4 a If $x\dfrac{\mathrm{d}y}{\mathrm{d}x} - 2y = 3\dfrac{\mathrm{d}y}{\mathrm{d}x}$

show that $\dfrac{\mathrm{d}y}{\mathrm{d}x} = \dfrac{2y}{x-3}$.

b Solve $x\dfrac{\mathrm{d}y}{\mathrm{d}x} - 2y = 3\dfrac{\mathrm{d}y}{\mathrm{d}x}$.

c Solve $y\dfrac{\mathrm{d}y}{\mathrm{d}x} = 2x + \dfrac{\mathrm{d}y}{\mathrm{d}x}$.

d Solve $x\dfrac{\mathrm{d}y}{\mathrm{d}x} = y + xy$.

5 Solve

$$\frac{\mathrm{d}y}{\mathrm{d}x} = \frac{x^2}{y+3}$$

given that $y = 1$ when $x = 2$.

6 By finding a particular solution of the differential equation

$$\frac{\mathrm{d}y}{\mathrm{d}x} = \frac{3}{10}$$

obtain the equation of the straight line of gradient $\frac{3}{10}$ which passes through $(5, -2)$.

7 A family of parabolas has the differential equation

$$\frac{\mathrm{d}y}{\mathrm{d}x} = 2x - 3$$

Find the equation of the member of the family that passes through $(4, 5)$.

8 a Find the general solution of the differential equation

$$6t\frac{\mathrm{d}s}{\mathrm{d}t} + 1 = 0$$

b Find the particular solution given by the condition $s = 0$ when $t = 2$.

9 a A curve is such that

$$\frac{dy}{dx} = 4xy$$

If the curve passes through $(0, 5)$, find its equation.

b A curve passes through $(4, 0)$ and is such that

$$\frac{dy}{dx} = -\frac{x}{y}$$

Find its equation and describe the curve.

10 Given that

$$\frac{dy}{dx} = e^{x-y}$$

and $y = 1$ when $x = 0$, find y when $x = 1$.

11 Given that

$$(x^2 + 4)\frac{dy}{dx} = 2xy$$

and that $y = 15$ when $x = 1$, show that y is of the form $y = ax^2 + b$ and find the values of a and b.

12 a Show that

$$\int \cos^2 x \, dx = \frac{1}{2}x + \frac{1}{4}\sin 2x + c$$

b Find the general solution of the differential equation

$$\sec^2 x \frac{dy}{dx} = \sec y$$

c Given that $y = \frac{\pi}{2}$ when $x = \frac{\pi}{2}$, find an expression for $\sin y$.

13 a Find the displacement s m of a particle t s after leaving O, where

$$t\frac{ds}{dt} = t^2 + 4$$

b Given that $s = 4\ln 2$ when $t = 2$ and $s = a + b\ln 2$ when $t = 4$, find a and b.

14 a Use integration by parts to find

$$\int x \cos x \, dx$$

b Find the general solution of the differential equation

$$\sec x \frac{dy}{dx} = \frac{x}{y}$$

c Find the particular solution, given that $y = 1$ when $x = 0$.

15 A family of curves is defined by the differential equation

$$\frac{dy}{dx} = -\frac{y}{x}$$

a Find the gradients of the curves at $(2, 2)$ and at $(-1, 0)$.

b Copy and complete the diagram below showing the tangent field at points with integer coefficients between -2 and $+2$, excluding the origin.

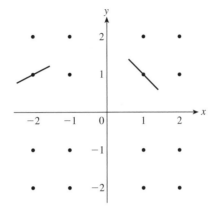

c Solve the differential equation assuming that x and y are both positive and give, in simplified form, the equation of the curve from this family which passes through the point $(1, 1)$. Sketch this curve on the same diagram as the tangent field.

d Another family of curves is defined by

$$\frac{dy}{dx} = -\frac{(y-2)}{(x+1)}$$

Draw another sketch showing a typical member of this family for $x > -1$, labelling any important features.

MEI

14.2 Forming differential equations

Differential equations often occur when a mathematical model is used to describe a physical situation.

Example 9 A circular inkblot, with radius r cm, is enlarging such that the rate of increase of the radius at time t seconds is given by

$$\frac{dr}{dt} = \frac{0.1}{r}$$

Initially the radius of the inkblot is 0.4 cm. Find the area of the inkblot after 2 seconds.

Solution

$$\frac{dr}{dt} = \frac{0.1}{r}$$

> Separate the variables.

$$\int r \, dr = \int 0.1 \, dt$$

> Integrate to obtain the general solution.

$$\frac{r^2}{2} = 0.1t + c$$

> Multiply throughout by 2 to obtain a simpler format.

$$r^2 = 0.2t + d$$

When $t = 0$, $r = 0.4$

> Substitute the initial condition to find d.

so $0.4^2 = d$

$$d = 0.16$$

$\therefore$ $r^2 = 0.2t + 0.16$

When $t = 2$ $r^2 = 0.4 + 0.16$

> There is no need to find r, since the value of r^2 is required to find the area.

$$= 0.56$$

so   area $= \pi r^2$

$$= 0.56\pi$$

$$= 1.76 \, \text{cm}^2 \; (3 \text{ s.f.})$$

Example 10   A spherical balloon, which is being inflated, has radius r cm at time t seconds. It takes 3 seconds to inflate the balloon to a radius of 16 cm from its initial value of 1 cm. In a simple model, the rate of increase of r is taken to be proportional to $1/r^2$. Express this statement as a differential equation, and find the time it would take to inflate the balloon to a radius of 20 cm. Give your answer in seconds, to two significant figures.

Solution

$$\frac{dr}{dt} \propto \frac{1}{r^2}$$

$$\frac{dr}{dt} = \frac{k}{r^2}, \text{ where } k > 0$$

> k is the proportionality constant and since $\dfrac{dr}{dt}$ is increasing, $k > 0$.

$$\int r^2 \, dr = \int k \, dt$$

> Separate the variables.

$$\frac{r^3}{3} = kt + c$$

> Multiply by 3 to obtain a simpler format. Note that another letter, say d, can be used for the integration constant, but the proportionality constant k should not be altered.

$$r^3 = 3kt + d$$

When $t = 0$, $r = 1$, so $d = 1$

> Use the given conditions to find k and d.

When $t = 3$, $r = 16$, so

$$16^3 = 9k + 1$$

> Substitute the value found for d.

$$k = \frac{4095}{9} = 455$$

$\therefore$ $r^3 = 1365t + 1$

When $r = 20$

$$20^3 = 1365t + 1$$
$$t = \frac{7999}{1365}$$
$$= 5.9 \ (2 \ \text{s.f.})$$

It would take 5.9 seconds to inflate the balloon to a radius of $20\,\text{cm}$.

14.3   Exponential growth and decay

Exponential growth and decay (page 269) are two special situations that often arise, especially in natural sciences, economics and social science.

Exponential growth

Example 11   A population, y, is increasing at a rate proportional to the total population at the time and is such that when $t = 0$, $y = y_0$.

a   Show that $y = y_0 e^{kt}$, where k is a positive constant.

b   If the population doubles in 10 years, find by what factor the initial population has been multiplied after a further 20 years.

Solution   **a** $\dfrac{dy}{dt} \propto y \Rightarrow \dfrac{dy}{dt} = ky$, where $k > 0$. k is the proportionality constant.

$$\int \frac{1}{y} \, dy = \int k \, dt$$ Separate the variables.

$$\ln y = kt + c$$ c is the integration constant.

$$y = A e^{kt}$$ $e^{kt+c} = e^{kt} \times e^c = A e^{kt}$ (writing e^c as A)

When $t = 0$, $y = y_0$, so   See Example 5c on page 340.

$$y_0 = A e^0 = A$$

$\therefore \qquad y = y_0 e^{kt}$

b   When $t = 10$, $y = 2y_0$, so   This condition allows k to be found.

$$2y_0 = y_0 e^{k \times 10}$$

$$e^{10k} = 2$$ Use $e^a = b \Leftrightarrow a = \ln b$

$$10k = \ln 2$$

$$k = \frac{1}{10} \ln 2$$

$\therefore \qquad y = y_0 e^{\left(\frac{1}{10} \ln 2 \right) t}$

After a further 20 years, $t = 30$, so

$$y = y_0 e^{\left(\frac{1}{10}\ln 2\right) \times 30}$$

$$= y_0 e^{3\ln 2}$$

| $3\ln 2 = \ln 2^3 = \ln 8$ and $e^{\ln 8} = 8$ |

$$= 8y_0$$

After 30 years, the population has increased by a factor of 8.

 In general, the differential equation

$$\frac{dy}{dt} = ky$$

where $k > 0$, represents exponential growth.
If $y = y_0$ when $t = 0$, then $y = y_0 e^{kt}$.

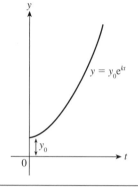

Example 12   A mathematical model for a number of bacteria, n, in a culture, states that n is increasing at a rate proportional to the number present.

At 11.00 a.m., there are 500 bacteria; at 11.30 a.m., there are 735 bacteria. According to the model, at what time will the number of bacteria have increased to 1000?

Solution

$$\frac{dn}{dt} \propto n \Rightarrow \frac{dn}{dt} = kn, \text{ where } k > 0.$$

| Separate the variables. |

$$\int \frac{1}{n} \, dn = \int k \, dt$$

$$\ln n = kt + c$$

$$n = n_0 e^{kt}$$

| This standard result for exponential growth can usually be quoted. |

where n_0 is the initial number.

| n_0 is the number of bacteria present when $t = 0$, i.e. at 11.00 a.m. |

When $t = 0$, $n = 500$, so

$$n_0 = 500$$

$$\therefore \quad n = 500e^{kt}$$

Considering t measured in minutes, when $t = 30$, $n = 735$, so

$$735 = 500e^{30k}$$

$$1.47 = e^{30k}$$

$$30k = \ln 1.47$$

$$k = \frac{1}{30}\ln 1.47$$

$$\therefore \quad n = 500e^{\left(\frac{1}{30}\ln 1.47\right)t}$$

| This is the exact value of k. Although it can be calculated to a particular degree of accuracy (e.g. 0.0128 to 3 s.f.), it is usually better to leave it in the exact form. |

When $n = 1000$

$$1000 = 500e^{\left(\frac{1}{30}\ln 1.47\right)t}$$

$$2 = e^{\left(\frac{1}{30}\ln 1.47\right)t}$$

$$\ln 2 = \frac{1}{30}\ln(1.47)t$$

$$t = \frac{30\ln 2}{\ln 1.47} = 53.97\ldots$$

According to the model, the number of bacteria will have increased to 1000 by 11.54 a.m.

Exponential decay

Example 13   Radium decays at a rate proportional to the amount, y, present at time t. If it takes 1600 years for half the original amount to decay, find the percentage of the original amount that remains after 200 years.

Solution

$$\frac{dy}{dt} \propto y \Rightarrow \frac{dy}{dt} = -ky, \text{ where } k > 0$$

Note: With $k > 0$, the negative is included to indicate decay, rather than growth.

$$\int \frac{1}{y}\,dy = -\int k\,dt$$

$$\ln y = -kt + c$$

$$y = Ae^{-kt}$$

This is similar to the general solution for exponential growth, but note the negative, indicating decay.

If $y = y_0$ when $t = 0$, $y_0 = A$

$$\therefore \qquad y = y_0e^{-kt}$$

When $t = 1600$, $y = \frac{1}{2}y_0$ so

$$\frac{1}{2}y_0 = y_0e^{-1600k}$$

Cancel y_0.

$$2 = e^{1600k}$$

Rewriting $\frac{1}{2} = e^{-1600k}$ avoids negatives.

$$\ln 2 = 1600k$$

$$\frac{1}{2} = \frac{1}{e^{1600k}} \Rightarrow 2 = e^{1600k}$$

$$k = \frac{\ln 2}{1600}$$

$$\therefore \qquad y = y_0e^{-\left(\frac{\ln 2}{1600}\right)t}$$

When $t = 200$

$$y = y_0e^{-\left(\frac{\ln 2}{1600}\right) \times 200}$$

$$= 0.917\ldots y_0$$

After 200 years, approximately 92% of the original radioactive radium still remains.

> In general, the differential equation
>
> $$\frac{dy}{dt} = -ky$$
>
> where $k > 0$, represents exponential decay.
>
> If $y = y_0$ when $t = 0$, then $y = y_0 e^{-kt}$.

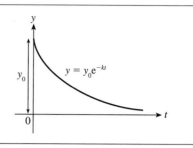

The rate of decay of radioactive material is usually described in terms of **half-life**, where half-life is the time it takes for half the material to decay.

Note The half-life of radium is 1600 years.

In general, if the half-life is T, then when $y = \frac{1}{2}y_0$, $t = T$, so

$$\frac{1}{2}y_0 = y_0 e^{-kT}$$

$$\frac{1}{2} = \frac{1}{e^{kT}}$$

$$\therefore \quad 2 = e^{kT}$$

$$T = \frac{1}{k}\ln 2$$

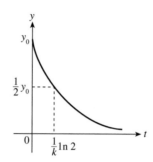

Note: The half-life does not depend on the original amount. The half-life is the same, irrespective of when observations begin.

Example 14 According to Newton's law of cooling, the rate at which the temperature of a body falls is proportional to the amount by which the temperature exceeds that of its surroundings.

A room is at a constant temperature of 20°C. An object has temperature 80°C when it is brought into the room and 5 minutes later its temperature is 65°C. What will its temperature be after a further interval of 5 minutes?

Solution Let $T°C$ be the temperature of the object at time t minutes after being brought into the room.

> Form the differential equation, using $k > 0$. The negative sign indicates that the temperature is falling.

$$\frac{dT}{dt} = -k(T - 20)$$

> Separate the variables.

$$\therefore \quad \int \frac{1}{T - 20}\, dT = -\int k\, dt$$

$$\ln(T - 20) = -kt + c$$

$$T - 20 = Ae^{-kt}$$

> $T - 20 = e^{-kt+c} = e^{-kt} \times e^c$. Then write $A = e^c$.

When $t = 0$, $t = 80$, so

> Use the conditions to find the values of the constants.

$$80 - 20 = A$$

$$A = 60$$

$$\therefore \quad T - 20 = 60e^{-kt}$$

When $t = 5$, $T = 65$, so

$$45 = 60e^{-5k}$$

$$e^{-5k} = \frac{3}{4}$$

$$e^{5k} = \frac{4}{3}$$

$$5k = \ln\left(\frac{4}{3}\right)$$

$$k = \frac{1}{5}\ln\left(\frac{4}{3}\right)$$

Write $e^{-5k} = \frac{3}{4}$ as $e^{5k} = \frac{4}{3}$ to avoid negatives.

When $t = 10$

$$T - 20 = 60e^{-\left(\frac{1}{5}\ln\frac{4}{3}\right) \times 10}$$

$$T = 20 + 60e^{-2\ln\frac{4}{3}}$$

$$= 20 + 60 \times \frac{9}{16}$$

$$= 53.75$$

Take care with the manipulation:
Use $e^{\ln a} = a$ and $-2\ln\left(\frac{4}{3}\right) = \ln\left(\frac{4}{3}\right)^{-2} = \ln\left(\frac{9}{16}\right)$

After a further 5 minutes the temperature of the object is 53.75°C.

Exercise 14B

1 In a scientific study, the size, P of a particular population at time t hours is being studied. Initially $P = 560$ and after 6 hours, P is found to be 1218. In a simple model of population growth, the rate of increase of the population is taken to be proportional to the population at that time.
Using this model, predict

  a the size of the population 24 hours after the start of the experiment

  b how long it will take for the population to increase tenfold

2 The mass, m g, of a radioactive substance at time t seconds, is such that

$$\frac{dm}{dt} = -m$$

When $t = 0$, $m = 4$.

  a Find the mass when $t = 0.5$.

  b At what value of t is the mass 2 g?

  c What is the half-life of the radioactive substance?

3 **a** Express in partial fractions:

$$\frac{1}{(2-x)(1-x)}$$

  b Find

$$\int \frac{1}{(2-x)(1-x)}\,dx$$

c In a chemical reaction, at time t seconds, a solution contains x mg of a particular substance. The rate of increase of the mass of the substance, per second, is given by the equation

$$\frac{dx}{dt} = (2 - x)(1 - x)$$

and $x = 0$ when $t = 0$

i Find x when $t = 0.5$.

ii Express x in terms of t.

iii As $t \to \infty$, x tends to a limiting value. Find this limiting value.

4 When a spherical mint is sucked, a simple model gives the rate of decrease of its radius as inversely proportional to the square of the radius. Initially, the radius of the mint is 5 mm and after 5 minutes the radius is 4 mm.

a Using this model, find the radius of the mint 8 minutes after being put in the mouth.

b At what time does the mint dissolve completely?

5 a Using partial fractions, find

$$\int \frac{1}{4 - x^2} \, dx$$

b The amount of chemical present in a particular reaction at time t is x. If

$$\frac{dx}{dt} = 10(4 - x^2)$$

and $x = 0$ when $t = 0$, express x in terms of t.

6 An economist studying the price changes of a particular commodity uses the model for which

$$\frac{dP}{dt} = kPt$$

where £P is the price at time t months and k is a positive constant.

a Given that $P = 100$ when $t = 0$, show that $P = 100e^{0.5kt^2}$.

b Find the price of the commodity, predicted by the model, at time 9 months, given that the price after 2 months is £102.

7 The surface area of a pond is 500 m². The area of weed on the surface of the pond is increasing at a rate proportional to its area at that instant. The area of the weed when observations first began was 30 m² and it was 51 m² after 3 days.

a Find the area of weed 9 days after observations began.

b When is 50% of the pond's surface covered by weed?

8 A sphere of radius R melts such that the radius after time t is r and the rate of change of the radius is given by

$$\frac{dr}{dt} = -r$$

If T is the time taken for the radius of the sphere to decrease to $\frac{1}{4}R$, show that $T = \ln 4$.

9 According to Newton's law of cooling, the rate at which the temperature of a body falls is proportional to the amount by which its temperature exceeds that of the surroundings. Suppose that the temperature of an object falls from 200°C to 100°C in 40 minutes, in a surrounding temperature of 10°C. Show that after t minutes the temperature, T°C of the body is given by

$$T = 10 + 190e^{-kt} \quad \text{where} \quad k = \frac{1}{40}\ln\left(\frac{19}{9}\right)$$

Calculate the time it takes to reach 50°C.

10 A simple model suggests that the rate at which the value of a car is depreciating is proportional to the value of the car at that instant. If a car costs £14 000 when new and is worth £10 000 after two years, predict how much it will be worth, according to the model, after a further three years. Give your answer to the nearest £10.

11 In a chemical reaction, hydrogen peroxide is converted into water and oxygen. At time t after the start of the reaction, the quantity of hydrogen peroxide that has **not** been converted is x and the rate at which x is decreasing is proportional to x.

a Write down a differential equation involving x and t.

b Given that $x = x_0$ initially, show that

$$\ln\frac{x}{x_0} = -kt$$

where k is a positive constant.

c In an experiment, the time taken for the hydrogen peroxide to be reduced to half of its original quantity was 3 minutes. Find, to the nearest minute, the time that would be required to reduce the hydrogen peroxide to one-tenth of its original quantity.

d i Express x in terms of x_0 and t.

   ii Sketch a graph showing how x varies with t.   *AQA*

12 A cylindrical tank with a horizontal circular base is leaking. At time t minutes, the depth of oil in the tank is h metres. It is known that $h = 10$ when $t = 0$ and that $h = 5$ when $t = 40$.

Student A assumes that an appropriate model is to take the rate of change of h with respect to t to be constant.

a Find a relation between h and t for this model.

Student B uses the model that the rate of change of h with respect to t is proportional to h.

b Form a differential equation and, using the conditions given, solve it to find Student B's relation between h and t.

c Find, for each model, the value of h when $t = 60$.

d Briefly explain which model you would use.   *London*

1 Find the general solution of each of these differential equations.

a $y\dfrac{dy}{dx} = e^{3x} + 2$

b $\dfrac{dy}{dx} + 3y = x\dfrac{dy}{dx}$

c $t^2\dfrac{dx}{dt} = x + 1$

d $\dfrac{dy}{dx} + \cos x - 1 = 0$

e $\sqrt{y}\dfrac{dy}{dx} = \dfrac{y}{x}$

f $\dfrac{d\theta}{dt} = \theta(t + 1)$

2 Given that $y = 1$ when $x = 1$, find, in the form $y = f(x)$, the particular solutions of these differential equations.

a $\dfrac{dy}{dx} = \dfrac{x}{y^2}$

b $\dfrac{dy}{dx} = xy^2$

c $\dfrac{dy}{dx} = \dfrac{y^2}{x}$

d $\dfrac{dy}{dx} = \dfrac{1}{xy^2}$

3 The gradient of the curve $y = f(x)$ is inversely proportional to the square root of x and the curve passes through $(0, 3)$ and $(4, 23)$. Find the equation of the curve.

4 Given that $y = 1$ when $x = 0$, express y in terms of x if

a $\dfrac{dy}{dx} = \dfrac{e^x}{e^y}$

b $\dfrac{dx}{dy} = \dfrac{e^x}{e^y}$

5 Solve, for y in terms of x, the differential equation

$$x\dfrac{dy}{dx} = 2(3 - y)$$

given that $y = 2$ when $x = 2$.

6 Express

$$\dfrac{1}{(1 + x)(3 + x)}$$

in partial fractions. Hence find the solution of the differential equation

$$\dfrac{dy}{dx} = \dfrac{y}{(1 + x)(3 + x)}$$

where $x > -1$, given that $y = 2$ when $x = 1$. Express your answer in the form $y = f(x)$.

7 Solve the differential equation $\dfrac{dy}{dx} = xye^{3x}$, given that $y = 1$ when $x = 0$.

8 A metal rod is 60 cm long and is heated at one end. The temperature at a point on the rod at distance x cm from the heated end is denoted by $T°C$. At a point half way along the rod, $T = 290$ and $\dfrac{dT}{dx} = -6$.

a In a simple model for the temperature of the rod, it is assumed that $\dfrac{dT}{dx}$ has the same value at all points on the rod. For this model, express T in terms of x and hence determine the temperature difference between the ends of the rod.

b In a more refined model, the rate of change of T with respect to x is taken to be proportional to x. Set up a differential equation for T, involving a constant of proportionality k.

Solve the differential equation and hence show that, in this refined model, the temperature along the rod is predicted to vary from 380°C to 20°C.

Cambridge

9 In a certain pond, the rate of increase of the number of fish is proportional to the number of fish, n, present at time t. Assuming that n can be regarded as a continuous variable, write down a differential equation relating n and t, and hence show that

$$n = Ae^{kt}$$

where A and k are constants.

In a revised model, it is assumed also that fish are removed from the pond, by anglers and by natural wastage, at the constant rate of p per unit time, so that

$$\frac{\mathrm{d}n}{\mathrm{d}t} = kn - p$$

Given that $k = 2$, $p = 100$ and that initially there were 500 fish in the pond, solve this differential equation, expressing n in terms of t.

Give a reason why this revised model is not satisfactory for large values of t.

Cambridge

There is a 'Test yourself' exercise (Ty14) and an 'Extension exercise' (Ext14) on the CD-ROM.

➤ **Key points**

Differential equations

The **general solution** of a first-order differential equation contains one arbitrary integration constant. Geometrically it is represented by a family of curves.

A **particular solution** is a specific member of the family of curves and is obtained by finding the value of the arbitrary integration constant using additional information.

Solving first-order differential equations by separating the variables
A differential equation that can be written in the form

$$f(y)\frac{dy}{dx} = g(x)$$

can be solved by separating the variables, where

$$\int f(y)\,dy = \int g(x)\,dx$$

When $f(y) = 1$, i.e. $\dfrac{dy}{dx} = g(x)$, then $y = \int g(x)\,dx$

When $g(x) = 1$, i.e. $f(y)\dfrac{dy}{dx} = 1$, then $\int f(y)\,dy = \int 1\,dx \Rightarrow \int f(y)\,dy = x + c$

Exponential growth and decay

$\dfrac{dy}{dt} = ky$, where $k > 0$, represents **exponential growth**.

If $y = y_0$ when $t = 0$, then $y = y_0 e^{kt}$.

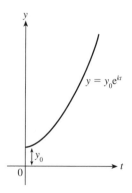

$\dfrac{dy}{dt} = -ky$, where $k > 0$, represents **exponential decay**.

If $y = y_0$ when $t = 0$, then $y = y_0 e^{-kt}$.

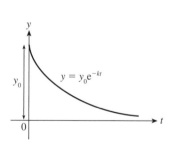

15 Vectors

Vectors have their origin in the middle of the nineteenth century. In recent years they have become an essential tool for engineers, physicists and other scientists. They provide a concise notation for dealing with geometrical problems and also assist in the visualisation of physical ideas.

This chapter introduces the idea of vector quantities and their use in geometry.

After working through the chapter you should

- understand the difference between a scalar and a vector quantity
- be able to find the magnitude of a vector
- be able to apply the rules of vector algebra and interpret them geometrically
- understand the idea of position and displacement vectors and use them to prove geometrical results
- be able to find the distance between two points
- be able to find the scalar product of two vectors
- know how to find the vector equation of a line in both two and three dimensions
- be able to use the scalar product to find the angle between two lines
- be able to determine whether a pair of lines is intersecting, parallel or skew and, in the case of intersecting lines, find the coordinates of the point of intersection
- be able to find the perpendicular distance from a point to a line.

15.1 Vector geometry

A quantity which has both **magnitude** (i.e. size) and **direction** is called a **vector quantity** or **vector**. A quantity which is completely specified by its magnitude alone is called a **scalar quantity** or **scalar**. Velocity is an example of a vector quantity; the velocity of an aircraft might be $750 \, \text{km h}^{-1}$ on a bearing of $274°$. In contrast, speed is a scalar quantity; the speed of the aircraft is $750 \, \text{km h}^{-1}$.

Vectors are particularly important in mechanics, where they are used to represent many quantities including velocity, acceleration, force and momentum. Vectors are also used in geometry to represent displacements.

> **A quantity which has both magnitude and direction is called a vector.**

Vector notation

Arrows are used to represent vectors.

The vector representing the movement from A to B is called a **displacement vector** and is written $\overrightarrow{AB}$. Alternatively, this vector may be represented by a single lower case letter, for example **u**. In print a vector is represented by bold type, **u**, in handwriting an underlined, lower case letter is used, ṵ.

Suppose that A is the point (2, 1) and B is the point (5, 7), then the vector $\overrightarrow{AB}$ representing the displacement from A to B can be written as the column vector

$$\overrightarrow{AB} = \begin{pmatrix} 3 \\ 6 \end{pmatrix} \quad \text{or} \quad \overrightarrow{AB} = \begin{bmatrix} 3 \\ 6 \end{bmatrix}$$

(either parentheses or square brackets can be used).

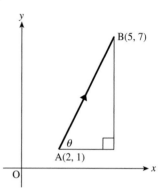

The upper number is the *increase* in the x-coordinate and the lower number is the *increase* in the y-coordinate.

Notice that

$$\overrightarrow{BA} = \begin{pmatrix} -3 \\ -6 \end{pmatrix} = -\overrightarrow{AB}$$

In general, for any points P and Q,

$$\overrightarrow{QP} = -\overrightarrow{PQ}$$

The **magnitude**, or **modulus**, of the vector $\overrightarrow{AB}$ is written as $|\overrightarrow{AB}|$ or AB. In the case of a single letter representing a vector, **u** say, its magnitude is written as $|\mathbf{u}|$ or u.

In the example above, the magnitude of the vector $\overrightarrow{AB}$ can be found using Pythagoras' theorem

$$|\overrightarrow{AB}|^2 = 3^2 + 6^2 = 45$$

$$\therefore \quad |\overrightarrow{AB}| = \sqrt{45} = 3\sqrt{5}$$

The direction of a vector, in two dimensions, is usually given as an angle to the positive x-axis. In this case,

$$\tan\theta = \frac{6}{3} = 2$$

$$\theta = 63.4° \text{ (to 1 d.p.)}$$

The vector $\overrightarrow{AB}$ has magnitude $3\sqrt{5}$ and it is inclined at an angle of 63.4° to the positive x-axis.

Scalar multiplication

Consider the two vectors

$$\mathbf{a} = \begin{pmatrix} 2 \\ -1 \end{pmatrix} \qquad \text{and} \qquad \mathbf{b} = \begin{pmatrix} 4 \\ -2 \end{pmatrix}$$

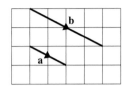

From the diagram, it can be seen that $\mathbf{b}$ is parallel to $\mathbf{a}$ and twice as long. This can be written as

$$\mathbf{b} = 2\mathbf{a}$$

In general, the vector $k\mathbf{a}$ is a vector with k times the magnitude of $\mathbf{a}$, and parallel to $\mathbf{a}$. If k is positive, $k\mathbf{a}$ is in the same direction as $\mathbf{a}$; if k is negative, $k\mathbf{a}$ is in the opposite direction to $\mathbf{a}$. Note also that

$$|\lambda\mathbf{a}| = \lambda|\mathbf{a}|$$

A vector with magnitude zero, representing no displacement, is called the **zero vector** and is written **0**. The zero vector has no direction.

A **unit vector** has magnitude 1. The notation $\hat{\mathbf{a}}$ is used for a unit vector in the direction of the vector $\mathbf{a}$. The magnitude of the vector $\mathbf{a}$ is $|\mathbf{a}|$ and so a unit vector in the direction of $\mathbf{a}$ is

$$\hat{\mathbf{a}} = \frac{\mathbf{a}}{|\mathbf{a}|}$$

Any vector in two dimensions can be written in terms of the standard unit **base vectors**

$$\mathbf{i} = \begin{pmatrix} 1 \\ 0 \end{pmatrix} \qquad \text{and} \qquad \mathbf{j} = \begin{pmatrix} 0 \\ 1 \end{pmatrix}$$

For example, $\overrightarrow{OA}$ can be written as

the column vector $\begin{pmatrix} 3 \\ 2 \end{pmatrix}$ or in terms

of the base vectors as

$$\overrightarrow{OA} = 3\mathbf{i} + 2\mathbf{j}$$

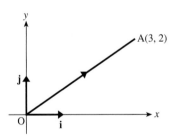

In three dimensions, the standard base vectors are

$$\mathbf{i} = \begin{pmatrix} 1 \\ 0 \\ 0 \end{pmatrix} \qquad \mathbf{j} = \begin{pmatrix} 0 \\ 1 \\ 0 \end{pmatrix} \qquad \mathbf{k} = \begin{pmatrix} 0 \\ 0 \\ 1 \end{pmatrix}$$

and, again, any other vector can be expressed in terms of these base vectors.

For example, in the diagram shown,

$$\mathbf{p} = \begin{pmatrix} 5 \\ 2 \\ -3 \end{pmatrix} \qquad \text{or} \qquad \mathbf{p} = 5\mathbf{i} + 2\mathbf{j} - 3\mathbf{k}$$

Notice that Pythagoras' theorem can still be used to find the magnitude of $\mathbf{p}$.

$$p = \sqrt{5^2 + 2^2 + 3^2} = \sqrt{38}$$

In general, if $\mathbf{a} = x\mathbf{i} + y\mathbf{j} + z\mathbf{k}$ then

$$a = |\mathbf{a}| = \sqrt{x^2 + y^2 + z^2}$$

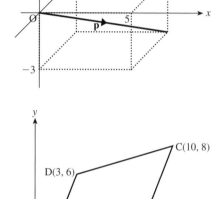

It follows that if the distance between two points (x_1, y_1, z_1) and (x_2, y_2, z_2) is d then
$$d^2 = (x_1 - x_2)^2 + (y_1 - y_2)^2 + (z_1 - z_2)^2$$

Consider the parallelogram ABCD shown in the diagram.

Notice that

$$\overrightarrow{AB} = \overrightarrow{DC} = \begin{pmatrix} 7 \\ 2 \end{pmatrix}$$

$\overrightarrow{AB}$ and $\overrightarrow{DC}$ are **equal vectors**, i.e. they have the same magnitude and direction.

Since ABCD is a parallelogram $\overrightarrow{AD}$ and $\overrightarrow{BC}$ are also equal.

In general, any vectors that are equal to each other can be represented by the same single letter.

Vectors which have a specific line of action are called **localised vectors**. In the example above, $\overrightarrow{AD}$ and $\overrightarrow{BC}$ are both localised vectors. In contrast, the vector $\mathbf{p} = 7\mathbf{i} + 2\mathbf{j}$ is a **free vector**. Its line of action is not specified and it could be drawn anywhere parallel to $\overrightarrow{AB}$.

Two vectors are parallel if one is a multiple of the other. This is best determined by considering the ratios of their separate components.

If $x_1\mathbf{i} + y_1\mathbf{j} + z_1\mathbf{k}$ and $x_2\mathbf{i} + y_2\mathbf{j} + z_2\mathbf{k}$ are parallel then $\dfrac{x_1}{x_2} = \dfrac{y_1}{y_2} = \dfrac{z_1}{z_2}$.

Addition and subtraction of vectors

Suppose that A, B and C are the points (1, 1), (8, 2) and (3, 5) respectively, as shown in the diagram.

Consider the paths that can be taken to travel from A to C.

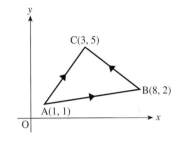

Travelling directly gives the vector

$$\overrightarrow{AC} = \begin{pmatrix} 2 \\ 4 \end{pmatrix}$$

Alternatively it is possible to go from A to B and then from B to C.

Vector addition

Vector addition is defined to mean one displacement followed by another and so, in this case

$$\overrightarrow{AC} = \overrightarrow{AB} + \overrightarrow{BC}$$

$\overrightarrow{AC}$ is called the **resultant** of the vectors $\overrightarrow{AB}$ and $\overrightarrow{BC}$.

Notice that this definition makes sense when the column vectors are considered.

$$\begin{pmatrix} 2 \\ 4 \end{pmatrix} = \begin{pmatrix} 7 \\ 1 \end{pmatrix} + \begin{pmatrix} -5 \\ 3 \end{pmatrix}$$

Notice also that, it does not matter in what order two vectors are added. From the diagram

$$\mathbf{c} = \mathbf{a} + \mathbf{b}$$

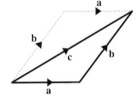

This is often called the **triangle law** for vector addition.

However, if the dotted vectors are used the result obtained is $\mathbf{c} = \mathbf{b} + \mathbf{a}$ and so,

$$\mathbf{a} + \mathbf{b} = \mathbf{b} + \mathbf{a}$$

Because of this important result, the rule for adding vectors is sometimes called the **parallelogram law** for vector addition.

Vector subtraction

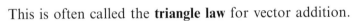

Subtraction of vectors, $\mathbf{a} - \mathbf{b}$, can be thought of as $\mathbf{a} + (-\mathbf{b})$. Hence, if the vectors $\mathbf{a}$ and $\mathbf{b}$ are as shown, the vector $\mathbf{a} - \mathbf{b}$ is formed by adding the vector $-\mathbf{b}$ to the vector $\mathbf{a}$.

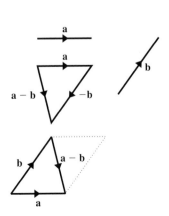

Notice that the vector $\mathbf{a} - \mathbf{b}$ can also be thought of as the other diagonal of the parallelogram formed when $\mathbf{a}$ and $\mathbf{b}$ are added.

Example 1   Find a unit vector in the direction of the vector $5\mathbf{i} - 4\mathbf{j} + 3\mathbf{k}$.

Let $\mathbf{p} = 5\mathbf{i} - 4\mathbf{j} + 3\mathbf{k}$

$\therefore$ $p = \sqrt{5^2 + 4^2 + 3^2} = \sqrt{50} = 5\sqrt{2}$

> $\sqrt{50} = \sqrt{25}\sqrt{2}$

Hence, a unit vector in the direction of $\mathbf{p}$ is

$$\hat{\mathbf{p}} = \frac{\mathbf{p}}{|\mathbf{p}|} = \frac{1}{5\sqrt{2}}(5\mathbf{i} - 4\mathbf{j} + 3\mathbf{k})$$

$$= \frac{1}{\sqrt{2}}\mathbf{i} - \frac{4}{5\sqrt{2}}\mathbf{j} + \frac{3}{5\sqrt{2}}\mathbf{k}$$

> The answer may be left in this form or written as $\frac{\sqrt{2}}{2}\mathbf{i} - \frac{2\sqrt{2}}{5}\mathbf{j} + \frac{3\sqrt{2}}{10}\mathbf{k}$.

Example 2   Given that $\mathbf{p} = 2\mathbf{i} - 3\mathbf{j} + 4\mathbf{k}$ and $\mathbf{q} = -6\mathbf{i} + 5\mathbf{j} + 2\mathbf{k}$, find

a $|\mathbf{p} + \mathbf{q}|$ **b** $|\mathbf{p} - \mathbf{q}|$

Solution   **a** $\mathbf{p} + \mathbf{q} = (2\mathbf{i} - 3\mathbf{j} + 4\mathbf{k}) + (-6\mathbf{i} + 5\mathbf{j} + 2\mathbf{k})$

$$= -4\mathbf{i} + 2\mathbf{j} + 6\mathbf{k}$$

> It is not necessary to bracket $\mathbf{p}$ or $\mathbf{q}$ here, although it is conventional to do so.

$|\mathbf{p} + \mathbf{q}| = \sqrt{4^2 + 2^2 + 6^2} = \sqrt{56} = 2\sqrt{14}$

> *Note:* $|\mathbf{p} + \mathbf{q}| \neq |\mathbf{p}| + |\mathbf{q}|$

b $\mathbf{p} - \mathbf{q} = (2\mathbf{i} - 3\mathbf{j} + 4\mathbf{k}) - (-6\mathbf{i} + 5\mathbf{j} + 2\mathbf{k})$

$$= 8\mathbf{i} - 8\mathbf{j} + 2\mathbf{k}$$

> Here $\mathbf{q}$ *must* be bracketed because it is to be subtracted.

$|\mathbf{p} - \mathbf{q}| = \sqrt{8^2 + 8^2 + 2^2} = \sqrt{132} = 2\sqrt{33}$

> *Note:* $|\mathbf{p} - \mathbf{q}| \neq |\mathbf{p}| - |\mathbf{q}|$

Example 3   Given that $\overrightarrow{AB} = 3\mathbf{i} - 5\mathbf{j} - 7\mathbf{k}$ and $\overrightarrow{CB} = \mathbf{i} - 10\mathbf{j} + 2\mathbf{k}$ find $\overrightarrow{AC}$.

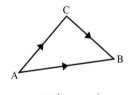

> While not necessary, a diagram can help to show how the vectors *relate* to each other. The diagram does *not* show the actual positions of the points in space.

$$\overrightarrow{AC} = \overrightarrow{AB} + \overrightarrow{BC}$$

$$= \overrightarrow{AB} - \overrightarrow{CB}$$

$$= (3\mathbf{i} - 5\mathbf{j} - 7\mathbf{k}) - (\mathbf{i} - 10\mathbf{j} + 2\mathbf{k})$$

$$= 2\mathbf{i} + 5\mathbf{j} - 9\mathbf{k}$$

Example 4   *In this example components are compared to determine whether pairs of vectors are parallel.*

Determine whether the following pairs of vectors are parallel.

a $\mathbf{p} = 4\mathbf{i} + 3\mathbf{j} - 5\mathbf{k}$ and $\mathbf{q} = 12\mathbf{i} + 9\mathbf{j} - 15\mathbf{k}$

b $\mathbf{p} = 6\mathbf{i} + 8\mathbf{j} - 4\mathbf{k}$ and $\mathbf{q} = 9\mathbf{i} + 12\mathbf{j} - 7\mathbf{k}$

Solution   **a** Compare the ratios of the **i**, **j** and **k**

components: $\dfrac{12}{4} = \dfrac{9}{3} = \dfrac{-15}{-5} = 3$ and

so the vectors are parallel.

> For parallel vectors the three ratios must be equal.

b Compare the ratios of the **i**, **j** and **k**

components: $\dfrac{9}{6} = \dfrac{12}{8} = 1.5$ but $\dfrac{-7}{-4} = 1.75$

and so the vectors are not parallel.

Exercise 15A

1 Find the magnitude of each of these vectors.

  a $3\mathbf{i} + 4\mathbf{j}$ **b** $5\mathbf{i} - 12\mathbf{j}$ **c** $\mathbf{i} + \mathbf{j}$ **d** $-6\mathbf{j}$

  e $\begin{pmatrix} 4 \\ 0 \end{pmatrix}$ **f** $\begin{pmatrix} 2 \\ -3 \end{pmatrix}$ **g** $\begin{pmatrix} -4 \\ 5 \end{pmatrix}$ **h** $4\mathbf{i} + 6\mathbf{j}$

  i $2\mathbf{i} - 2\mathbf{j} + \mathbf{k}$ **j** $4\mathbf{i} - 3\mathbf{j} + 5\mathbf{k}$ **k** $\begin{pmatrix} 1 \\ 1 \\ 1 \end{pmatrix}$ **l** $\begin{pmatrix} 2 \\ -3 \\ 6 \end{pmatrix}$

2 Find the magnitude and direction of each of these vectors.

  a $12\mathbf{i} + 5\mathbf{j}$ **b** $2\mathbf{i} - 2\mathbf{j}$ **c** $\mathbf{i} - \sqrt{3}\mathbf{j}$

  d $-8\mathbf{i} + 15\mathbf{j}$ **e** $-8\mathbf{i}$ **f** $16\mathbf{j}$

3 The vector **p** has magnitude 10 units and is inclined at $150°$ to the x-axis. Express **p** as a column vector.

4 The vector **q** has magnitude 16 units and is inclined at $300°$ to the x-axis. Express **q** in the form $a\mathbf{i} + b\mathbf{j}$.

5 The vector **r** has magnitude $6\sqrt{2}$ units and is inclined at $225°$ to the x-axis. Express **r** in the form $a\mathbf{i} + b\mathbf{j}$.

6 Find unit vectors in the direction of these vectors.

  a $\begin{pmatrix} 5 \\ -12 \end{pmatrix}$ **b** $\begin{pmatrix} 3 \\ 4 \\ -12 \end{pmatrix}$ **c** $3\mathbf{i} - 6\mathbf{j} + 12\mathbf{k}$

  d $2\mathbf{i} + 2\mathbf{j} + 2\mathbf{k}$ **e** $15\mathbf{i} + 20\mathbf{j}$ **f** $\mathbf{i} - 2\mathbf{j} + 5\mathbf{k}$

7 Find a vector of magnitude 12 in the direction of the vector $2\mathbf{i} - \mathbf{j} + 2\mathbf{k}$.

8 Find a vector of magnitude $\sqrt{2}$ in the direction of the vector $3\mathbf{i} - 4\mathbf{j} + 5\mathbf{k}$.

9 Given that $\overrightarrow{AB} = 2\mathbf{i} + 3\mathbf{j} - 4\mathbf{k}$ and $\overrightarrow{BC} = 5\mathbf{i} + 6\mathbf{j} + 8\mathbf{k}$, find $\overrightarrow{AC}$.

10 Given that $\overrightarrow{CD} = -5\mathbf{i} + 6\mathbf{j} + 3\mathbf{k}$ and $\overrightarrow{DE} = 5\mathbf{i} + 3\mathbf{j} - 3\mathbf{k}$, find $\overrightarrow{CE}$.

11 Given that

$$\overrightarrow{PQ} = \begin{pmatrix} 4 \\ 2 \\ -1 \end{pmatrix} \quad \text{and} \quad \overrightarrow{QR} = \begin{pmatrix} -6 \\ -5 \\ 10 \end{pmatrix}$$

find $\overrightarrow{PR}$.

12 Given that $\overrightarrow{EF} = 4\mathbf{i} + 10\mathbf{j} + 12\mathbf{k}$ and $\overrightarrow{EG} = 15\mathbf{i} + 6\mathbf{j} + 18\mathbf{k}$, find $\overrightarrow{FG}$.

13 Given that $\overrightarrow{LN} = 4\mathbf{i} - 5\mathbf{j} + 7\mathbf{k}$ and $\overrightarrow{LM} = 20\mathbf{i} - 14\mathbf{j} + 13\mathbf{k}$, find $\overrightarrow{NM}$.

14 Given that

$$\overrightarrow{XY} = \begin{pmatrix} 2 \\ 4 \\ 6 \end{pmatrix} \quad \text{and} \quad \overrightarrow{ZX} = \begin{pmatrix} -10 \\ -8 \\ 6 \end{pmatrix}$$

find $\overrightarrow{YZ}$.

15 Given that $\mathbf{a} = 3\mathbf{i} + 4\mathbf{j} - 7\mathbf{k}$ and $\mathbf{b} = -2\mathbf{i} - 2\mathbf{j} + 5\mathbf{k}$, find

  a $2\mathbf{a}$ **b** $-3\mathbf{b}$ **c** $\mathbf{a} + \mathbf{b}$ **d** $2\mathbf{a} + \mathbf{b}$

  e $\mathbf{a} + 2\mathbf{b}$ **f** $2\mathbf{a} + 3\mathbf{b}$ **g** $\mathbf{a} - \mathbf{b}$ **h** $2\mathbf{a} - 3\mathbf{b}$

  i $-5\mathbf{a} + 4\mathbf{b}$ **j** $-\mathbf{a} - \mathbf{b}$ **k** $4\mathbf{b} - 5\mathbf{a}$ **l** $6\mathbf{b} - 3\mathbf{a}$

16 Given that $\mathbf{a} = 4\mathbf{i} - 3\mathbf{j} + \alpha\mathbf{k}$ and $\mathbf{b} = \beta\mathbf{i} - 6\mathbf{j} - 10\mathbf{k}$, and that $\mathbf{b} = 2\mathbf{a}$, find the values of α and β.

17 Given that $\mathbf{v} = \lambda\mathbf{i} - 2\mathbf{j} + 6\mathbf{k}$, find λ such that $|\mathbf{v}| = 2\sqrt{14}$.

18 Given that $\mathbf{a} = 5\mathbf{i} + 9\mathbf{j} - 2\mathbf{k}$ and $\mathbf{b} = -3\mathbf{i} - 7\mathbf{j} + 4\mathbf{k}$, find

  a $|\mathbf{a} + \mathbf{b}|$ **b** $|\mathbf{a} - \mathbf{b}|$ **c** $|2\mathbf{a} + \mathbf{b}|$ **d** $|4\mathbf{a} + 2\mathbf{b}|$

19 Given that

$$\mathbf{a} = \begin{pmatrix} 5 \\ 6 \\ -3 \end{pmatrix} \quad \text{and} \quad \mathbf{b} = \begin{pmatrix} 1 \\ 2 \\ -1 \end{pmatrix}$$

find a unit vector in the direction of $7\mathbf{b} - 2\mathbf{a}$.

20 In this diagram, each set of parallel lines is equally spaced and $\overrightarrow{HI} = \mathbf{a}$, $\overrightarrow{HN} = \mathbf{b}$.

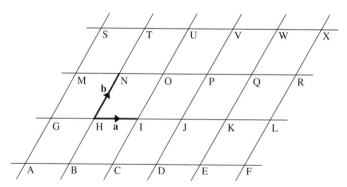

Express these vectors in terms of $\mathbf{a}$ and $\mathbf{b}$.

  a $\overrightarrow{HJ}$ **b** $\overrightarrow{FX}$ **c** $\overrightarrow{DK}$ **d** $\overrightarrow{WP}$

  e $\overrightarrow{OW}$ **f** $\overrightarrow{AR}$ **g** $\overrightarrow{SJ}$ **h** $\overrightarrow{EN}$

  i $\overrightarrow{VH}$ **j** $\overrightarrow{UW} + \overrightarrow{WI}$ **k** $\overrightarrow{HD} + \overrightarrow{DQ} + \overrightarrow{QS}$ **l** $\overrightarrow{TK} + \overrightarrow{KM} + \overrightarrow{MT}$

15.2 Position vectors and geometrical applications

The vector, $\overrightarrow{OA}$, from a fixed origin O to a point A is called the **position vector** of the point A.

The corresponding lower case letter is often used to represent the position vector of a point. i.e. **a** for $\overrightarrow{OA}$, **b** for $\overrightarrow{OB}$, etc.

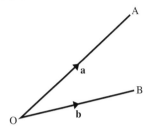

Thus, for example, if A is the point $(3, 2, -1)$, the position vector of A is

$$\mathbf{a} = \overrightarrow{OA} = \begin{pmatrix} 3 \\ 2 \\ -1 \end{pmatrix}$$

or $\quad \mathbf{a} = 3\mathbf{i} + 2\mathbf{j} - \mathbf{k}$

Note It is important to distinguish between $(3, 2, -1)$, meaning the point A, and $\begin{pmatrix} 3 \\ 2 \\ -1 \end{pmatrix}$ the displacement vector from the origin to A.

Example 5 Suppose A and B are two points with position vectors **a** and **b** with respect to some fixed origin O.

Find expressions, in terms of **a** and **b**, for the vector $\overrightarrow{AB}$, and the position vector of the mid-point of the line segment AB.

Solution

$$\begin{aligned} \overrightarrow{AB} &= \overrightarrow{AO} + \overrightarrow{OB} \\ &= -\overrightarrow{OA} + \overrightarrow{OB} \\ &= -\mathbf{a} + \mathbf{b} \\ &= \mathbf{b} - \mathbf{a} \end{aligned}$$

> There are two paths from A to B: either direct, or via O.

> Vectors can be added in either order.

Let M be the mid-point of AB.

$$\begin{aligned} \overrightarrow{OM} &= \overrightarrow{OA} + \overrightarrow{AM} \\ &= \overrightarrow{OA} + \tfrac{1}{2}\overrightarrow{AB} \\ &= \mathbf{a} + \tfrac{1}{2}(\mathbf{b} - \mathbf{a}) \\ &= \mathbf{a} + \tfrac{1}{2}\mathbf{b} - \tfrac{1}{2}\mathbf{a} \\ &= \tfrac{1}{2}\mathbf{a} + \tfrac{1}{2}\mathbf{b} \end{aligned}$$

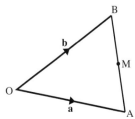

> Use the result $\overrightarrow{AB} = \mathbf{b} - \mathbf{a}$.

> Or write as $\overrightarrow{OM} = \tfrac{1}{2}(\mathbf{a} + \mathbf{b})$.

Note   It is important to distinguish between

$\frac{1}{2}(\mathbf{a} + \mathbf{b})$, the *position vector* of the mid-point of AB and

$\frac{1}{2}(\mathbf{b} - \mathbf{a})$, the vector from A to the *mid-point* of AB.

The two results from Example 5 are particularly important. They apply for any two points, regardless of the position of the origin. (The position vector of *any* point on the line AB can be found by using the Ratio Theorem; see Extension Exercise 15, Question 5 on the CD-ROM.)

Geometrical applications of vectors

Vectors give two pieces of information at once, i.e. the magnitude and the direction of a displacement. This makes them particularly useful in solving geometrical problems as they are able to show clearly whether line segments are parallel or of the same length.

Example 6   In triangle OAB, $\overrightarrow{OA} = \mathbf{a}$ and $\overrightarrow{OB} = \mathbf{b}$.

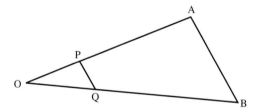

Given that P and Q are points on OA and OB such that OP:PA and OQ:QB are both 1:2, express $\overrightarrow{PQ}$ and $\overrightarrow{AB}$ in terms of $\mathbf{a}$ and $\mathbf{b}$. Hence state the geometrical relationship between PQ and AB.

Solution $\mathbf{p} = \overrightarrow{OP} = \frac{1}{3}\mathbf{a}$ and $\mathbf{q} = \overrightarrow{OQ} = \frac{1}{3}\mathbf{b}$

> P and Q lie one third of the way along OA and OB respectively.

$$\overrightarrow{PQ} = \mathbf{q} - \mathbf{p} \quad \text{and} \quad \overrightarrow{AB} = \mathbf{b} - \mathbf{a}$$

> Using the result from Example 5.

$$= \tfrac{1}{3}\mathbf{b} - \tfrac{1}{3}\mathbf{a}$$

$$= \tfrac{1}{3}(\mathbf{b} - \mathbf{a})$$

Hence $\overrightarrow{PQ} = \frac{1}{3}\overrightarrow{AB}$

PQ is one third the length of AB and parallel to it.

Example 7   ABCD is a skew quadrilateral. Show that the figure formed by joining the mid-points of the four sides is a parallelogram.

Let the mid-points of the sides be E, F, G and H.

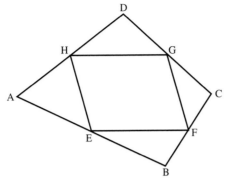

Note: A skew quadrilateral is a quadrilateral which is not a 'special' shape, such as a parallelogram or a kite.

Assume some arbitrary origin O, and let $\overrightarrow{OA} = \mathbf{a}$, $\overrightarrow{OB} = \mathbf{b}$, etc. To show that EFGH is a parallelogram, it is sufficient to show, for example, that EF and HG are parallel and of equal length.

Or EH and FG

$$\overrightarrow{EF} = \mathbf{f} - \mathbf{e}$$

$$= \tfrac{1}{2}\mathbf{b} + \tfrac{1}{2}\mathbf{c} - \left(\tfrac{1}{2}\mathbf{a} + \tfrac{1}{2}\mathbf{b}\right)$$

$$= \tfrac{1}{2}\mathbf{b} + \tfrac{1}{2}\mathbf{c} - \tfrac{1}{2}\mathbf{a} - \tfrac{1}{2}\mathbf{b}$$

$$= \tfrac{1}{2}\mathbf{c} - \tfrac{1}{2}\mathbf{a}$$

$$\overrightarrow{HG} = \mathbf{g} - \mathbf{h}$$

$$= \tfrac{1}{2}\mathbf{c} + \tfrac{1}{2}\mathbf{d} - \left(\tfrac{1}{2}\mathbf{d} + \tfrac{1}{2}\mathbf{a}\right)$$

$$= \tfrac{1}{2}\mathbf{c} + \tfrac{1}{2}\mathbf{d} - \tfrac{1}{2}\mathbf{d} - \tfrac{1}{2}\mathbf{a}$$

$$= \tfrac{1}{2}\mathbf{c} - \tfrac{1}{2}\mathbf{a}$$

$\overrightarrow{AB} = \mathbf{b} - \mathbf{a}$ holds true for any two points.

Using the result from Example 5.

Hence $\overrightarrow{EF} = \overrightarrow{HG}$ and so EF and HG are parallel and of equal length. This is sufficient to prove that EFGH is a parallelogram.

The working for Example 7 could have been simplified a little if the original quadrilateral had been labelled OABC taking one of the vertices as the origin, O. This is a technique that can be used in many similar problems in both two, and three, dimensions.

Example 8   The points P, Q and R have coordinates $(3, -5, 7)$, $(9, -1, -1)$ and $(12, 1, -5)$ respectively. Prove that the three points are collinear.

$$\overrightarrow{PQ} = \begin{pmatrix} 6 \\ 4 \\ -8 \end{pmatrix} = 2\begin{pmatrix} 3 \\ 2 \\ -4 \end{pmatrix} \quad \text{and} \quad \overrightarrow{PR} = \begin{pmatrix} 9 \\ 6 \\ -12 \end{pmatrix} = 3\begin{pmatrix} 3 \\ 2 \\ -4 \end{pmatrix}$$

$$\therefore \qquad \overrightarrow{PR} = \tfrac{3}{2}\overrightarrow{PQ}$$

Hence P, Q and R are collinear.

This proves that PQ and PR are parallel. But PQ and PR also have the point P in common and so the three points must be collinear.

1   In the triangle ABC, $\overrightarrow{AB} = \mathbf{p}$ and $\overrightarrow{AC} = \mathbf{q}$. D is a point on BC such that
    BD:DC = 1:3. Find, in terms of **p** and **q**

    a $\overrightarrow{BC}$ **b** $\overrightarrow{BD}$ **c** $\overrightarrow{AD}$

2   Repeat Question 1 parts **b** and **c**, given that the ratio BD:DC is

    i 1:4   **ii** 2:3   **iii** 4:1   **iv** *l:m*

3   ABCDEF is a regular hexagon. The vectors **p** and **q** are as shown in the
    diagram.

    a Find, in terms of **p** and **q**

    i $\overrightarrow{AC}$

    ii $\overrightarrow{FD}$

    iii $\overrightarrow{FC}$

    b What can be deduced from part **a** about

    i AC and FD

    ii FC and AB?

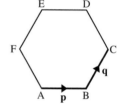

4   OACB is a parallelogram, as shown in the diagram. The position vectors of A
    and B, with respect to the origin O, are **a** and **b**.

    a Find, in terms of **a** and **b**

    i the position vector of C

    ii the position vector of the mid-point of OC

    iii the position vector of the mid-point of AB

    b What fact about a parallelogram can be deduced
    from the answers to part **a**?

5   OACB is a parallelogram, as shown in the diagram. $\overrightarrow{OA} = \mathbf{a}$ and $\overrightarrow{OB} = \mathbf{b}$.
    The point M lies one quarter of the way along OA and N lies three-quarters of
    the way along AC.

    a Find, in terms of **a** and **b**

    i the position vector of M

    ii the vector $\overrightarrow{AC}$

    iii $\overrightarrow{AN}$

    iv the position vector of N

    v $\overrightarrow{MN}$

    vi $\overrightarrow{OC}$

    b What can you deduce about OC and MN from the answers to part **a**?

6   In the triangle OAB, M and N are the mid-points of OA and OB respectively.
    Given that $\overrightarrow{OA} = \mathbf{a}$ and $\overrightarrow{OB} = \mathbf{b}$, find an expression for $\overrightarrow{MN}$ in terms of **a** and
    b. What can be deduced about MN and AB?

7   In the triangle OAB, M and N lie on OA and OB respectively, and
    OM:MA = $1:\lambda$, ON:NB = $1:\lambda$. Given that $\overrightarrow{OA} = \mathbf{a}$ and $\overrightarrow{OB} = \mathbf{b}$, find an
    expression for $\overrightarrow{MN}$ in terms of **a**, **b** and λ. What can be deduced about MN and
    AB?

8 In each of these cases determine, giving reasons, whether the three points given are collinear.

a $(4, 6, -8)$, $(14, 12, -16)$, $(-1, 3, -4)$

b $(2, 3, 5)$, $(6, 9, 15)$, $(8, 12, 20)$

c $(4, 0, -3)$, $(11, 5, 1)$, $(25, 15, 8)$

d $(0, 0, 0)$, $(18, -27, 45)$, $(-10, 15, -25)$

e $(17, 19, 36)$, $(50, 70, 0)$, $(-5, -15, 60)$

9 The object shown in the diagram is a parallelepiped i.e. all the faces are parallelograms. Taking O as the origin, the position vectors of the vertices A, B and C are **a**, **b** and **c** respectively.
Find, in terms of **a**, **b** and **c**

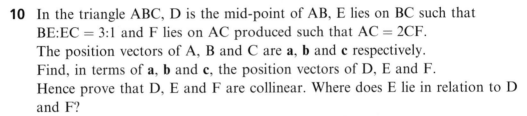

a the position vectors of D, E, F and G

b the position vectors of the mid-points of OG, AF, BE and CD

What fact can be deduced about a parallelepiped from these results?

10 In the triangle ABC, D is the mid-point of AB, E lies on BC such that BE:EC = 3:1 and F lies on AC produced such that AC = 2CF.
The position vectors of A, B and C are **a**, **b** and **c** respectively.
Find, in terms of **a**, **b** and **c**, the position vectors of D, E and F.
Hence prove that D, E and F are collinear. Where does E lie in relation to D and F?

15.3 | Scalar product

There are two main ways of 'multiplying' vectors: the scalar (or dot) product gives a scalar result; the vector (or cross) product gives a vector result. This book considers only the scalar product.

The **scalar** (or **dot**) **product** of two vectors is a way of combining them to give a *scalar* result. The scalar product of the vectors **a** and **b** is written as **a.b** (read as '**a** dot **b**').

> **The scalar product of the vectors a and b is defined as**
>
> $$\mathbf{a.b} = |\mathbf{a}||\mathbf{b}| \cos \theta$$
>
> **where θ is the angle between the two vectors when they are placed together as shown.**

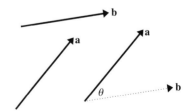

Clearly the angle, θ, must lie between $0°$ and $180°$. If θ is acute, the scalar product will be positive; if θ is obtuse, the scalar product will be negative.

There are a number of important points to note concerning the scalar product.

■ Scalar multiplication is *commutative*.

$$\mathbf{b}.\mathbf{a} = \mathbf{a}.\mathbf{b}$$

This result is straightforward to prove since

$$\mathbf{b}.\mathbf{a} = |\mathbf{b}||\mathbf{a}|\cos\theta = |\mathbf{a}||\mathbf{b}|\cos\theta = \mathbf{a}.\mathbf{b}$$

■ Scalar multiplication is *distributive* over vector addition.

$$\mathbf{a}.(\mathbf{b} + \mathbf{c}) = \mathbf{a}.\mathbf{b} + \mathbf{a}.\mathbf{c}$$

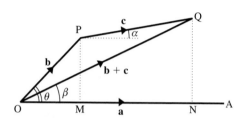

$$
\begin{aligned}
\mathbf{a}.(\mathbf{b} + \mathbf{c}) &= |\mathbf{a}||\mathbf{b} + \mathbf{c}|\cos\beta \\
&= |\mathbf{a}| \times OQ \times \cos\beta \\
&= |\mathbf{a}| \times ON \\
&= |\mathbf{a}| \times (OM + MN) \\
&= |\mathbf{a}| \times (OP\cos\theta + PQ\cos\alpha) \\
&= |\mathbf{a}||\mathbf{b}|\cos\theta + |\mathbf{a}||\mathbf{c}|\cos\alpha \\
&= \mathbf{a}.\mathbf{b} + \mathbf{a}.\mathbf{c}
\end{aligned}
$$

■ $\mathbf{a}.(\lambda\mathbf{b}) = (\lambda\mathbf{a}).\mathbf{b} = \lambda(\mathbf{a}.\mathbf{b})$

The proof of this follows easily from the fact that $|\lambda\mathbf{a}| = \lambda|\mathbf{a}|$

■ If $\mathbf{a}$ and $\mathbf{b}$ are *perpendicular* then, since $\cos 90° = 0$,

$$\mathbf{a}.\mathbf{b} = 0$$

In particular

$$\mathbf{i}.\mathbf{j} = \mathbf{j}.\mathbf{k} = \mathbf{k}.\mathbf{i} = 0$$

■ If $\mathbf{a}.\mathbf{b} = 0$ then either $|\mathbf{a}| = 0$, $|\mathbf{b}| = 0$ or $\mathbf{a}$ and $\mathbf{b}$ are perpendicular.

■ If $\mathbf{a}$ and $\mathbf{b}$ are *parallel* then, since $\cos 0° = 1$,

$$\mathbf{a}.\mathbf{b} = |\mathbf{a}||\mathbf{b}|$$

It follows that, for any vector $\mathbf{a}$,

$$\mathbf{a}.\mathbf{a} = |\mathbf{a}|^2$$

In particular,

$$\mathbf{i}.\mathbf{i} = \mathbf{j}.\mathbf{j} = \mathbf{k}.\mathbf{k} = 1$$

Scalar products may also be calculated using components. Suppose that

$$\mathbf{a} = a_1\mathbf{i} + a_2\mathbf{j} + a_3\mathbf{k} \qquad \text{and} \qquad \mathbf{b} = b_1\mathbf{i} + b_2\mathbf{j} + b_3\mathbf{k}$$

then

$$
\begin{aligned}
\mathbf{a}.\mathbf{b} &= (a_1\mathbf{i} + a_2\mathbf{j} + a_3\mathbf{k}).(b_1\mathbf{i} + b_2\mathbf{j} + b_3\mathbf{k}) \\
&= a_1 b_1 \mathbf{i}.\mathbf{i} + a_1 b_2 \mathbf{i}.\mathbf{j} + a_1 b_3 \mathbf{i}.\mathbf{k} + a_2 b_1 \mathbf{j}.\mathbf{i} + a_2 b_2 \mathbf{j}.\mathbf{j} + a_2 b_3 \mathbf{j}.\mathbf{k} \\
&\quad + a_3 b_1 \mathbf{k}.\mathbf{i} + a_3 b_2 \mathbf{k}.\mathbf{j} + a_3 b_3 \mathbf{k}.\mathbf{k} \\
&= a_1 b_1 + a_2 b_2 + a_3 b_3
\end{aligned}
$$

All terms are zero except for those involving $\mathbf{i}.\mathbf{i}$, $\mathbf{j}.\mathbf{j}$ and $\mathbf{k}.\mathbf{k}$.

Using this result, with the definition of scalar product, allows angles between vectors to be calculated since

$$\mathbf{a.b} = |\mathbf{a}||\mathbf{b}|\cos\theta \quad \Rightarrow \quad \cos\theta = \frac{\mathbf{a.b}}{|\mathbf{a}||\mathbf{b}|}$$

Example 9

Prove that the vectors $3\mathbf{i} - 5\mathbf{j}$ and $10\mathbf{i} + 6\mathbf{j}$ are perpendicular.

$(3\mathbf{i} - 5\mathbf{j}).(10\mathbf{i} + 6\mathbf{j}) = 3 \times 10 - 5 \times 6 = 30 - 30 = 0$

Hence, since clearly neither vector is zero, the two vectors are perpendicular.

Example 10

Given that $\mathbf{a} = 3\mathbf{i} - 2\mathbf{j} + 4\mathbf{k}$ and $\mathbf{b} = \mathbf{i} - 3\mathbf{j} + 2\mathbf{k}$, find the angle between the two vectors, to the nearest tenth of a degree.

$$\mathbf{a.b} = 3 + 6 + 8 = 17$$

$$|\mathbf{a}| = \sqrt{9 + 4 + 16} = \sqrt{29}$$

$$|\mathbf{b}| = \sqrt{1 + 9 + 4} = \sqrt{14}$$

$$\mathbf{a.b} = |\mathbf{a}||\mathbf{b}|\cos\theta$$

$$17 = \sqrt{29}\sqrt{14}\cos\theta$$

$$\cos\theta = \frac{17}{\sqrt{29}\sqrt{14}}$$

$$\theta = 32.5° \text{ (to 1 d.p.)}$$

> Always calculate the dot product first. If it is zero, then the vectors are perpendicular.

> Or, start directly from this result.

Example 11

Find non-zero values of a and b which make the vector $\begin{pmatrix} a \\ -5 \\ b \end{pmatrix}$ perpendicular to the vector $\begin{pmatrix} 6 \\ 2 \\ 1 \end{pmatrix}$.

Since the two vectors are perpendicular

$$\begin{pmatrix} a \\ -5 \\ b \end{pmatrix}.\begin{pmatrix} 6 \\ 2 \\ 1 \end{pmatrix} = 0$$

$$6a - 10 + b = 0$$

$$6a + b = 10$$

One possible solution is $a = 1$, $b = 4$.

> There are infinitely many solutions to this equation, as there are infinitely many vectors perpendicular to $6\mathbf{i} + 2\mathbf{j} + \mathbf{k}$.

Exercise 15C

1. Find **a.b** for each of these pairs of vectors.

   a $\mathbf{a} = 5\mathbf{i} + 4\mathbf{j}, \mathbf{b} = 2\mathbf{i} + 3\mathbf{j}$ **b** $\mathbf{a} = -3\mathbf{i} + 2\mathbf{j}, \mathbf{b} = 6\mathbf{i} + 9\mathbf{j}$

   c $\mathbf{a} = 2\mathbf{i} + 7\mathbf{j} + 6\mathbf{k}, \mathbf{b} = \mathbf{i} - 4\mathbf{j} - 3\mathbf{k}$ **d** $\mathbf{a} = 10\mathbf{i} + 5\mathbf{j} - 4\mathbf{k}, \mathbf{b} = 3\mathbf{i} + 7\mathbf{j} - 2\mathbf{k}$

   e $\mathbf{a} = 5\mathbf{i} + 3\mathbf{j}, \mathbf{b} = 4\mathbf{j} + 8\mathbf{k}$ **f** $\mathbf{a} = -3\mathbf{i} + 8\mathbf{j} + 2\mathbf{k}, \mathbf{b} = -4\mathbf{i} + 2\mathbf{j} + 6\mathbf{k}$

   g $\mathbf{a} = 4\mathbf{i} - 3\mathbf{j} + 2\mathbf{k}, \mathbf{b} = \mathbf{i} + \mathbf{j} + \mathbf{k}$ **h** $\mathbf{a} = 3\mathbf{i} - 4\mathbf{j} + 5\mathbf{k}, \mathbf{b} = 3\mathbf{i} - 4\mathbf{j} + 5\mathbf{k}$

   i $\mathbf{a} = \begin{pmatrix} 7 \\ 3 \\ -1 \end{pmatrix}, \mathbf{b} = \begin{pmatrix} -1 \\ 3 \\ 6 \end{pmatrix}$ **j** $\mathbf{a} = \begin{pmatrix} -10 \\ -5 \\ -8 \end{pmatrix}, \mathbf{b} = \begin{pmatrix} -15 \\ 6 \\ 10 \end{pmatrix}$

2. Given that $\mathbf{a} = 3\mathbf{i} + 6\mathbf{j} - 2\mathbf{k}$, $\mathbf{b} = 4\mathbf{i} - 2\mathbf{j} - 5\mathbf{k}$ and $\mathbf{c} = \mathbf{i} + \mathbf{j} + 3\mathbf{k}$, find

   a $\mathbf{a.b}$ **b** $\mathbf{a.c}$ **c** $\mathbf{a.(b + c)}$

   d $\mathbf{a.b + a.c}$ **e** $\mathbf{a.(b - c)}$ **f** $\mathbf{(a + b).c}$

3. Given that $\mathbf{p} = 4\mathbf{i} + 6\mathbf{j} - 2\mathbf{k}$, $\mathbf{q} = 4\mathbf{i} + 6\mathbf{j} - 3\mathbf{k}$ and $\mathbf{r} = 2\mathbf{i} + 3\mathbf{j}$, find

   a $\mathbf{p.q}$ **b** $\mathbf{p.r}$ **c** $\mathbf{q.r}$ **d** $\mathbf{(p - q).r}$

   e $\mathbf{p.(q + r)}$ **f** $\mathbf{(p - 2r).q}$ **g** $\mathbf{(3p - 2q).r}$

4. Given that $|\mathbf{a}| = 7$, $|\mathbf{b}| = 4$ and that the angle between the vectors **a** and **b** is 150°, find the exact value of **a.b**.

5. Given that $\mathbf{p.q} = -12\sqrt{2}$, $|\mathbf{p}| = 6$ and that the angle between the vectors **p** and **q** is 135°, find the value of $|\mathbf{q}|$.

6. Identify which of these pairs of vectors are parallel, perpendicular or neither.

   a $\mathbf{p} = 5\mathbf{i} + 3\mathbf{j} - 4\mathbf{k}, \mathbf{q} = 7\mathbf{i} - 9\mathbf{j} + 2\mathbf{k}$ **b** $\mathbf{p} = -5\mathbf{i} + 3\mathbf{j} + 7\mathbf{k}, \mathbf{q} = -20\mathbf{i} + 12\mathbf{j} + 28\mathbf{k}$

   c $\mathbf{p} = 4\mathbf{i} - 3\mathbf{j} - \mathbf{k}, \mathbf{q} = 4\mathbf{i} + 5\mathbf{j} - \mathbf{k}$ **d** $\mathbf{p} = 9\mathbf{i} - 5\mathbf{j} + 2\mathbf{k}, \mathbf{q} = 10\mathbf{i} + 16\mathbf{j} - 5\mathbf{k}$

7. Find, in degrees correct to 1 decimal place, the angle between each of these pairs of vectors:

   a $\mathbf{p} = \mathbf{i} + \mathbf{j} + \mathbf{k}, \mathbf{q} = \mathbf{i} - \mathbf{j} + \mathbf{k}$ **b** $\mathbf{p} = \mathbf{i} + 3\mathbf{j} + \mathbf{k}, \mathbf{q} = 3\mathbf{i} + 2\mathbf{j} + 5\mathbf{k}$

   c $\mathbf{p} = 4\mathbf{i} - 3\mathbf{j}, \mathbf{q} = 3\mathbf{i} + 4\mathbf{j}$ **d** $\mathbf{p} = 5\mathbf{i} - \mathbf{j} + 7\mathbf{k}, \mathbf{q} = 12\mathbf{i} + 20\mathbf{j} - 5\mathbf{k}$

   e $\mathbf{p} = \mathbf{i} - \mathbf{k}, \mathbf{q} = \mathbf{j} - \mathbf{k}$ **f** $\mathbf{p} = -6\mathbf{i} - 3\mathbf{j} - 5\mathbf{k}, \mathbf{q} = 4\mathbf{i} + 2\mathbf{j} - 3\mathbf{k}$

   g $\mathbf{p} = 2\mathbf{i} + 5\mathbf{j} - 7\mathbf{k}, \mathbf{q} = -3\mathbf{i} + 6\mathbf{j} + 5\mathbf{k}$ **h** $\mathbf{p} = 3\mathbf{i}, \mathbf{q} = 4\mathbf{i} + 4\mathbf{j} - 2\mathbf{k}$

   i $\mathbf{p} = \begin{pmatrix} 5 \\ -2 \\ 1 \end{pmatrix}, \mathbf{q} = \begin{pmatrix} 5 \\ 2 \\ 1 \end{pmatrix}$ **j** $\mathbf{p} = \begin{pmatrix} 2 \\ 3 \\ -6 \end{pmatrix}, \mathbf{q} = \begin{pmatrix} 2 \\ -2 \\ 1 \end{pmatrix}$

8. Given that the vectors $\mathbf{c} = \lambda\mathbf{i} + 4\mathbf{j} - 3\mathbf{k}$ and $\mathbf{d} = 2\mathbf{i} + 2\lambda\mathbf{j} + 5\mathbf{k}$ are perpendicular, find the value of λ.

9. Given that the vectors $\mathbf{l} = 2\lambda\mathbf{i} + 5\mathbf{j} - \mathbf{k}$ and $\mathbf{m} = \lambda\mathbf{i} - 3\mathbf{j} + \lambda\mathbf{k}$ are perpendicular, find the possible values of λ.

10 Given that the vectors

$$\mathbf{u} = \begin{pmatrix} \lambda + 1 \\ 4 \\ 1 \end{pmatrix} \quad \text{and} \quad \mathbf{v} = \begin{pmatrix} \lambda - 5 \\ \lambda \\ -11 \end{pmatrix}$$

are perpendicular, find the possible values of λ.

11 $\mathbf{p} = \mu\mathbf{i} + 6\mathbf{j} - 14\mathbf{k}$, $\mathbf{q} = 10\mathbf{i} + 15\mathbf{j} - 35\mathbf{k}$. Find the value of μ, if

    a $\mathbf{p}$ and $\mathbf{q}$ are perpendicular

    b $\mathbf{p}$ and $\mathbf{q}$ are parallel

12 Simplify $\mathbf{a}.(\mathbf{b} + \mathbf{c}) - \mathbf{a}.(\mathbf{b} - \mathbf{c})$.

13 Given that $\mathbf{a} = 3\mathbf{i} + 4\mathbf{j} + 12\mathbf{k}$, $\mathbf{b} = 3\mathbf{i} + 4\mathbf{j} - 5\mathbf{k}$, and that θ is the angle between the vectors $\mathbf{a}$ and $\mathbf{b}$, find $\cos\theta$.

14 Find, to the nearest $0.1°$, the angle between the vectors $\mathbf{a}$ and $\mathbf{b}$ if

    a $|\mathbf{a}| = 3$, $|\mathbf{b}| = 6$ and $\mathbf{a}.\mathbf{b} = -9$ **b** $|\mathbf{a}| = 5$, $|\mathbf{b}| = 10$ and $\mathbf{a}.\mathbf{b} = 7$

    c $|\mathbf{a}| = 4$, $|\mathbf{b}| = 7$ and $\mathbf{a}.\mathbf{b} = -10$ **d** $|\mathbf{a}| = 10$, $|\mathbf{b}| = 12$ and $\mathbf{a}.\mathbf{b} = 0$

    e $|\mathbf{a}| = \sqrt{50}$, $|\mathbf{b}| = 2$ and $\mathbf{a}.\mathbf{b} = 10$ **f** $|\mathbf{a}| = \sqrt{6}$, $|\mathbf{b}| = \sqrt{8}$ and $\mathbf{a}.\mathbf{b} = -6$

15 A, B and C are the points $(1, 2, -7)$, $(3, 0, 5)$ and $(2, -1, 5)$ respectively. Find $\angle BAC$ to the nearest $0.1°$.

16 C is the point $(1, 1)$, $D(6, 0)$ and $E(2, 6)$. Show that the triangle CDE is right-angled. Which angle of the triangle is the right angle?

17 The vectors $\mathbf{a}$ and $\mathbf{b}$ are given by

$$\mathbf{a} = \begin{pmatrix} 1 \\ 2 \\ -2 \end{pmatrix} \quad \text{and} \quad \mathbf{b} = \begin{pmatrix} 2 \\ 1 \\ 2 \end{pmatrix}$$

    a Verify that $\mathbf{a}$ and $\mathbf{b}$ are perpendicular.

    b Find a vector $\mathbf{c}$ that is perpendicular to both $\mathbf{a}$ and $\mathbf{b}$.   *NEAB*

18 The points P and Q have position vectors

$$\mathbf{p} = 2\mathbf{i} + m\mathbf{j} - 7\mathbf{k} \quad \text{and} \quad \mathbf{q} = m\mathbf{i} + 6\mathbf{j} + 4\mathbf{k}$$

respectively, where m can take different values.

    a Determine the value of m for which $\mathbf{p}$ and $\mathbf{q}$ are perpendicular.

    b In the case when $m = 4$, find the acute angle between the vectors $\mathbf{p}$ and $\mathbf{q}$, giving your answer to the nearest $0.1°$.   *AEB*

15.4   The vector equation of a line

Consider a line through a point A which is parallel to a vector **b**, as shown in the diagram.

The position vector of A, with respect to a fixed origin O, is **a**. Let R be a general point on the line, with position vector **r**. Then, since the line is parallel to **b**

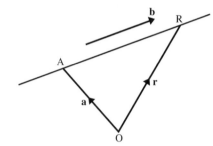

$$\overrightarrow{AR} = t\mathbf{b}$$

where t is some scalar

$$\therefore \quad \mathbf{r} = \overrightarrow{OR} = \overrightarrow{OA} + \overrightarrow{AR}$$

$$= \mathbf{a} + t\mathbf{b}$$

> As different values of the parameter, t, are taken, all points on the line can be obtained.

Note   The equation of the line is not unique. Any other point on the line can be used in place of A and any multiple of **b** can be used for the direction of the line.

> **A vector equation of the line through a point A and parallel to a vector b is**
>
> $$\mathbf{r} = \mathbf{a} + t\mathbf{b}$$

The vector **b**, in the direction of the line, is called the **direction vector** of the line.

Example 12   Find a vector equation for the line that passes through the point $(2, 3, -8)$ and is parallel to the vector $5\mathbf{i} - 10\mathbf{j} + 15\mathbf{k}$.

The equation of the line is

$$\mathbf{r} = 2\mathbf{i} + 3\mathbf{j} - 8\mathbf{k} + t(5\mathbf{i} - 10\mathbf{j} + 15\mathbf{k})$$

> This could be written as
> $$\mathbf{r} = (2 + 5t)\mathbf{i} + (3 - 10t)\mathbf{j} + (-8 + 15t)\mathbf{k}$$

One alternative equation for the line is

$$\mathbf{r} = 2\mathbf{i} + 3\mathbf{j} - 8\mathbf{k} + t(\mathbf{i} - 2\mathbf{j} + 3\mathbf{k})$$

> Since $5\mathbf{i} - 10\mathbf{j} + 15\mathbf{k} = 5(\mathbf{i} - 2\mathbf{j} + 3\mathbf{k})$, $\mathbf{i} - 2\mathbf{j} + 3\mathbf{k}$ can also be used as the direction vector.

Example 13   Find a vector equation for the line that passes through the points A and B with coordinates $(2, 4, 5)$ and $(5, 8, 0)$ respectively.

Since A and B both lie on the line its direction can be taken as $\overrightarrow{AB}$

> Or $\overrightarrow{BA}$, or any other multiple of $\overrightarrow{AB}$.

$$\overrightarrow{AB} = \mathbf{b} - \mathbf{a}$$

$$= 5\mathbf{i} + 8\mathbf{j} - (2\mathbf{i} + 4\mathbf{j} + 5\mathbf{k})$$

$$= 3\mathbf{i} + 4\mathbf{j} - 5\mathbf{k}$$

Hence, as the line passes through A,
a vector equation of the line is

$$\mathbf{r} = 2\mathbf{i} + 4\mathbf{j} + 5\mathbf{k} + \lambda(3\mathbf{i} + 4\mathbf{j} - 5\mathbf{k})$$

> If B is used instead of A, another form
> of the equation of the line is
> $\mathbf{r} = 5\mathbf{i} + 8\mathbf{j} + \mu(3\mathbf{i} + 4\mathbf{j} - 5\mathbf{k})$

> **A vector equation of the straight line
> through two points C and D is**
>
> $$\mathbf{r} = \mathbf{c} + t(\mathbf{d} - \mathbf{c})$$

Pairs of lines

In two dimensions, a pair of lines must be either parallel, or intersecting. In three
dimensions, there is a third possibility: the lines do not intersect but are not parallel.
Such lines are called **skew**. In the case of both intersecting and skew lines, it is
possible to find the angle between the lines. If three, or more, lines intersect at a
point, the lines are said to be **concurrent**.

Example 14   *In this example the components of two vectors are used to obtain simultaneous equations.
Two of these equations are then solved and checked for concurrency by comparing the
third components.*

For each of these pairs of lines determine whether the lines are parallel, intersecting
or skew. If the lines intersect, find the coordinates of their point of intersection. If
the lines intersect, or are skew, find the acute angle between them.

a $\mathbf{r} = 13\mathbf{i} + 5\mathbf{j} - 3\mathbf{k} + s(2\mathbf{i} - 4\mathbf{j} + 7\mathbf{k})$ and $\mathbf{r} = 4\mathbf{i} - 5\mathbf{j} - 7\mathbf{k} + t(-6\mathbf{i} + 12\mathbf{j} - 21\mathbf{k})$

b $\mathbf{r} = \mathbf{i} + \mathbf{j} - 3\mathbf{k} + \lambda(-4\mathbf{i} + \mathbf{j})$ and $\mathbf{r} = 9\mathbf{i} + 2\mathbf{j} + \mathbf{k} + \mu(2\mathbf{i} + \mathbf{j} + 2\mathbf{k})$

c $\mathbf{r} = \begin{pmatrix} 3 \\ 4 \\ 2 \end{pmatrix} + \lambda \begin{pmatrix} 2 \\ 0 \\ -1 \end{pmatrix}$ and $\mathbf{r} = \begin{pmatrix} -2 \\ 0 \\ -2 \end{pmatrix} + \mu \begin{pmatrix} 3 \\ 4 \\ 6 \end{pmatrix}$

Solution   **a** $\mathbf{r} = 13\mathbf{i} + 5\mathbf{j} - 3\mathbf{k} + s(2\mathbf{i} - 4\mathbf{j} + 7\mathbf{k})$

and $\mathbf{r} = 4\mathbf{i} - 5\mathbf{j} - 7\mathbf{k} + t(-6\mathbf{i} + 12\mathbf{j} - 21\mathbf{k})$

> $-6\mathbf{i} + 12\mathbf{j} - 21\mathbf{k} = -3(2\mathbf{i} - 4\mathbf{j} + 7\mathbf{k})$

Since the direction vectors of the two lines are multiples of each other the lines
are parallel.

b $\mathbf{r} = \mathbf{i} + \mathbf{j} - 3\mathbf{k} + \lambda(-4\mathbf{i} + \mathbf{j})$

and $\mathbf{r} = 9\mathbf{i} + 2\mathbf{j} + \mathbf{k} + \mu(2\mathbf{i} + \mathbf{j} + 2\mathbf{k})$

> The direction vectors of the lines
> are not multiples of each other.

The lines are not parallel.

> To find if they intersect,
> solve them simultaneously.

$$\mathbf{r} = (1 - 4\lambda)\mathbf{i} + (1 + \lambda)\mathbf{j} - 3\mathbf{k}$$

$$\mathbf{r} = (9 + 2\mu)\mathbf{i} + (2 + \mu)\mathbf{j} + (1 + 2\mu)\mathbf{k}$$

> First rewrite the equations in this form.

Equating the coefficients of **i**, **j** and **k** gives

> If the lines intersect then, for
> some value of λ and μ, the
> position vectors will be the same.

$$1 - 4\lambda = 9 + 2\mu \qquad \text{①}$$
$$1 + \lambda = 2 + \mu \qquad \text{②}$$
$$-3 = 1 + 2\mu \qquad \text{③}$$

Rearranging ① and ② gives

$$4\lambda + 2\mu = -8$$

Cancel the common factor 2.

$$2\lambda + \mu = -4 \qquad ④$$

$$\lambda - \mu = 1 \qquad ⑤$$

$$④ + ⑤ \qquad 3\lambda = -3$$

$$\lambda = -1$$

Substituting in ④

$$-2 + \mu = -4$$

$$\mu = -2$$

Substituting $\mu = -2$ in the RHS of ③ gives

$$1 - 4 = -3$$

and so this equation is also satisfied by the solutions of ① and ②.

Thus the two lines do intersect, at the point with position vector $\mathbf{r} = 5\mathbf{i} - 3\mathbf{k}$.

This is obtained either by substituting $\lambda = -1$ in the equation of the first line, or $\mu = -2$ in the equation of the second line.

The coordinates of the point of intersection are $(5, 0, -3)$.

The direction vectors of the two lines are

$$\mathbf{d}_1 = -4\mathbf{i} + \mathbf{j}$$

and

$$\mathbf{d}_2 = 2\mathbf{i} + \mathbf{j} + 2\mathbf{k}$$

$$\mathbf{d}_1.\mathbf{d}_2 = -8 + 1 + 0 = -7$$

$$|\mathbf{d}_1| = \sqrt{16 + 1} = \sqrt{17}$$

$$|\mathbf{d}_2| = \sqrt{4 + 1 + 4} = \sqrt{9} = 3$$

$$\mathbf{d}_1.\mathbf{d}_2 = |\mathbf{d}_1||\mathbf{d}_2| \cos\theta$$

$$\cos\theta = \frac{\mathbf{d}_1.\mathbf{d}_2}{|\mathbf{d}_1||\mathbf{d}_2|}$$

$$\cos\theta = \frac{-7}{3\sqrt{17}}$$

$$\theta = 124.5° \text{ (1 d.p.)}$$

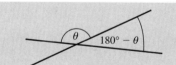

In this case, θ is the obtuse angle between the lines. The acute angle is $180° - \theta$.

So, the acute angle between the two lines is $55.5°$ (1 d.p.).

c

$$r = \begin{pmatrix} 3 \\ 4 \\ 2 \end{pmatrix} + \lambda \begin{pmatrix} 2 \\ 0 \\ -1 \end{pmatrix} \text{ and } r = \begin{pmatrix} -2 \\ 0 \\ -2 \end{pmatrix} + \mu \begin{pmatrix} 3 \\ 4 \\ 6 \end{pmatrix}$$

Rewrite the equations using single column vectors.

$$\mathbf{r} = \begin{pmatrix} 3 + 2\lambda \\ 4 \\ 2 - \lambda \end{pmatrix} \qquad \text{and } \mathbf{r} = \begin{pmatrix} -2 + 3\mu \\ 4\mu \\ -2 + 6\mu \end{pmatrix}$$

Equating **i**, **j** and **k** components gives

$$3 + 2\lambda = -2 + 3\mu \qquad \textcircled{1}$$

$$4 = 4\mu \qquad \textcircled{2}$$

$$2 - \lambda = -2 + 6\mu \qquad \textcircled{3}$$

From $\textcircled{2}$

$$\mu = 1$$

Substituting in $\textcircled{1}$ gives

$$3 + 2\lambda = -2 + 3$$

$$2\lambda = -2$$

$$\lambda = -1$$

However, checking in $\textcircled{3}$ gives

$$\text{LHS} = 2 - (-1) = 3$$

and

$$\text{RHS} = -2 + 6 = 4$$

> The third equation must be used to check the solution obtained. A solution that satisfies two equations can *always* be found (unless the lines are parallel).

So equation $\textcircled{3}$ is not satisfied and so the lines do not intersect. Clearly the lines are not parallel and so they must be skew.

The direction vectors of the two lines are

$$\mathbf{d}_1 = \begin{pmatrix} 2 \\ 0 \\ -1 \end{pmatrix} \qquad \text{and} \qquad \mathbf{d}_2 = \begin{pmatrix} 3 \\ 4 \\ 6 \end{pmatrix}$$

$$\mathbf{d}_1 . \mathbf{d}_2 = 6 + 0 - 6 = 0$$

So the two lines are perpendicular.

The distance of a point from a line

Consider a line through a point A which is parallel to a vector **b**. The equation of this line is $\mathbf{r} = \mathbf{a} + t\mathbf{b}$.

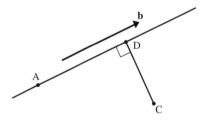

Now consider a point C, not on the line, as shown in the diagram. Suppose that D is the point on the line such that the vector $\overrightarrow{CD}$ is perpendicular to the line. Then the distance of the point C from the line is the length of the vector $\overrightarrow{CD}$.

The point D is called the foot of the perpendicular from C to the line.

Example 15   A line has vector equation $\mathbf{r} = \begin{pmatrix} 9 \\ -1 \\ 20 \end{pmatrix} + t \begin{pmatrix} 2 \\ -1 \\ 3 \end{pmatrix}$ and the point A is $(6, -3, 12)$.

  a Find the coordinates of the foot of the perpendicular from A to the line.

  b Hence find the distance of A from the line.

Solution   **a** A general point R on the line has position vector

$$\mathbf{r} = \begin{pmatrix} 9 + 2t \\ -1 - t \\ 20 + 3t \end{pmatrix}$$

> Rewrite the equation of the line as a single vector.

The displacement vector $\overrightarrow{AR}$ is

$$\overrightarrow{AR} = \mathbf{r} - \mathbf{a}$$

$$= \begin{pmatrix} 9 + 2t \\ -1 - t \\ 20 + 3t \end{pmatrix} - \begin{pmatrix} 6 \\ -3 \\ 12 \end{pmatrix}$$

$$= \begin{pmatrix} 3 + 2t \\ 2 - t \\ 8 + 3t \end{pmatrix}$$

Since R is the foot of the perpendicular from A to the line, AR is perpendicular to the line, so

$$\mathbf{d}.\overrightarrow{AR} = 0 \text{ where } \mathbf{d} \text{ is the direction vector of the line.}$$

Hence, $\begin{pmatrix} 2 \\ -1 \\ 3 \end{pmatrix}.\begin{pmatrix} 3 + 2t \\ 2 - t \\ 8 + 3t \end{pmatrix} = 0$

> The direction vector of the line is $\begin{pmatrix} 2 \\ -1 \\ 3 \end{pmatrix}$.

$$6 + 4t - 2 + t + 24 + 9t = 0$$

$$28 + 14t = 0$$

$$t = -2$$

When $t = -2$, $\mathbf{r} = \begin{pmatrix} 9 - 4 \\ -1 + 2 \\ 20 - 6 \end{pmatrix} = \begin{pmatrix} 5 \\ 1 \\ 14 \end{pmatrix}$

> Substitute the value for t into the expression for the position vector of R.

Hence the foot of the perpendicular from A to the line has coordinates $(5, 1, 14)$.

  b $\overrightarrow{AR} = \begin{pmatrix} -1 \\ 4 \\ 2 \end{pmatrix}$ and so

> Evaluate $\mathbf{r} - \mathbf{a}$, or substitute $t = -2$ into the expression $\overrightarrow{AR}$ obtained earlier.

$$|\overrightarrow{AR}| = \sqrt{1^2 + 4^2 + 2^2} = \sqrt{21}$$

Exercise 15D

1 Find a vector equation for the line that passes through the point A and is parallel to the vector **b**, where

   a A is $(2, 3, -4)$, $\mathbf{b} = \mathbf{i} + \mathbf{j} + \mathbf{k}$ **b** A is $(2, -4, 6)$, $\mathbf{b} = 7\mathbf{i} + \mathbf{j}$

   c A is $(0, 5, 3)$, $\mathbf{b} = 2\mathbf{i} - 5\mathbf{j} + 7\mathbf{k}$ **d** A is $(0, 0, 0)$, $\mathbf{b} = 5\mathbf{i} + 30\mathbf{j} - 5\mathbf{k}$

   e A is $(7, 0, 1)$, $\mathbf{b} = \begin{pmatrix} 0 \\ 3 \\ 0 \end{pmatrix}$ **f** A is $(5, 3, -4)$, $\mathbf{b} = \begin{pmatrix} 2 \\ 6 \\ -4 \end{pmatrix}$

2 Find vector equations for the lines passing through these points.

   a $(3, 2, -1)$ and $(5, 7, 3)$ **b** $(4, -2, 7)$ and $(0, 0, 0)$

   c $(3, 8, -5)$ and $(-1, -2, 5)$ **d** $(6, 4, 2)$ and $(2, 4, 6)$

   e $(10, 20, -15)$ and $(20, 30, -5)$ **f** $(-1, 1, 6)$ and $(2, -1, 8)$

3 Given that the points A and B are $(3, 5)$ and $(1, 10)$ respectively, find a vector equation for the line through the point A and perpendicular to AB.

4 Find a vector equation for the perpendicular bisector of AB where A is the point $(-3, 7)$ and B is the point $(7, 13)$.

5 The lines l_1 and l_2 have equations

   $$l_1 : \mathbf{r} = \mathbf{i} - \mathbf{j} - 5\mathbf{k} + \lambda(2\mathbf{i} + 3\mathbf{j} + 4\mathbf{k})$$
   $$l_2 : \mathbf{r} = 16\mathbf{i} + 5\mathbf{j} - 2\mathbf{k} + \mu(7\mathbf{i} - 6\mathbf{j} - 10\mathbf{k})$$

   a Show that l_1 and l_2 do not intersect.

   b Find the acute angle, to the nearest $0.1°$, between l_1 and l_2.

6 Determine which of these pairs of lines are parallel, intersecting or skew. In the case of intersecting lines, find the coordinates of the point of intersection. In the case of intersecting, or skew, lines, find the angle between the lines, giving your answer to the nearest tenth of a degree.

   a $\mathbf{r} = 2\mathbf{i} - 3\mathbf{j} - 7\mathbf{k} + s(\mathbf{i} + \mathbf{j} + \mathbf{k})$ and $\mathbf{r} = 4\mathbf{i} - 6\mathbf{j} - 14\mathbf{k} + t(5\mathbf{j} + 9\mathbf{k})$

   b $\mathbf{r} = s(3\mathbf{i} + 7\mathbf{j} + 4\mathbf{k})$ and $\mathbf{r} = 20\mathbf{i} + 15\mathbf{j} - 5\mathbf{k} + t(4\mathbf{i} + 3\mathbf{j} - \mathbf{k})$

   c $\mathbf{r} = 7\mathbf{i} - 4\mathbf{j} + 6\mathbf{k} + s(-2\mathbf{i} + 3\mathbf{j} - \mathbf{k})$ and $\mathbf{r} = 6\mathbf{i} + 15\mathbf{j} + 20\mathbf{k} + t(2\mathbf{i} + 4\mathbf{j} + 6\mathbf{k})$

   d $\mathbf{r} = 2\mathbf{i} - \mathbf{j} - \mathbf{k} + s(6\mathbf{i} - 9\mathbf{j} + 3\mathbf{k})$ and $\mathbf{r} = \mathbf{i} + \mathbf{j} + \mathbf{k} + t(-4\mathbf{i} + 6\mathbf{j} - 2\mathbf{k})$

   e $\mathbf{r} = \begin{pmatrix} 1 \\ 2 \\ 3 \end{pmatrix} + s\begin{pmatrix} 3 \\ 2 \\ -1 \end{pmatrix}$ and $\mathbf{r} = \begin{pmatrix} 9 \\ 2 \\ 5 \end{pmatrix} + t\begin{pmatrix} -1 \\ -2 \\ 1 \end{pmatrix}$

7 Referred to a fixed origin O, the points P, Q and R have position vectors $(2\mathbf{i} + \mathbf{j} + \mathbf{k})$, $(5\mathbf{j} + 3\mathbf{k})$ and $(5\mathbf{i} - 4\mathbf{j} + 2\mathbf{k})$ respectively.

   a Find in the form $\mathbf{r} = \mathbf{a} + t\mathbf{b}$, an equation of the line PQ.

   b Show that the point S with position vector $(4\mathbf{i} - 3\mathbf{j} - \mathbf{k})$ lies on PQ.

   c Show that the lines PQ and RS are perpendicular.

   d Find the size of $\angle PQR$, giving your answer to $0.1°$.   *London*

8 With respect to a fixed origin O, the lines l_1 and l_2 are given by the equations

$$l_1 : \mathbf{r} = (2\mathbf{i} + 3\mathbf{j} - 2\mathbf{k}) + \lambda(-2\mathbf{i} + 4\mathbf{j} + \mathbf{k}),$$

$$l_2 : \mathbf{r} = (-6\mathbf{i} - 3\mathbf{j} + \mathbf{k}) + \mu(5\mathbf{i} + \mathbf{j} - 2\mathbf{k}),$$

where λ and μ are scalar parameters.

a Show that l_1 and l_2 meet and find the position vector of their point of intersection.

b Find, to the nearest $0.1°$, the acute angle between l_1 and l_2. *London*

9 Find the distance of each of these points from the line given.

a $(7, 10, 8)$, $\mathbf{r} = \begin{pmatrix} 2 \\ 1 \\ 3 \end{pmatrix} + t \begin{pmatrix} -3 \\ 1 \\ -4 \end{pmatrix}$ **b** $(18, 4, 6)$, $\mathbf{r} = \begin{pmatrix} 6 \\ 3 \\ 2 \end{pmatrix} + t \begin{pmatrix} 5 \\ -2 \\ 0 \end{pmatrix}$

c $(4, 5, 0)$, $\mathbf{r} = \begin{pmatrix} 8 \\ 15 \\ 0 \end{pmatrix} + t \begin{pmatrix} -1 \\ 4 \\ 1 \end{pmatrix}$ **d** $(1, 0, 13)$, $\mathbf{r} = \begin{pmatrix} 5 \\ -8 \\ 7 \end{pmatrix} + t \begin{pmatrix} -2 \\ 4 \\ 3 \end{pmatrix}$

---

1 OABC and OPQR are parallelograms and

$$\overrightarrow{OA} = \mathbf{a} \quad \overrightarrow{OC} = \mathbf{c} \quad \overrightarrow{OP} = \tfrac{2}{3}\mathbf{a} \quad \overrightarrow{OR} = \tfrac{1}{2}\mathbf{c}$$

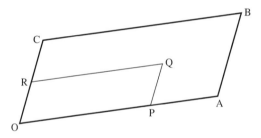

Express these vectors in terms of $\mathbf{a}$ and $\mathbf{c}$.

a $\overrightarrow{OB}$ **b** $\overrightarrow{AC}$ **c** $\overrightarrow{OQ}$ **d** $\overrightarrow{PR}$ **e** $\overrightarrow{RC}$

f $\overrightarrow{AQ}$ **g** $\overrightarrow{QC}$ **h** $\overrightarrow{PB}$ **i** $\overrightarrow{PC}$ **j** $\overrightarrow{BQ}$

2 Given that $\mathbf{a} = 2\mathbf{i} + \mathbf{j} - 5\mathbf{k}$ and $\mathbf{b} = 5\mathbf{i} - \mathbf{j} + 8\mathbf{k}$, find

a $|\mathbf{a} + \mathbf{b}|$ **b** $|\mathbf{a} - \mathbf{b}|$ **c** $|2\mathbf{a} + 2\mathbf{b}|$

3 Find a vector of magnitude 39, in the opposite direction to the vector $-3\mathbf{i} + 4\mathbf{j} - 12\mathbf{k}$.

4 In the diagram, $\overrightarrow{OB} = \mathbf{b}$, $\overrightarrow{OC} = \frac{4}{3}\mathbf{b}$ and $\overrightarrow{AP} = \frac{2}{3}\overrightarrow{AB}$.

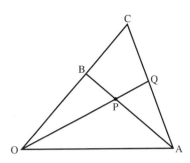

Given that $\overrightarrow{AQ} = m\overrightarrow{AC}$ and that $\overrightarrow{OQ} = n\overrightarrow{OP}$, calculate the values of m and n, and the ratio AQ:QC.

5 The vectors $\mathbf{a}$ and $\mathbf{b}$ are $4\mathbf{i} - 6\mathbf{j} + m\mathbf{k}$ and $6\mathbf{i} + n\mathbf{j} + 15\mathbf{k}$ respectively.

    a Find the values of m and n if $\mathbf{a}$ and $\mathbf{b}$ are parallel.

    b Find n, in terms of m, if $\mathbf{a}$ and $\mathbf{b}$ are perpendicular.

6 Evaluate the scalar product of the vectors

$$\mathbf{a} = \begin{pmatrix} -4 \\ 3 \end{pmatrix} \quad \text{and} \quad \mathbf{b} = \begin{pmatrix} 8 \\ 19.2 \end{pmatrix}$$

and hence find the angle between $\mathbf{a}$ and $\mathbf{b}$.

7 The points A, B and C have coordinates $(7, 1, 4)$, $(2, -5, 7)$ and $(-4, 4, 15)$ respectively. Prove that angle ABC is a right angle.

8 Find a unit vector that is perpendicular to both $4\mathbf{i} + 10\mathbf{j} - 3\mathbf{k}$ and $5\mathbf{i} - 9\mathbf{j} + 7\mathbf{k}$.

9 Given that $|\mathbf{a}| = 4$, $|\mathbf{b}| = 5$ and $|\mathbf{a} - \mathbf{b}| = 3$ use the cosine rule to find the cosine of the angle between $\mathbf{a}$ and $\mathbf{b}$. Hence find the value of $\mathbf{a.b}$.

10 A and B are the points $(1, -3)$ and $(7, 9)$ respectively.
Find a vector equation for the perpendicular bisector of AB.

11 Find the point of intersection of the lines

$$\mathbf{r} = \begin{pmatrix} 7 \\ 0 \\ 1 \end{pmatrix} + s\begin{pmatrix} -2 \\ 1 \\ -4 \end{pmatrix} \quad \text{and} \quad \mathbf{r} = \begin{pmatrix} 8 \\ 7 \\ 3 \end{pmatrix} + t\begin{pmatrix} 3 \\ -4 \\ 6 \end{pmatrix}$$

What is the acute angle between the lines?

12 Show that the lines

$$\mathbf{r} = 8\mathbf{i} + 14\mathbf{j} + 6\mathbf{k} + \lambda(7\mathbf{i} - 3\mathbf{j} - 2\mathbf{k})$$

and

$$\mathbf{r} = 4\mathbf{i} - 40\mathbf{j} + 20\mathbf{k} + \mu(5\mathbf{i} + 9\mathbf{j} + 4\mathbf{k})$$

are both skew and perpendicular.

13 Show that the lines

$$\mathbf{r} = \begin{pmatrix} 7 \\ 10 \\ 1 \end{pmatrix} + \lambda \begin{pmatrix} 3 \\ 4 \\ 0 \end{pmatrix} \quad \text{and} \quad \mathbf{r} = \begin{pmatrix} 5 \\ 14 \\ 4 \end{pmatrix} + \mu \begin{pmatrix} 4 \\ 12 \\ 3 \end{pmatrix}$$

intersect, and find their point of intersection.

14 OABC is a rectangle. With respect to the origin, O, the position vectors of **a** and **b** are $2\mathbf{i} - 3\mathbf{j} + 5\mathbf{k}$ and $s\mathbf{i} + 3\mathbf{j} + 7\mathbf{k}$ respectively.

   a Find the value of s.

   b Find vector equations for the diagonals AC and OB.

   c Find the cosine of the acute angle between these diagonals.

15 Find a vector equation for the line l that passes through the points $(2, 5, 3)$ and $(6, 8, 5)$.

   The point P lies on l and is such that OP is perpendicular to l.

   a Find the coordinates of P.

   b Hence find the shortest distance from O to l.

There is a 'Test yourself' exercise (Ty15) and an 'Extension exercise' (Ext15) on the CD-ROM.

➤ Key points

Vector geometry

■ A quantity which has both magnitude and direction is called a **vector**.

■ A **unit vector** is a vector of magnitude 1.

■ The vector $-\mathbf{a}$ has the same magnitude as $\mathbf{a}$, but is in the opposite direction.

■ The vector $k\mathbf{a}$ has k times the magnitude of $\mathbf{a}$, and is parallel to it.
If k is positive $k\mathbf{a}$ is in the same direction as $\mathbf{a}$, if k is negative $k\mathbf{a}$ is in the opposite direction to $\mathbf{a}$.

■ Vectors are **added** using the triangle, or parallelogram, law of vector addition:

$$\mathbf{c} = \mathbf{a} + \mathbf{b}$$

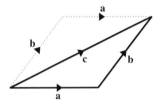

■ Vectors are **subtracted** by adding a negative vector. $\mathbf{a} - \mathbf{b}$ is formed by adding the vector $-\mathbf{b}$ to the vector $\mathbf{a}$.

$$\mathbf{a} - \mathbf{b} = \mathbf{a} + (-\mathbf{b})$$

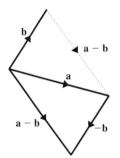

■ The **base vectors i**, **j** and **k** are defined as

$$\mathbf{i} = \begin{pmatrix} 1 \\ 0 \\ 0 \end{pmatrix} \qquad \mathbf{j} = \begin{pmatrix} 0 \\ 1 \\ 0 \end{pmatrix} \qquad \mathbf{k} = \begin{pmatrix} 0 \\ 0 \\ 1 \end{pmatrix}$$

In two dimensions

$$\mathbf{i} = \begin{pmatrix} 1 \\ 0 \end{pmatrix} \qquad \mathbf{j} = \begin{pmatrix} 0 \\ 1 \end{pmatrix}$$

■ If $\mathbf{a} = x\mathbf{i} + y\mathbf{j} + z\mathbf{k}$, then the **magnitude** of $\mathbf{a}$ is $a = |\mathbf{a}| = \sqrt{x^2 + y^2 + z^2}$.

■ If the distance between two points (x_1, y_1, z_1) and (x_2, y_2, z_2) is d then
$d^2 = (x_1 - x_2)^2 + (y_1 - y_2)^2 + (z_1 - z_2)^2$.

■ A **unit vector** in the direction of a vector $\mathbf{a}$ is $\hat{\mathbf{a}} = \dfrac{\mathbf{a}}{|\mathbf{a}|}$.

■ The **position vector** of a point A is the vector from the origin to A.

■ If A and B have position vectors $\mathbf{a}$ and $\mathbf{b}$ respectively, then $\overrightarrow{AB} = \mathbf{b} - \mathbf{a}$ and the position vector of the mid-point of AB is $\frac{1}{2}(\mathbf{a} + \mathbf{b})$.

Scalar product

- The scalar (or dot) product of two vectors **a** and **b** is defined as $\mathbf{a}.\mathbf{b} = |\mathbf{a}||\mathbf{b}| \cos \theta$.

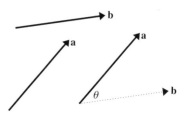

- If $\mathbf{a} = a_1\mathbf{i} + a_2\mathbf{j} + a_3\mathbf{k}$ and $\mathbf{b} = b_1\mathbf{i} + b_2\mathbf{j} + b_3\mathbf{k}$ then $\mathbf{a}.\mathbf{b} = a_1b_1 + a_2b_2 + a_3b_3$,
- If **a** and **b** are perpendicular then $\mathbf{a}.\mathbf{b} = 0$.
- If $\mathbf{a}.\mathbf{b} = 0$ then either $|\mathbf{a}| = 0$, $|\mathbf{b}| = 0$ or **a** and **b** are perpendicular.
- If $\mathbf{a}.\mathbf{b} = |\mathbf{a}||\mathbf{b}|$ then **a** and **b** are parallel (because $\cos \theta = 1 \Rightarrow \theta = 0°$).
- $\mathbf{a}.\mathbf{a} = |\mathbf{a}|^2$. In particular $\mathbf{i}.\mathbf{i} = \mathbf{j}.\mathbf{j} = \mathbf{k}.\mathbf{k} = 1$.

Lines

- A vector equation of the line through a point A and parallel to a vector **b** is $\mathbf{r} = \mathbf{a} + t\mathbf{b}$.

- A vector equation of the line through two points C and D is $\mathbf{r} = \mathbf{c} + t(\mathbf{d} - \mathbf{c})$.

- To find the angle between two lines $\mathbf{r} = \mathbf{a} + \lambda\mathbf{b}$ and $\mathbf{r} = \mathbf{c} + \mu\mathbf{d}$, use scalar product to find the angle between the vectors **b** and **d**, the direction vectors of the lines.

- To find whether two lines intersect, compare their components to produce simultaneous equations. Then solve two of the equations and check the solution in the third equation.

1 A curve is defined parametrically by

$$x = t^2 + t, \quad y = t^2 - t, \quad t > 0,$$

and P is the point where $t = 2$.

a Show that $\dfrac{dy}{dx} = 1 - \dfrac{2}{2t + 1}$.

b Determine an equation of the normal to the curve at P.

c Find a cartesian equation of the curve.

AQA June 2001

2

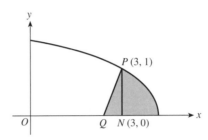

The diagram shows the curve C which is defined parametrically by

$$x = 4\sin^2 t, \qquad y = 2\cos t, \qquad 0 \leqslant t \leqslant \frac{\pi}{2}$$

The point $P(3, 1)$ lies on the curve C and the foot of the perpendicular from P to the x-axis is $N(3, 0)$. The normal to the curve C at P intersects the x-axis at the point Q.

i Obtain an expression for $\dfrac{dy}{dx}$ in terms of t.

ii Find the value of t at the point P.

iii Show that the equation of the normal PQ is $y = 2x - 5$.

AQA Jan 2004

3 A curve is defined parametrically by the equations

$$x = 3\ln t, \quad y = t^3 + 6t$$

where $t > 0$.

a Find $\dfrac{dy}{dx}$ in terms of t, simplifying your answer as much as possible.

b Determine a cartesian equation of the normal to the curve at the point where $t = 1$.

AQA (AEB) Jan 1999

(Not from the live examinations for the current specification.)

4 The diagram shows the sketch of part of the curve C which is defined parametrically by

$$x = t^2, \quad y = \sin t, \quad t \geqslant 0.$$

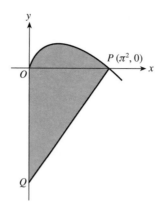

The curve cuts the positive x-axis for the first time at the point $P(\pi^2, 0)$.

The normal to the curve C at P intersects the y-axis at the point Q.

Show that the equation of the normal PQ is

$$y = 2\pi x - 2\pi^3.$$

AQA June 2002

5 A curve is defined by the parametric equations

$$x = t + \frac{3}{t}, \quad y = t - \frac{3}{t}, \quad t \neq 0.$$

a i Express $x + y$ and $x - y$ in terms of t.

ii Hence verify that the cartesian equation of the curve is

$$x^2 - y^2 = 12.$$

b i By finding $\dfrac{\mathrm{d}x}{\mathrm{d}t}$ and $\dfrac{\mathrm{d}y}{\mathrm{d}t}$, find $\dfrac{\mathrm{d}y}{\mathrm{d}x}$ in terms of t.

ii Differentiate $x^2 - y^2 = 12$ implicitly to obtain $\dfrac{\mathrm{d}y}{\mathrm{d}x}$ in terms of x and y.

c Verify that the expressions you obtained in parts **b i** and **b ii** are equivalent.

6 a Use integration by parts to find

$$\int x e^x \mathrm{d}x.$$

b Hence find the solution of the differential equation

$$\frac{\mathrm{d}y}{\mathrm{d}x} = yx e^x$$

given that $y = \mathrm{e}$ when $x = 1$.

AQA June 2003

7 A pond covers an area of $300\,\text{m}^2$. A specimen of pondweed grows on the surface of the pond. At time t days after the weed is first discovered, it covers an area of $A\,\text{m}^2$.

The area of the pond covered by the weed increases at a rate which is proportional to the square root of the area of the pond already covered by the weed.

Initially, the area covered by the weed is $0.25\,\text{m}^2$ and its rate of growth is $1\,\text{m}^2$ per day.

  a Show that
  $$\frac{\mathrm{d}A}{\mathrm{d}t} = 2A^{\frac{1}{2}}.$$

  b Find the relationship between t and A.

  c Deduce, to the nearest day, the time taken for the pond's surface to be completely covered by this weed.

  AQA Jan 2002

8 **a** Use the substitution $u = 3 + \cos x$, or otherwise, to determine

  $$\int \frac{\sin x}{(3 + \cos x)^2}\,\mathrm{d}x.$$

  b Hence find the solution of the differential equation

  $$(3 + \cos x)^2 \frac{\mathrm{d}y}{\mathrm{d}x} = y \sin x,$$

  given that $y = \mathrm{e}$ when $x = 0$.

  AQA June 2001

9 Solve the differential equation

  $$\frac{\mathrm{d}y}{\mathrm{d}x} = 2y \sec^2 2x$$

given that $y = 2$ when $x = 0$, expressing your answer in the form $y = \mathrm{f}(x)$.

10 **a** Express $\dfrac{x + 1}{(x - 1)(x - 2)}$ in partial fractions.

  Given that $x > 2$, find $\displaystyle\int \frac{x + 1}{(x - 1)(x - 2)}\,\mathrm{d}x.$

  b Given that $x > 2$, solve the differential equation

  $$\frac{\mathrm{d}y}{\mathrm{d}x} = \frac{(x + 1)(y + 4)}{(x - 1)(x - 2)}, \text{ given that } y = 4 \text{ when } x = 3.$$

  Express y in terms of x in a form not involving logarithms.

  AQA (AEB) Jan 1998

  (Not from the live examinations for the current specification.)

11 a Use integration by parts to find $\int 4x\,e^{-2x}\,dx$.

b Solve the differential equation

$$e^{2x}\frac{dy}{dx} = 4x\,y^{\frac{1}{2}}, \quad (y > 0),$$

given that $y = 9$ when $x = 0$.

Express y in terms of x.

AQA (AEB) 1999

(Not from the live examinations for the current specification.)

12 a The quantity N varies with t so that $\dfrac{dN}{dt} = 4N$, and $N = 200$ when $t = 0$.

i Write down an expression for $\dfrac{dt}{dN}$ in terms of N.

ii Given that

$$4t = \int \frac{1}{N}\,dN$$

and using the fact $N = 200$ when $t = 0$, find an expression for t in terms of N.

iii Hence, show that $N = 200\,e^{4t}$.

b The number of bacteria, N, in a colony is such that the **hourly** rate of increase of N is equal to 4 times the number of bacteria present. Initially there are 200 bacteria present. Use the results from part **a** to find the time taken for the number of bacteria to grow to 700, giving your answer to the nearest minute.

AQA June 2003

13 At each point (x, y) on a curve the gradient of the curve is equal to the sum of the x-coordinate and the y-coordinate of that point.

a Express this information in the form of differential equation.

b Given that $y = z - x$

i find $\dfrac{dy}{dx}$ in terms of $\dfrac{dz}{dx}$

ii hence show that your differential equation can be written as

$$\frac{dz}{dx} = (1 + z).$$

c i Solve the differential equation $\dfrac{dz}{dx} = (1 + z)$.

ii Given that the curve passes through the point $(0, 0)$, show that the equation of the curve is

$$y = e^x - x - 1.$$

d Find an equation of the normal to the curve $y = e^x - x - 1$ at the point $(1,\ e - 2)$.

AQA June 2003

14 The points A and B have coordinates $(2, 1, 3)$ and $(4, 2, 3)$ respectively. The line l_1 has vector equation $\mathbf{r} = \mathbf{i} - \mathbf{j} - 3\mathbf{k} + s(\mathbf{i} + \mathbf{j} + 2\mathbf{k})$, where s is a scalar parameter.

  a Write down a vector equation for the line l_2 which passes through the points A and B, giving the equation in terms of a scalar parameter t.

  b Show that the lines l_1 and l_2 intersect and find the coordinates of the point of intersection.

  c Calculate the acute angle between the lines l_1 and l_2, giving your answer in degrees to one decimal place.

15 For some value of the scalar constant a, the lines with equations

$$\mathbf{r} = (3 + \lambda)\mathbf{i} + (5 - 2\lambda)\mathbf{j} + (1 + 3\lambda)\mathbf{k}$$

and

$$\mathbf{r} = (6 - \mu)\mathbf{i} + (1 + \mu)\mathbf{j} + (a - 4\mu)\mathbf{k}$$

meet at the point P. Determine the value of a and the position vector of P.

16 a Given that p is a constant, and that the lines with equations

$$\mathbf{r} = \begin{pmatrix} 24 \\ 8 \\ p \end{pmatrix} + \lambda \begin{pmatrix} 3 \\ 2 \\ -2 \end{pmatrix} \quad \text{and} \quad \mathbf{r} = \begin{pmatrix} 13 \\ -6 \\ 11 \end{pmatrix} + \mu \begin{pmatrix} 1 \\ 4 \\ -7 \end{pmatrix}$$

  intersect at the point X, determine the value of p and the position vector of X.

  b Calculate the acute angle between the two lines, giving your answer to the nearest degree.

AQA June 2001

17 The line l has equation $\begin{pmatrix} x \\ y \\ z \end{pmatrix} = \begin{pmatrix} 4 \\ -3 \\ 0 \end{pmatrix} + t \begin{pmatrix} 3 \\ 2 \\ 1 \end{pmatrix}$.

The point P on the line l is where $t = p$, and the point Q has coordinates $(8, 3, 4)$.

  a Show that $\overrightarrow{QP} . \begin{pmatrix} 3 \\ 2 \\ 1 \end{pmatrix} = 14p - 28$.

  b Hence find the coordinates of the foot of the perpendicular from the point Q to the line l.

  c Hence find the distance of the point Q from the line l.

- **Time $1\frac{1}{2}$ hours.**

- **You are advised to show all your working.**

- **Calculators may be used.**

---

1. Express $\dfrac{7x}{x^2 - 3x - 10} + \dfrac{5}{5 - x}$ as a single fraction in its simplest terms.

   (6 marks)

---

2. The curve C has equation $3x^2 - 2xy + y^2 = 6$. The point P on the curve C has coordinates $(1, -1)$.

   (a) Find the gradient of the curve at P. *(5 marks)*

   (b) Find an equation of the normal to C at P in the form $ax + by + c = 0$, where a, b and c are integers. *(3 marks)*

---

3. (a) Expand $(4 - x)^{-\frac{1}{2}}$, $|x| < 4$, in ascending powers of x up to and including the term in x^3, simplifying each term. *(5 marks)*

   (b) In the series expansion of $\dfrac{1 + ax}{\sqrt{4 - x}}$, the coefficient of x^3 is $-\dfrac{67}{2048}$.

   Find the value of the constant a. *(3 marks)*

---

4. The value, £V, of a car at age t months is modelled by the formula $V = Ae^{-kt}$, where k and A are positive constants. The value of the car when new was £9000. The value of the car is expected to decrease to £4500 after 36 months.

   (a) Write down the value of A. *(1 mark)*

   (b) Show that $k \approx 0.019254$. *(3 marks)*

   (c) Use the model to

   (i) calculate the value, to the nearest pound, of the car when it is 18 months old; *(2 marks)*

   (ii) find the age of the car, to the nearest month, when its value first falls below £1800. *(4 marks)*

---

5. The diagram shows the plan of a rectangular garden $ACEF$.

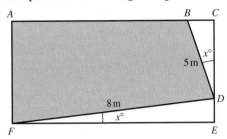

The shaded area $ABDF$ represents the lawn, which has a perimeter of 26 m.

It is given that $BD = 5$ m, $DF = 8$ m and angle DFE = angle $BDC = x°$, $x \neq 0$.

(a) Show that
$$13 \cos x° + 3 \sin x° = 13.$$
(5 marks)

(b) Express $13 \cos x° + 3 \sin x°$ in the form $R \cos(x° - \alpha°)$, where $R > 0$ and $0 < \alpha < 90$, giving your values of R and α to two decimal places. (3 marks)

(c) Hence find the value of x, giving your answer to one decimal place.
(3 marks)

---

6. Relative to a fixed origin O, the points A and B have position vectors $4\mathbf{i} + 3\mathbf{j} - \mathbf{k}$ and $\mathbf{i} + 4\mathbf{j} + 4\mathbf{k}$ respectively.

(a) Find a vector equation of the line l_1 which passes through A and B.
(2 marks)

(b) The line l_2 has equation $\mathbf{r} = 22\mathbf{i} + a\mathbf{j} + 4\mathbf{k} + \mu(b\mathbf{i} - \mathbf{j} + 2\mathbf{k})$, where a and b are constants. The lines l_1 and l_2 are perpendicular and intersect.
  (i) Find the values of a and b. (6 marks)
  (ii) Find the position vector of the point of intersection of l_1 and l_2.
  (2 marks)

---

7. (a) Express
$$\frac{x^2 + 1}{(x - 1)(x - 3)}$$
in the form
$$A + \frac{B}{x - 1} + \frac{C}{x - 3}.$$
(4 marks)

(b) Hence show that the exact value of $\displaystyle\int_4^5 \frac{x^2 + 1}{(x - 1)(x - 3)}\, \mathrm{d}x$ is $1 + \ln 24$.
(5 marks)

---

8. (a) By using the identity
$$\sin(A+B) = \sin A \cos B + \cos A \sin B$$
show that
$$\sin 3x = 3\sin x - 4\sin^3 x. \qquad \textit{(5 marks)}$$

(b) Solve the differential equation
$$\frac{dy}{dx} = 4y\sin^3 x$$
given that $y = 1$ when $x = \dfrac{\pi}{3}$. $\qquad \textit{(5 marks)}$

(c) Find, in an exact from, the value of $\dfrac{dy}{dx}$ when $x = \dfrac{\pi}{2}$. $\qquad \textit{(3 marks)}$

END OF QUESTIONS

Answers

1 Functions

Exercise 1A (p. 8)

1 a $\{0, 4, 8, 16\}$; one–one
  b $y \in \mathbb{R}, y \geqslant 1$; many–one
  c $y \in \mathbb{R}, -7 < y < 5$; one–one
  d $y \in \mathbb{R}, y \neq 0$; one–one
  e $y \in \mathbb{R}, -2 \leqslant y \leqslant 128$; one–one
  f $y \in \mathbb{R}, -2 \leqslant y \leqslant 1$; many–one
  g $y \in \mathbb{R}^+$; many–one

2 a $\{3, 5\frac{1}{2}, 8, 10\frac{1}{2}\}$; one–one
  b $x \in \mathbb{R}, 4 < x < 49$; one–one
  c $x \in \mathbb{R}, x < -2$ and $x > 2$;
    many–one
  d $x \in \mathbb{R}, 7 \leqslant x \leqslant 12$; one–one
  e $x \in \mathbb{R}, 1 < x \leqslant 3$ and
    $-3 \leqslant x < -1$; many–one

3 a $f(x) = (x - 1)^2 + 8$
  b $y \in \mathbb{R}, y \geqslant 8$

4 a $y \in \mathbb{R}, y \geqslant -6\frac{1}{4}$
  b $x = 1$ or $x = -4$

5 a $p = 7, q = 1$
  b $y \in \mathbb{R}, y \leqslant 7$

6 a $y \in \mathbb{R}, y \geqslant -11$
  b $y \in \mathbb{R}, y \geqslant -\frac{7}{2}$
  c $y \in \mathbb{R}, y \geqslant -\frac{5}{4}$
  d $y \in \mathbb{R}, y \leqslant \frac{33}{4}$

7 a $(2, 5)$
  c $x = 4$

8 b $y \in \mathbb{R}, y > 4$
  c $x = 5$

9 a $f(x) = 2 + \dfrac{5}{x - 1}$
  b $A\left(-\frac{3}{2}, 0\right)$; $B(0, -3)$; $C(1, 2)$;
    $x = 1, y = 2$
  c $y \in \mathbb{R}, y \neq 2$
  d $x = 1 \pm \sqrt{5}$

Exercise 1B (p. 13)

1 a 4 **b** 16
  c $3x^2 + 4$ **d** 16

2 a $x + 3$; $x + 3$ **b** $6x + 3$; $6x - 1$
  c x; x **d** $9 - 6x$; $1 - 6x$
  e $x^2 + 1$; $(x + 1)^2$
  f $\dfrac{1}{2x}$; $\dfrac{2}{x}$ **g** $\dfrac{1}{1 + x}$; $\dfrac{1}{x} + 1$
  h $2(2x + 1)(x - 1)$; $2(2x + 1)(x - 2)$

3 a $\dfrac{1}{x^2}$ **b** $\dfrac{1}{x^2}$
  c $\dfrac{1}{3x - 1}$ **d** x^4

e x **f** $9x - 4$
g $\dfrac{3}{x^2} - 1$ **h** $\dfrac{1}{(3x - 1)^2}$
i $\left(\dfrac{3}{x} - 1\right)^2$ **j** $\dfrac{3}{x^2} - 1$
k $\dfrac{1}{3x^2 - 1}$ **l** $\dfrac{1}{(3x - 1)^2}$

4 a $k = 2$

5 a $x = -2$ or $x = \frac{2}{3}$
  b $x = -\frac{1}{2}$
  c $x = \frac{1}{3}$ or $x = -1$
  d $x = \frac{1}{9}$ or $x = -1$

6 a $2x^2 + 7$, $y \in \mathbb{R}, y \geqslant 7$
  b $2x^3 + 7$, $y \in \mathbb{R}$
  c $(2x + 7)^2$, $y \in \mathbb{R}, y \geqslant 0$
  d $(2x + 7)^3$, $y \in \mathbb{R}$

7 a $fg: x \rightarrow (x + 4)^2 - 3$, $gf: x \rightarrow x^2 + 1$
  b $y \geqslant -3, y \geqslant 1$
  c $k = -\frac{3}{2}$
  d $l = \frac{1}{2} \pm \frac{1}{2}\sqrt{7}$

8 a $x + 10$; $x + 15$; $x + 5n$
  b $g(x) = 2x$
  c $h(x) = \dfrac{x}{a}$

9 a $gf(x) = 12x^2 - 12x + 5$
  b $a = 1, b = 1$
  c $x = -4$ or $x = 2$

10 a $x + 4$
  b $8x^4 + 24x^2 + 21$
  c x
  d $\dfrac{1}{2(x + 2)^2 + 3}$
  e $\dfrac{1}{2x^2 + 5}$
  f $\dfrac{2}{x^2} + 5$
  g $\dfrac{1}{2x^2 + 3} + 2$
  h $\dfrac{x^2}{2 + 3x^2}$

11 a $x \geqslant -2$ **b** $x \leqslant 4$
  c $-2 \leqslant x \leqslant 2$ **d** $0 < x \leqslant \frac{2}{3}$

Exercise 1C (p. 19)

1 a $\dfrac{x}{3}$ **b** $x - 4$
  c $x + 5$ **d** $6x$
  e $\dfrac{1}{x}$, self-inverse   **f** $\sqrt{x}$
  g $\dfrac{x + 4}{3}$ **h** $\dfrac{3 - x}{2}$

i $4 - x$, self-inverse   **j** $5x - 6$
k $\dfrac{5}{x}$, self-inverse   **l** $\dfrac{x - 1}{x}$

2 a $\dfrac{3x - 5}{2}$, $\mathbb{R}$
  b $\frac{3}{2}(x - 5)$, $\mathbb{R}$
  c $3\left(\dfrac{x}{2} - 5\right)$, $\mathbb{R}$
  d $\dfrac{x}{2} - \dfrac{5}{6}$, $\mathbb{R}$
  e $\dfrac{3x}{2} - 5$, $\mathbb{R}$
  f $x^2 - 4$, $x \geqslant 0$
  g $\dfrac{x}{7} + 4$, $\mathbb{R}$
  h $\dfrac{4 - x}{7}$, $\mathbb{R}$
  i $\sqrt{\dfrac{x - 2}{3}}$, $x \geqslant 2$
  j $\dfrac{4}{7x}$, $x \neq 0$, self-inverse
  k $\dfrac{1}{x} - 1$, $x \neq 0$
  l $\dfrac{1}{1 - \frac{1}{x}} = \dfrac{x}{x - 1}$, $x \neq 0$, $x \neq 1$,
    self-inverse
  m $\dfrac{1}{x - 1}$, $x \neq 1$
  n $\dfrac{3}{5x}$, $x \neq 0$, self-inverse
  o $\dfrac{a}{bx}$, $x \neq 0$, self-inverse
  p $4x^2$, $x \in \mathbb{R}^+$
  q $2 - x^2$, $x \geqslant 0$
  r $\sqrt[3]{x + 1}$, $\mathbb{R}$
  s $\dfrac{\sqrt{x + 4} - 1}{3}$, $x \geqslant -4$
  t $\dfrac{(x - 5)^2 + 1}{2}$, $x \geqslant 5$

3 a $y > -9$; $\sqrt{x + 9} - 3$; $x > -9$,
    $y > -3$
  b $y > 3$; $\sqrt{x - 3} + 2$; $x > 3, y > 2$
  c $y \leqslant 3$; $\sqrt{3 - x} - 1$; $x \leqslant 3, y \geqslant -1$

4 a i $\dfrac{x - 1}{3}$, $\mathbb{R}$ **iv** $x = -\frac{1}{2}, \left(-\frac{1}{2}, -\frac{1}{2}\right)$
  b i $\dfrac{2 - x}{4}$, $\mathbb{R}$ **iv** $x = \frac{2}{5}, \left(\frac{2}{5}, \frac{2}{5}\right)$

5 a $f^{-1}(x) = \dfrac{x - 2}{3}$; $g^{-1}(x) = \dfrac{1}{x}$;
    $gf(x) = \dfrac{1}{3x + 2}$

7 a $x \geqslant 0, y \geqslant 0; \sqrt{x}$
  b $x \geqslant -2, y \geqslant 0; \sqrt[4]{x} - 2$
  c $x \geqslant -1, y \leqslant 4; \sqrt{4 - x} - 1$

8 b Symmetrical in $y = x$

9 b $f^{-1}(x)$: $(-1, 0), (0, 2)$
  $f(x - 2)$: $(4, 0), (0, -2)$

10 a $f^{-1}(x) = \dfrac{4}{x} + 1, \mathbb{R}^+$
  c $x = 2.56$

11 b and **c**

12 a $f(6) = \frac{1}{2}$
  b $f^{-1}(2) = 2$
  c $k = 2$ or $k = -5$

13 a $g^{-1}(x) = \dfrac{2x - 1}{2 - x}, x \in \mathbb{R}, x \neq 2$
  b $x = \pm 1$

14 a $h^{-1}(x) = -\dfrac{1}{x} - 1, x \in \mathbb{R}, x \neq 0$

15 a $f^{-1}(x) = \dfrac{ax - 1}{x + 2}, x \neq -2$
  b $a = 7$
  c $a = -\frac{1}{2}$

Exercise 1D (p. 25)

1 a $-2 < x < 2$
  b $x \leqslant -3$ or $x \geqslant 3$
  c $-4 < x < 6$

3 b $y \in \mathbb{R}, 0 \leqslant y \leqslant 9$
  c Horizontal line cuts twice
  d $x = \frac{1}{2}$ or $x = \frac{9}{2}$

Exercise 1E (p. 27)

1 a $x = -5, x = 3$ **b** $x = -2, x = 3$
  c $x = 3$ **d** $x = -1, x = 2$
  e $x = -\frac{1}{3}, x = 2$ **f** $x = -1$
  g $x = -\frac{1}{3}, x = 2$ **h** $x = 0, x = 1$
  i $x = -1, x = 3$ **j** $x = \frac{3}{5}, x = \frac{3}{7}$

2 a $-5 < x < 3$
  b $x \leqslant -2$ or $x \geqslant 3$
  c $x \in \mathbb{R}, x \neq 3$
  d $-1 \leqslant x \leqslant 2$
  e $-\frac{1}{3} < x < 2$
  f $x \leqslant -1$
  g $-\frac{1}{2} \leqslant x \leqslant 2$
  h $x \leqslant 0$ or $x \geqslant 1$
  i $x < -1$ or $x > 3$
  j $\frac{3}{7} < x < \frac{3}{5}$

Exercise 1F (p. 31)

2 a $y = x^2 - 1$
  b $y = (x + 1)^2$
  c $y = 4x^2$
  d $y = x^2 + 6x + 11$

3 a $y = \sqrt{x} - 1, x \geqslant 0$
  b $y = \sqrt{x + 1}, x \geqslant -1$
  c $y = -\sqrt{x}, x \geqslant 0$
  d $y = 2 + \sqrt{x + 3}, x \geqslant -3$

7 A: $(2, -3), (-2, 3), (-2, -3)$
  B: $(3, 0), (-3, 0), (-3, 0)$

8 $f(x) = \dfrac{1}{x^2}$

9 b $y = |f(x)|$: $(-3, 0), (-1, 0), (3, 0),$
  $(0, 2)$
  $y = f(|x|)$: $(-3, 0), (3, 0), (0, -2)$

10 a Translation $\binom{30°}{0}$ followed by
  stretch parallel to y-axis, s.f. 3,
  followed by translation $\binom{0}{-1}$
  b i $y = \tan\left(\frac{1}{3}x - 10°\right) - 2$
    ii $y = 2(\tan(x - 60°) + 4)$

11 a $0 \leqslant y \leqslant 2$; neither
  b $-1 \leqslant y \leqslant 5$; even
  c $5 \leqslant y \leqslant 15$; neither
  d $0 \leqslant y \leqslant 2$; even
  e $-1 \leqslant y \leqslant 1$; odd
  f $-2 \leqslant y \leqslant 2$; even

12 a $-2\frac{1}{2} \leqslant y \leqslant -1\frac{1}{2}$; translation of
  $\binom{\frac{\pi}{4}}{-4}$, followed by stretch parallel
  to y-axis, s.f. $\frac{1}{2}$; 2π
  b $y \in \mathbb{R}$; stretch parallel to x-axis, s.f. 2,
  followed by reflection in x-axis,
  followed by translation $\binom{0}{1}$; 2π
  c $-1 \leqslant y \leqslant 1$; stretch parallel to
  x-axis, s.f. $\frac{1}{2}$, followed by reflection
  in y-axis, followed by translation
  $\binom{\frac{\pi}{6}}{0}$; π
  d $y \in \mathbb{R}$; stretch parallel to x-axis, s.f. $\frac{1}{3}$,
  followed by reflection in y-axis; $\frac{\pi}{3}$

Exercise 1G (Review) (p. 33)

1 a 7   **b** $x^3 - 1$
  c $(x - 3)^3 + 2$ **d** 1

2 a 77   **b** 30
  c 1   **d** $\frac{1}{77}$

3 a $5x$; $7 - x$; $\dfrac{7 - x}{5}$; $7 - 5x$
  b $x = \frac{7}{6}$

4 a $\dfrac{6 - x}{5}$ **b** $x = 1$

5 a hgf   **b** ghf   **c** gfh
  d fgh   **e** fhg   **f** hfg

6 a f and g: one–one; h: many–one
  b fgh: $x \to \dfrac{2x^2 + 20}{x^2 + 5}$
  c Yes
  d $2 < y \leqslant 4$
  e $x = \pm\sqrt{5}$

7 a $a = 5, b = 2$
  b $\binom{3}{2}$, $(0, \frac{1}{3})$, $(\frac{1}{2}, 0)$
  c $x = -2$ or $x = 4$

11 a $x = -\frac{1}{3}, x = 5$
  b $x = \frac{4}{3}, x = 4$
  c $x = -2, x = 0$
  d $x = 0, x = 2$

12 a $-\frac{1}{3} < x < 5$
  b $\frac{4}{3} < x < 4$
  c $x \leqslant -2$ or $x \geqslant 0$
  d $0 \leqslant x \leqslant 2$

15 a $y \geqslant 1, y \geqslant 4$
  b $ff(x) = |x| + 2, y \geqslant 2$;
  $gg(x) = (x^2 + 4)^2 + 4, y \geqslant 20$;
  $fg(x) = x^2 + 5, y \geqslant 5$

2 Trigonometry

Exercise 2A (p. 42)

1 a $a \cot\theta$ **b** $b^2 \operatorname{cosec}^2\theta$
  c $c^2 \tan^2\theta$ **d** $\dfrac{1}{a}\tan\theta\sec\theta$
  e $\dfrac{1}{b}\sin\theta\cos\theta$ **f** $\sin\theta$

2 a $-\frac{8}{15}$ **b** $\frac{17}{15}$

3 a $-\frac{25}{7}$ **b** $-\frac{25}{24}$

4 a $\dfrac{y^2}{b^2} - \dfrac{x^2}{a^2} = 1$ **b** $\dfrac{x^2}{a^2} - \dfrac{y^2}{b^2} = 1$
  c $\dfrac{b^2}{y^2} - \dfrac{x^2}{a^2} = 1$ **d** $\dfrac{b^2}{y^2} - \dfrac{x^2}{a^2} = 1$

5 a $\theta = 30°, 150°$
  b $\theta = \pm35.3°, \pm144.7°$
  c $\theta = -135°, 45°$
  d $\theta = -30°, 90°$
  e $\theta = 90°$
  f $\theta = -45°, 135°$
  g $\theta = -170°, -80°, 10°, 100°$
  h $\theta = -86.6°, 93.4°$
  i $\theta = -134.5°, 74.5°$
  j $\theta = -176.2°, -123.8°, -56.2°,$
    $-3.8°, 63.8°, 116.2°$
  k $\theta = -111\frac{1}{2}°, -51\frac{1}{2}°, 68\frac{1}{2}°, 128\frac{1}{2}°$
  l $\theta = -58.4°, 121.6°$

6 a $\theta = \pm\dfrac{\pi}{4}, \pm\dfrac{3\pi}{4}$
  b $\theta = 0.20, 2.94$
  c $\theta = -\dfrac{\pi}{4}, \dfrac{3\pi}{4}; \theta = -1.82, 1.32$
  d $\theta = 0.41, 2.73$
  e $\theta = \dfrac{\pi}{6}, \dfrac{5\pi}{6}$
  f $\theta = \pm\dfrac{2\pi}{3}, \pm1.16$

7 a $\theta = \dfrac{\pi}{12}, \dfrac{5\pi}{12}, \dfrac{7\pi}{12}, \dfrac{11\pi}{12}, \dfrac{13\pi}{12}, \dfrac{17\pi}{12},$
    $\dfrac{19\pi}{12}, \dfrac{23\pi}{12}$

b $\theta = 1.62, 4.66$

c $\theta = \dfrac{\pi}{4}, \dfrac{5\pi}{4}$; $\theta = 1.11, 4.25$

d $\theta = \dfrac{3\pi}{4}, \dfrac{7\pi}{4}$; $\theta = 0.46, 3.61$

e $\theta = \dfrac{\pi}{3}, \dfrac{5\pi}{3}$

f $\theta = \dfrac{\pi}{6}, \dfrac{5\pi}{6}$; $\theta = 0.73, 2.41$

9 b $f(x): x \in \mathbb{R}, x \neq n\pi; y \leqslant -1, y \geqslant 1, 2\pi$
$g(x): x \in \mathbb{R}, x \neq n\pi; y \leqslant 0, y \geqslant 2, 2\pi$

11 a Translation $\begin{pmatrix} \frac{\pi}{4} \\ 0 \end{pmatrix}$,

stretch s.f. 2, parallel to y-axis

12 Reflection in x-axis, translation $\begin{pmatrix} 0 \\ 1 \end{pmatrix}$

13 a $x \in \mathbb{R}, -1 \leqslant y \leqslant 1; \dfrac{\pi}{2}$

b $x \in \mathbb{R}, x \neq (2n+1)\dfrac{\pi}{2};$
$y \leqslant 1, y \geqslant 3; 2\pi$

c $x \in \mathbb{R}, x \neq (2n+1)\dfrac{\pi}{2};$
$y \leqslant -3, y \geqslant 3; 2\pi$

d $x \in \mathbb{R}, x \neq (2n+1)\dfrac{\pi}{2};$
$y \leqslant -4, y \geqslant 4; 4\pi$

e $x \in \mathbb{R}, x \neq n\pi; y \in \mathbb{R}, \dfrac{\pi}{2}$

f $x \in \mathbb{R}, x \neq 2n\pi; y \in \mathbb{R}, 2\pi$

Exercise 2B (p. 48)

1 a $\dfrac{\pi}{3}$ **b** $\dfrac{\pi}{4}$ **c** $\dfrac{\pi}{4}$ **d** $-\dfrac{\pi}{6}$

e $\dfrac{5\pi}{6}$ **f** $-\dfrac{\pi}{4}$ **g** $-\dfrac{\pi}{2}$ **h** π

i 0 **j** $\dfrac{\pi}{2}$

4 a 0.7 **b** $\dfrac{\pi}{7}$ **c** $\dfrac{\sqrt{3}}{2}$

d 2 **e** $-\dfrac{\sqrt{3}}{3}$ **f** $\dfrac{\pi}{3}$

6 $\sqrt{1 - x^2}$

Exercise 2C (Review) (p. 48)

1 a $-\frac{4}{3}$ **b** $-\frac{5}{4}$

2 a $7.2°, 82.8°, 187.2°, 262.8°$
b $135°, 315°$
c $50°, 230°$
d $20°, 170°, 200°, 350°$
e $270°$
f $15°, 75°, 195°, 255°$

3 a $\pm\dfrac{5\pi}{6}, \pm\dfrac{\pi}{6}$ **b** $\pm\dfrac{3\pi}{4}, \pm\dfrac{\pi}{4}$

c $\pm\dfrac{3\pi}{4}, \pm\dfrac{\pi}{4}$ **d** $-\dfrac{3\pi}{4}, \dfrac{\pi}{4}, -2.68, 0.46$

e $0.34, 2.80$ **f** ± 1.98

4 Stretch parallel to y-axis s.f. 3 followed
by translation $\begin{pmatrix} 0 \\ 4 \end{pmatrix}$

3 The functions e^x and $\ln x$

Exercise 3A (p. 53)

1 a 20.1 **b** 1.39
c 0.0672 **d** -1.39

2 a 4 **b** x
c 3 **d** x
e $3 - x$ **f** $2x + 1$
g 4 **h** $\frac{1}{64}$

3 a $y \geqslant 1$ **b** $y \in \mathbb{R}$
c $y > 0$ **d** $y \in \mathbb{R}$

4 a $y \in \mathbb{R}^+$ **b** 1

c $f^{-1}(x) = \ln\left(\dfrac{x}{k}\right), x \in \mathbb{R}^+$ **e** $\sqrt{2}k$

5 a $f^{-1}(x) = e^{x-5}; y \in \mathbb{R};$
$x \in \mathbb{R}, y \in \mathbb{R}^+$
b $f^{-1}(x) = \ln(x - 4); y > 4;$
$x > 4, y \in \mathbb{R}$
c $f^{-1}(x) = \ln(2x); y \in \mathbb{R}^+;$
$x \in \mathbb{R}^+, y \in \mathbb{R}$

6 a $f^{-1}(x) = e^x + 2; y \in \mathbb{R}; x \in \mathbb{R}, y > 2$
b $f^{-1}(x) = \ln x - 1; y \in \mathbb{R}^+;$
$x \in \mathbb{R}^+, y \in \mathbb{R}$
c $f^{-1}(x) = 1 - e^{\frac{x}{3}}; y \in \mathbb{R}; x \in \mathbb{R}, y < 1$

Exercise 3B (p. 55)

1 a 1.82 **b** 2.80 **c** 3.74
d 1.25 **e** -0.693 **f** -1.95
g -0.520 **h** 0

2 a 0.369 **b** 2.20 **c** 0.489
d 2.70 **e** 2.95 **f** 0.323

3 a 7.39 **b** 1.95 **c** 7.72
d 12.9 **e** -3.39 **f** 1.98
g 7 **h** 3

4 a -3 **b** 0 **c** 1.39
d 0 **e** 1.10 **f** -1.39
g $0.693, 1.10$ **h** $-0.405, -0.28$

Exercise 3C (Review) (p. 56)

1 a $y > 4$ **b** $y \in \mathbb{R}$
c $y > -5$ **d** $y \in \mathbb{R}$

2 Translation $\begin{pmatrix} 2 \\ -4 \end{pmatrix}$; stretch parallel to
x-axis, s.f. $\frac{1}{6}$
or
Stretch parallel to x-axis, s.f. $\frac{1}{6}$;
translation $\begin{pmatrix} \frac{1}{3} \\ -4 \end{pmatrix}$

3 Stretch parallel to y-axis, s.f. 2;
translation $\begin{pmatrix} 3 \\ 5 \end{pmatrix}$

4 a 0.766 **b** 3.5
c 1.87 **d** 92.3

5 a 10 **b** 17.5
c 18.2 **d** -39.2

4 Differentiation

Exercise 4A (p. 65)

1 a $6(2x + 5)^2$ **b** $24(3x - 5)^3$

c $-12(4 - 3x)^3$ **d** $-\dfrac{2}{\sqrt{6 - 4x}}$

e $-\dfrac{3}{(3x+1)^{\frac{3}{2}}}$ **f** $\frac{8}{3}(4x + 5)^{-\frac{2}{3}}$

g $\dfrac{3}{2(1 - x)^{\frac{5}{2}}}$ **h** $-\dfrac{15}{(5x - 2)^4}$

i $\dfrac{9}{(2 - 5x)^4}$

2 a $36x(3x^2 + 2)^5$
b $2(6x + 5)(3x^2 + 5x)^3$
c $-\dfrac{5}{2}x(1 - 5x^2)^{-\frac{3}{4}}$
d $-\dfrac{18x^2}{(6x^3 - 2)^2}$
e $\dfrac{-4x}{(4x^2 - 9)^{\frac{3}{2}}}$
f $-\dfrac{6(4x + 1)}{(6x^2 + 3x - 1)^2}$

3 $\dfrac{(y + 2)(y + 4)}{(y + 3)^2}$

4 a $\dfrac{1}{3y^2}$ **b** $\dfrac{1}{8y^3}$ **c** $\dfrac{y^{\frac{1}{2}}}{4(5y^{\frac{9}{2}} - 1)}$

d $-2y^3$ **e** $\dfrac{1}{4y - 1}$ **f** $\dfrac{y^3}{2 - y}$

5 $160x + y + 144 = 0$

6 $(-1, 0), (-5, 0)$

7 a $a = 2$ **b** 0.5
c -2 **d** $y + 2x = 2$

8 a Min at $(1.5, 0)$
b Min at $(1, 2)$, max at $(-1, -2)$
c Min at $(-2, 9)$
d Point of inflexion at $(2, 0)$

9 a 10 **b** -80

10 $3y + x = 4$

11 a $A(-0.5, 0), P(0, 1)$
b $6y + x = 6$
c 3.25 square units

12 $-\frac{1}{16}$

14 $(0, 256), (-2, 0), (2, 0)$

17 a $b = \frac{1}{2}(3 + \sqrt{2})$ **c** $(2.5, -22)$

Exercise 4B (p. 70)

1 a e^x **b** $4e^{4x}$ **c** $-e^{-x}$

d $-2e^{-2x}$ **e** $15e^{-5x}$ **f** $\frac{3}{2}e^{\frac{3x}{2}}$

2 a $4e^{4x-5}$ **b** $-6e^{6-2x}$ **c** e^{3x}

d $3e^{3x+1}$ **e** $-\dfrac{3}{e^{6x+1}}$ **f** e^3

3 a $7e^{7x-1}$ **b** e^{x+1}

4 a a^2e^{ax+b} **b** a^ne^{ax+b}

5 a $a = 1.5$ **b** $y = 2x - 2$
c $b = e^{-3}$ **d** $y = -\frac{1}{2}e^3 x + e^{-3}$

6 Max at $(1, 0)$

7 $x > 0.5$

8 $y = -4x - 1$

9 a $3ey + x = 1 + 3e^2$

b $\dfrac{(1 + 3e^2)^2}{6e}$ square units

10 b $6e$, $y = 6ex + 4e$

12 a $(0, 2)$

b Min $\left(-\frac{1}{3}\ln 2, \ 3 \times 2^{-\frac{2}{3}}\right)$

13 $a = 5$, $b = -1$

14 a $8xe^{x^2 - 2}$ **b** $(6t^2 + 3)e^{2t^3 + 3t}$

c ue^{u^2} **d** $\dfrac{e^{\sqrt{x}}}{2\sqrt{x}}$

15 $(0, 2)$

16 a $6\,\mathrm{m}$ (1 s.f.)   **b** $12\,\mathrm{m\,s^{-1}}$
c $-24\,\mathrm{m\,s^{-2}}$

Exercise 4C (p. 75)

1 a $\dfrac{1}{x}$ **b** $\dfrac{1}{x}$ **c** $\dfrac{3}{x}$
d $\dfrac{2}{x}$ **e** $-\dfrac{7}{x}$ **f** $\dfrac{2}{x}$

2 a $\dfrac{2}{2x + 3}$ **b** $\dfrac{1}{x - 1}$

c $-\dfrac{8}{5 - 2x}$ **d** $-\dfrac{3}{2(2 - 3x)}$

e $\dfrac{12}{4x + 3}$

3 a $\dfrac{2}{x}$ **b** $-\dfrac{1}{x}$

c $\dfrac{1}{4x}$ **d** $\dfrac{2}{x + 4}$

e $\dfrac{3}{3x + 1} - \dfrac{2}{2x - 5}$ **f** $\dfrac{10}{5x + 6}$

g $\dfrac{4}{2x + 1}$ **h** $-\dfrac{6}{3x - 1}$

i 4

4 $y = 2x - 2 + \ln 2$

5 $y = \frac{1}{5}x + 3$

6 b $A\left(-\frac{1}{8}, 0\right)$, $B(0, \ln 2)$
c $T\left(\frac{1}{4}(\ln 2 - 1), 2\ln 2 - 1\right)$

7 a -3 **b** $(1, -1)$

8 $a = 6$, $b = 1$

10 $(1.5, \ln 3.5)$, $\frac{6}{7}, \frac{2}{7}$

11 a $\dfrac{4x}{(x^2 - 3)}$ **b** $\dfrac{3(x^2 - 1)}{x(x^2 - 3)}$

c $\dfrac{8}{x(5x - 2)}$

12 a $\dfrac{4x + 1}{x(2x + 1)}$ **b** $\dfrac{3}{x}$

c $\dfrac{30x + 7}{(3x + 2)(5x - 1)}$

d $\dfrac{12x}{3x^2 + 5}$ **e** $\dfrac{1}{x - 2}$

f $\dfrac{2(8x^3 + 9x^2 + 3)}{(2x^3 + 3)(2x + 3)}$

14 $\left(-\frac{1}{2}, \frac{1}{4} + \ln 2\right)$, $(-1, 1)$

15 $a = 0$, $k = 6$

Exercise 4D (p. 82)

1 a $x(5x + 2)(1 + x)^2$

b $(9x^2 + 1)(x^2 + 1)^3$

c $2(2x - 1)(x + 1)^2$

d $x^2(5x + 1)(25x + 3)$

e $(5x + 1)e^{5x}$

f $4(2x + 3)x^2 e^{2x}$

g $1 + \ln x$

h $\dfrac{x^2}{x + 3} + 2x\ln(x + 3)$

2 0.2, -1

3 $y = -13ex + 3e$

4 $a = 1$, $b = 8$, $c = 2$ (or multiples of these values)

5 a Max at $\left(\dfrac{1}{2}, \dfrac{1}{2e}\right)$

b Min at $(0, 0)$, max at $(2, 4e^{-2})$

6 $-\dfrac{1}{3e}$

7 $2(x + 1)(x - 1)(2x + 3)$

8 a $\dfrac{1}{(x + 1)^2}$ **b** $-\dfrac{1}{(x - 1)^2}$

c $\dfrac{1 - x^2}{(x^2 + 1)^2}$ **d** $\dfrac{8}{(4 - x)^2}$

e $\dfrac{26}{(5 - x)^2}$ **f** $\dfrac{1 - 2x}{(1 + 2x)^3}$

g $\dfrac{2(1 - x)}{(x + 2)^3}$ **h** $\dfrac{3(1 - x)}{(x + 3)^5}$

9 a $\dfrac{e^x(x - 1)}{x^2}$ **b** $\dfrac{2(4x - 5)e^{2x + 1}}{(4x - 3)^2}$

c $\dfrac{3(1 + 4x)}{e^{1 - 4x}}$ **d** $e^{x + 3}$

e $\dfrac{1 - \ln x}{x^2}$ **f** $\dfrac{2(1 - \ln x)}{x^2}$

g $\dfrac{1 - 2\ln x}{x^3}$ **h** $\dfrac{x(2\ln x - 1)}{(\ln x)^2}$

10 -40

11 $(0, 0)$, $\left(\dfrac{2}{3}, \dfrac{4}{9}\right)$

12 a $y = x$ **b** $(3, 3)$

13 $3x(3x^3 + 2)e^{x^3}$

14 a $a = 14$, $b = 5$ **b** $-1, \dfrac{1}{2}, -\dfrac{5}{14}$
c $y = -5x + 1$

15 $k = 48$

16 $\dfrac{1}{(1 + x^2)^{\frac{3}{2}}}$

Exercise 4E (p. 88)

1 a $-3\sin 3x$ **b** $5\cos 5x$
c $3\cos x$ **d** $-6\sin(3x - 1)$
e $-2\cos 4x$ **f** $-2\sin 4x$

g $-\cos\left(\dfrac{\pi}{2} - x\right)$

h $-6\sin 2x + 12\cos 6x$

2 $-2 + \dfrac{1}{\sqrt{2}}$

3 $y + 12x = 2\pi$

4 $\dfrac{\pi}{3}, \dfrac{2\pi}{3}$

5 a $2x\cos x^2$ **b** $-\dfrac{1}{2\sqrt{x}}\sin\sqrt{x}$

c $-\dfrac{1}{x^2}\cos\left(\dfrac{1}{x}\right)$ **d** $-\dfrac{\pi}{180}\sin x^\circ$

e $\dfrac{\pi^2}{180}\cos(\pi x)^\circ$

6 a $3\cos 3x\,e^{\sin 3x}$ **b** $-8\sin x\,e^{2\cos x}$
c $2\cot 2x$ **d** $-2\tan x$
e $-e^x\sin(e^x)$

7 a $\dfrac{dy}{dx} = \frac{1}{15}\sec 3y$

b $\dfrac{dy}{dx} = -2\operatorname{cosec}(2y - 1)$

c $\dfrac{dy}{dx} = \frac{1}{2}\operatorname{cosec} 4y$

d $\dfrac{dy}{dx} = \dfrac{1}{6(\cos 3y - \sin 2y)}$

8 $-3\sin x + 4\cos x$

9 a π **b** $10\,\mathrm{m\,s^{-1}}$

10 $-\dfrac{1}{\sqrt{2}}\,\mathrm{m\,s^{-1}}$ **b** $5\,\mathrm{m}$

c $-2 - 3\dfrac{\sqrt{3}}{2}\,\mathrm{m\,s^{-2}}$

11 $A(0, 1)$, $B\left(\frac{1}{2}\pi, 0\right)$, $C(\pi, -1)$,
$D\left(\frac{3}{2}\pi, 0\right)$, $E\left(-\frac{1}{2}\pi, 0\right)$

Exercise 4F (p. 92)

1 a $-x\sin x + \cos x$
b $2x\cos 2x + \sin 2x$
c $x(x\cos x + 2\sin x)$
d $\cos^2 x - \sin^2 x$

e $\dfrac{x\cos x - \sin x}{x^2}$

f $-\dfrac{2x\sin 2x + \cos 2x}{x^2}$

g $\dfrac{\sin x - x\cos x}{\sin^2 x}$

h $\dfrac{x(2\cos x + x\sin x)}{\cos^2 x}$

i $e^x(2\cos 2x + \sin 2x)$

j $-e^{-3x}(2\sin 2x + 3\cos 2x)$

k $\dfrac{e^x(\sin x - \cos x)}{\sin^2 x}$

l $-(\sin x + \cos x)e^{-x}$

3 $x^2 \sin x$

5 0.75π

6 a $\dfrac{2}{x} - \tan x$ **b** $\cot x - \tan x$

   c $-2\operatorname{cosec} x$

7 a $5\sec^2 5x$ **b** $\frac{1}{4}\sec^2\left(\frac{1}{2}x\right)$

   c $\sec^2 3x$ **d** $4\sec^2 4x$

   e $4\sec 4x \operatorname{cosec} 4x$ **f** $2\sec x \operatorname{cosec} x$

   g $x\sec^2 x + \tan x$ **h** $\dfrac{x\sec^2 x - \tan x}{x^2}$

   i $\dfrac{\sec^2 x}{2\sqrt{\tan x}}$ **j** $\pi\sec^2\left(\dfrac{\pi}{4} + \pi x\right)$

8 a $y = 2x + 1 - \frac{1}{2}\pi$ **b** $4y + x = 4 + \dfrac{\pi}{4}$

9 a $\dfrac{dy}{dx} = \frac{1}{2}\cos^2 2y$

   b $\dfrac{dy}{dx} = 2\sin^2 3y$

   c $\dfrac{dy}{dx} = -\sin 2y \tan 2y$

   d $\dfrac{dy}{dx} = \cos(4y+1)\cot(4y+1)$

10 2

11 a $-\operatorname{cosec}^2 x$ **b** $-\operatorname{cosec}^2 x$

12 a $-5\operatorname{cosec} 5x \cot 5x$

   b $-6\operatorname{cosec}^2 3x$

   c $-12\sec 3x \tan 3x$

   d $3\tan x \sec^3 x$

   e $\sec^2 x$

   f $e^x \sec^2(e^x)$

13 $y + x = 2$

Exercise 4G (Review) (p. 94)

2 $c = 50, n = 3$

3 $a = 10, b = 19$

4 a $y = -\dfrac{1}{16}x + \dfrac{1}{4}$ **b** -12.5

5 a $\left(-\ln 2, -\frac{1}{4}\right)$

6 $x > 3$

7 a $3e^6$

   b $(0, -5e^6), \left(\frac{5}{3}, 0\right)$

9 a $\min\left(\frac{1}{2}, 4\right), \max\left(-\frac{1}{2}, -4\right)$

   b $x = 0$

10 a $(-2, 0)$ and $(2, 0)$

   b $P(0, 0.5)$

11 Min at $(0, 0)$, max at $(2, 108)$

12 a $2x\sec^2(x^2)$ **b** $\dfrac{1}{2\sqrt{x}}\sec^2(\sqrt{x})$

13 $-\operatorname{cosec} x \cot x$

---

14 a $(14x+1)(x-1)^2(2x+3)^3$

   b $\dfrac{(2x-17)(2x+3)^3}{(x-1)^4}$

15 a $x^2(x\cos x + 3\sin x)$

   b $x^2 e^x(x+3)$

   c $(1 - 2x^2)e^{-x^2}$

16 $\left(-\frac{1}{6}, -\frac{1}{3}\right)$

17 a $y = -8x + \pi$

   b $y = 12x + 4\pi - 3\sqrt{3}$

18 2π

19 0.5

Examination questions 1 (p. 98)

1 $\dfrac{a}{4} < x < \dfrac{a}{2}$

2 a $(1, 0), (2, 0), (0, 2)$

   b Range $f(x) \geqslant -1$

   c $4, \frac{2}{3}$

3 a $(-1, 0), (2, 0), (0, 2)$

   b Range $f(x) \leqslant 3$

   c $x < -2, x > \frac{4}{3}$

4 a $\left(-\dfrac{3}{2}, 0\right), (0, 3); (0, -9), \left(\pm\dfrac{3}{\sqrt{2}}, 0\right)$

   b 2 roots; $3, -\frac{1}{2}\left(1 + \sqrt{13}\right)$

5 a $-0.10067, -1.4701$

   b ii   3

6 a Stretch scale factor $\frac{1}{3}$ parallel to x-axis

   b Translation by $\begin{pmatrix} 3 \\ 0 \end{pmatrix}$

   c Reflection in $y = x$

7 a Stretch scale factor $\frac{1}{2}$ parallel to x-axis followed by translation of $\begin{pmatrix} 0 \\ 4 \end{pmatrix}$

   b i -4 **ii** $4x + y = \dfrac{\pi}{2} + 5$

8 b $70.5°, 104.5°, 255.5°, 289.5°$

9 $198°, 342°$

11 a $\dfrac{4}{x} + \dfrac{1}{x^2}$ **b** $f'(x) > 0$ since $x > 0$

   c f^{-1} exists as f is one–one

12 a $\dfrac{1}{x} - 4x$ **c** -8 **d** Maximum

14 a i $2x - 3 + \dfrac{1}{x}$

   b ii $1, \frac{1}{2}$

     iii $2 - x^{-2}$

     iv $1, -2$

15 a ii 2 roots

   b i $1 - \dfrac{2}{x}$ **iii** $\dfrac{2}{x^2}$

   c ii $p = 9, q = 42$

---

16 a $(2x+5)\cos 3x + (x^2 + 5x + 4)$ $(-3\sin 3x)$

   b $y = 5x + 4$

5 Integration

Exercise 5A (p. 115)

1 a $12\frac{2}{3}$ **b** $10\frac{2}{3}$

   c 13 **d** $12\frac{2}{3}$

   e 15 **f** 4

2 a $P(-1, 1), Q(3, 9); 10\frac{2}{3}$

   b $P(-2, 8), Q(2, 0); 21\frac{1}{3}$

   c $P(-4, 9), Q(4, 9); 42\frac{2}{3}$

3 a $A(-1, 3), B(1, 3)$

   b $2\frac{2}{3}$

4 a $\frac{32}{5}\pi$ **b** $\frac{3}{4}\pi$

   c $\frac{1296}{5}\pi$ **d** 4π

5 a $\frac{9}{2}\pi$ **b** $\frac{96}{5}\pi$

   c $\frac{17}{6}\pi$ **d** $\frac{512}{15}\pi$

6 $\frac{3}{16}\pi$

7 a 144π **b** $\frac{28}{15}\pi$

   c 2π **d** $\frac{16}{15}\pi$

8 a $A(2, 4)$ **b** $\frac{48}{5}\pi$ **c** $\frac{24}{5}\pi$

9 a $10\frac{2}{3}$

10 a $\frac{9}{4}\pi$ **b** $\frac{224}{15}\pi$

   c $\frac{15}{2}\pi$ **d** 96π

11 $V = \frac{1}{3}\pi r^2 h$

12 b 18π **c** Hemisphere, radius 3

Exercise 5B (p. 121)

1 a $2e^6 - 2$ **b** $\ln 3$ **c** 1

2 $e^3 - 1$

3 $\ln 2$

4 $e - \dfrac{1}{e}$

5 $\ln 6$

6 a $\left(\frac{1}{2}, 3\right)$

   b $2\ln 2 + \frac{21}{2}$

Exercise 5C (p. 129)

1 a $\frac{1}{15}(3x+1)^5 + c$

   b $\frac{1}{20}(2+5x)^4 + c$

   c $-\frac{1}{12}(1-2x)^6 + c$

   d $-\frac{2}{7}\left(5 - \frac{1}{2}x\right)^7 + c$

   e $\frac{2}{3}(2+x)^{\frac{3}{2}} + c$

   f $\frac{1}{6}(4x+1)^{\frac{3}{2}}$

   g $-\dfrac{1}{6(2x-1)^3} + c$

   h $-\frac{1}{2}(2x-1)^{-1} + c$

   i $\frac{2}{3}\sqrt{3x-1} + c$

j $-\dfrac{1}{4(x+3)} + c$

k $-2(3+2x)^{-1} + c$

l $\dfrac{1}{\sqrt{1-2x}} + c$

2 a 156.2   **b** $\frac{1}{3}(7\sqrt{7} - 1)$
  c $\frac{15}{32}$

3 $y = \frac{1}{4}(x+2)^4 + 3$

4 $y = (2x-1)^3 + 1$

5 a $2\frac{2}{3}$ **b** 2   **c** A = 1.75, B = 0.5

6 $\frac{2}{3}\pi$

7 a $\frac{1}{30}(3x^2 + 1)^5 + c$

b $-\frac{1}{6}(2 - x^3)^2 + c$

c $\frac{1}{4}(x^2 - 5)^4 + c$

d $\frac{2}{3}(x^2 + 6)^{\frac{3}{2}} + c$

e $-\dfrac{1}{2(1+x^2)} + c$

f $\frac{1}{4}(x^2 - 3x + 7)^4 + c$

g $\sqrt{3 + x^2} + c$

h $-\frac{1}{6}(x^2 + 2x + 3)^{-3} + c$

i $\frac{2}{3}\sqrt{x^3 + 6x} + c$

8 a $\frac{1}{3}e^{3x+1} + c$ **b** $-\frac{1}{4}e^{1-4x} + c$

c $\frac{1}{4}e^{2x+3} + c$ **d** $\frac{1}{5}e^{5x+1} + c$

e $\frac{1}{4}e^{4x+2} + c$ **f** $-e^{-(x+1)} + c$

10 a $\dfrac{e^2 - 1}{2e}$ **b** $\left(\dfrac{e^4 - 1}{4e^2}\right)\pi$

11 $s = 5 - 2e$

12 a A$(-0.5, 0)$, B$(0, 1)$

b $y + x = 1$ **c** C$(1, 0)$

d $\frac{5}{6}$ **e** $\frac{7}{12}\pi$

13 b $(0, 2), (-2, 0)$

14 a $\frac{3}{2}e^{x^2} + c$ **b** $-e^{-x^3} + c$

c $-\frac{1}{2}e^{-x^2} + c$

15 a $(0.5, \sqrt{e})$ **b** 0.222 (3 s.f.)

16 b $\frac{512}{15}\pi$ **c** $\frac{256}{5}\pi$

17 a $2^x \ln 2$

b i $\dfrac{1}{\ln 2} 2^x + c$ **ii** $\dfrac{1}{2\ln 3} 3^{2x-1} + c$

Exercise 5D (p. 136)

3 a $\frac{1}{20}(6x + 1)(4x - 1)^{\frac{3}{2}} + c$

b $\frac{2}{375}(15x - 4)(5x + 2)^{\frac{3}{2}} + c$

c $\frac{1}{224}(14x + 1)(2x - 1)^7 + c$

d $\frac{2}{3}(x + 4)\sqrt{x - 2} + c$

e $\frac{1}{30}(5x + 13)(x - 1)^5 + c$

f $\frac{1}{168}(x - 2)^6(21x^2 + 156x + 304) + c$

g $x + \dfrac{4}{x - 2} + c$

h $\frac{1}{3}(x - 6)\sqrt{2x + 3} + c$

5 a $\frac{26}{15}$ **b** $\frac{1}{30}$

c $\sqrt{3} - \frac{2}{3}$ **d** $-\frac{7}{20}$

e $\frac{67}{48}$

6 a $a = 3$ **b** 24.3

7 4.16 (2 d.p.)

8 By recognition: a (ii), b (i) and c (ii)

a i $\frac{2}{135}(9x + 8)(3x - 4)^{\frac{3}{2}} + c$

ii $\frac{1}{9}(3x^2 - 4)^{\frac{3}{2}} + c$

b i $\frac{1}{14}(x^2 + 5)^7 + c$

ii $\frac{1}{56}(7x - 5)(x + 5)^7 + c$

c i $\frac{2}{3}(x + 2)\sqrt{x - 1} + c$

ii $\sqrt{x^2 - 1} + c$

9 2

Exercise 5E (p. 144)

1 a $\frac{1}{3}\ln(3x - 2) + c$

b $\frac{3}{5}\ln(2 + 5x) + c$

c $-\frac{1}{2}\ln(4 - 2x) + c$

d $-\frac{4}{5}\ln(3 - x) + c$

2 a $\frac{1}{3}\ln 3$ **b** $\frac{1}{2}\ln 1.4$ **c** $2\ln 1.4$

3 b ii, iv

c i $\ln\frac{2}{3}$ **iii** $\ln 2$

4 b i $\frac{1}{3}\ln\frac{10}{7}$ **ii** $\frac{1}{3}\ln\frac{2}{5}$

5 $a = \frac{1}{2}\pi$, $b = \ln 3$

6 a $\frac{3}{2}\ln(x^2 - 1) + c$

b $-\frac{1}{2}\ln(1 - x^2) + c$

c $\ln(x^2 + x - 2) + c$

d $\frac{1}{3}\ln(3x^2 - 9x + 4) + c$

e $\frac{1}{4}\ln(4 + 2e^{2x}) + c$

f $3\ln(e^x + 2) + c$

7 a $x - 2\ln(x + 2) + c$

b $\frac{3}{2}x - \frac{9}{4}\ln(2x + 3) + c$

c $-2x - 6\ln(3 - x) + c$

8 a $2 + 2\ln 2$ **b** $2 + 2\ln 2$

c $-\frac{1}{2} - \ln\frac{3}{2}$

9 $a = 13.5$

10 $\ln 7$

Exercise 5F (Review) (p. 145)

1 a e^x **b** $2e^{4x}$ **c** e^{-x}

d $-\frac{1}{12}e^{-4x}$ **e** $-e^{-5x}$ **f** $-2e^{\frac{3x}{2}}$

2 a $\ln x$ **b** $4\ln x$ **c** $\frac{1}{2}\ln x$

d $\frac{1}{3}\ln x$ **e** $-\frac{2}{3}\ln x$ **f** $-5\ln x$

3 $\frac{1}{3}e^3 - \frac{1}{3}e^{\frac{3}{2}} - \ln 2$

4 $\dfrac{1}{2} - \dfrac{1}{2e^4}$

5 a $(0, 0)$ **b** $e^{-4} + 3$

6 b $\frac{1}{3}(e^3 - 1)$

7 a $\frac{1}{12}(3x + 2)^4 + c$

b $\frac{1}{6}(2x + 3)^3 + c$

c $-\dfrac{1}{3(3x - 4)} + c$

d $\frac{2}{15}(5x - 2)^{\frac{3}{2}} + c$

e $\frac{1}{3}e^{3x-7}$

f $-\frac{1}{2}e^{2-4x} + c$

8 a $\frac{2}{3}\sqrt{3x + 2} + c$ **b** $\frac{3}{8}(4x - 2)^{\frac{2}{3}} + c$

9 a $60\frac{2}{3}$ **b** $\ln\frac{5}{6}$

c $\frac{2}{3}\ln\frac{2}{5}$ **d** $\frac{1}{2}e(e^4 - 1)$

e $3(e^2 - e^{-1})$ **f** $\frac{1}{6}$

10 $\ln 1.5$

11 $(e^2 - 1)\pi$

12 42π

13 a $(2, 4), (4, 2)$ **b** $6 - 8\ln 2$

c $2\frac{2}{3}\pi$

14 $\frac{16}{5}\pi$

15 a $1 + e^2$

b Minimum at $(1, 2e)$

c $e^2 - 1$

16 a $\frac{1}{378}(18x - 1)(3x + 1)^6 + c$

b $\frac{2}{375}(5x - 2)^{\frac{3}{2}}(15x + 4) + c$

c $\dfrac{y^2 - 8y + 32}{y - 4} + c$

17 a $\frac{1}{3}(x^2 + 3)^{\frac{3}{2}} + c$

b $\frac{1}{3}\sqrt{2 + 3x^2} + c$

c $-2e^{-x^2} + c$

18 $\frac{1}{2}\ln\frac{3}{4}$

19 2

6 Further integration

Exercise 6A (p. 158)

1 a $-\frac{1}{3}\cos 3x + c$

b $\frac{1}{3}\sin 3x + c$

c $-\frac{1}{2}\cos 4x + c$

d $\sin 2x + c$

e $\frac{1}{12}\cos 6x + c$

f $\frac{3}{2}\sin 4x + c$

g $-\frac{1}{2}\cos(2x + 1) + c$

h $\frac{3}{2}\sin(2x - 1) + c$

i $-\frac{4}{3}\cos(\frac{1}{2}x) + c$

2 a $\frac{1}{2}\tan 2x + c$

b $3\tan\left(x - \dfrac{\pi}{4}\right) + c$

c $\tan x + c$

d $2\tan(\frac{1}{2}x) - x + c$

e $\frac{1}{4}\tan 4x + c$

f $\tan x + c$

3 a 1   **b** $\frac{4}{3}$ **c** $\frac{4}{3}\sqrt{3}$

4 a 0

b Half the area is below the x-axis and half above it

5 $\frac{1}{2}$

6 b 2π

7 a $\sqrt{3} - \frac{1}{3}\pi$ **c** $2 - \frac{1}{8}\pi^2$

8 $\frac{8}{3}\pi^{\frac{2}{3}} - 2\sqrt{3}\pi$

9 a $V_1 = 2\pi$ **b** $V_2 = \pi$
   c π

10 $\frac{1}{2}\pi - 1$

11 12

12 a $\frac{1}{3}\sec 3x + c$ **b** $2\operatorname{cosec}\left(\frac{1}{2}x\right) + c$
   c $-\frac{2}{3}\cot 3x + c$ **d** $-\operatorname{cosec} x + c$

13 a $\cot x + c$ **b** $\tan x - x + c$
   c $-\cot x - x + c$ **d** $-\frac{1}{2}\cot 2x + c$

14 a $2\sin(x^2) + c$
   b $2\sin\sqrt{x} + c$
   c $e^{\tan x} + c$
   d $3\ln(2 + \sin x) + c$
   e $-\frac{1}{2}e^{-\sin 2x} + c$

15 a $-\frac{1}{3}\ln(\cos 3x) + c$
   b $\frac{1}{3}\ln(\sin 3x) + c$
   c $\ln(\sin x) + c$
   d $-\frac{1}{2}\ln(\cos 2x) + c$
   e $\frac{1}{4}\ln(\sin 4x) + c$

16 c $\left(\sqrt{3} - \frac{\pi}{3}\right)\pi$

Exercise 6B (p. 167)

1 a $x\sin x + \cos x + c$
   b $-\frac{1}{2}x\cos 2x + \frac{1}{4}\sin 2x + c$
   c $-3x\cos\left(x + \frac{\pi}{3}\right) + 3\sin\left(x + \frac{\pi}{3}\right) + c$
   d $xe^x - e^x + c$
   e $-\frac{1}{3}xe^{-3x} - \frac{1}{9}e^{-3x} + c$
   f $\frac{2}{3}xe^{3x+1} - \frac{2}{9}e^{3x+1} + c$

2 a and **b** $\frac{1}{56}(7x - 1)(1 + x)^7 + c$

3 $-\frac{1}{2}xe^{-x} - \frac{1}{2}e^{-x} + c$

4 a $\frac{1}{9}\pi$
   b $e + 1$

5 $A = \frac{1}{2}\pi$, $B = 2\pi$, $C = 2\pi$

6 a $x\tan x + \ln\cos x + c$
   b $x\tan x + \ln\cos x - \frac{1}{2}x^2 + c$
   c $-x\cot x + \ln\sin x + c$
   d $-x\cot x + \ln\sin x - \frac{1}{2}x^2 + c$

7 a $x^2\sin x + 2x\cos x - 2\sin x + c$
   b $-\frac{1}{2}x^2\cos 2x + \frac{1}{2}x\sin 2x$
     $+ \frac{1}{4}\cos 2x + c$
   c $-\frac{1}{3}x^2e^{-3x} - \frac{2}{9}xe^{-3x} - \frac{2}{27}e^{-3x} + c$
   d $e^x(x^3 - 3x^2 + 6x - 6) + c$

8 a $P(-1, -2e^{-1})$ **b** $4e^{-1} - 2$
   c $\pi(1 - 5e^{-2})$

9 a $\frac{1}{4}x^2(2\ln x - 1) + c$
   b $\frac{1}{9}x^3(3\ln x - 1) + c$

   c $\frac{1}{25}x^5(5\ln x - 1) + c$
   d $-\frac{1}{4}x^{-2}(2\ln x + 1) + c$

10 a $x(\ln 2x - 1) + c$
   b $2x(\ln x - 1) + c$
   c $(x - 1)\ln(x - 1) - x + c$

11 a $\frac{1}{2}e^x(\sin x + \cos x) + c$
   b $\frac{1}{13}e^{2x}(2\sin 3x - 3\cos 3x) + c$
   c $\frac{1}{13}e^{3x}(2\sin 2x + 3\cos 2x) + c$
   d $\frac{1}{25}e^{4x}(4\sin 3x - 3\cos 3x) + c$

12 a $2xe^{x^2}$, $\frac{1}{2}e^x(x^2 - 1) + c$
   b $\frac{1}{3}e^{x^3}(x^3 - 1) + c$
   c $-\frac{1}{2}e^{-x^2} + c$
   d $-\frac{1}{2}e^{-x^2}(x^2 + 1) + c$

13 a $a^x\ln a$, $\dfrac{a^x}{(\ln a)^2}(x\ln a - 1) + c$
   b $\dfrac{1}{x\ln a} + c$, $\dfrac{1}{2}x^2\log_a x - \dfrac{x^2}{4\ln a} + c$
   c $\dfrac{x}{\ln a}(\ln x - 1) + c$

Exercise 6C (p. 171)

1 a $\sin^{-1}\dfrac{x}{5} + c$ **b** $\sin^{-1}\dfrac{\sqrt{3}x}{3} + c$
   c $\sin^{-1}\dfrac{x}{8} + c$ **d** $\sin^{-1}\dfrac{\sqrt{3}x}{9} + c$

2 a $\dfrac{1}{5}\tan^{-1}\dfrac{x}{5} + c$
   b $\dfrac{\sqrt{7}}{7}\tan^{-1}\dfrac{\sqrt{7}x}{7} + c$
   c $\dfrac{\sqrt{2}}{8}\tan^{-1}\dfrac{\sqrt{2}x}{8} + c$
   d $\dfrac{1}{6}\tan^{-1}\dfrac{x}{6} + c$

3 a $\dfrac{\pi}{12}$ **b** $\dfrac{\pi}{6}$
   c $\dfrac{2\pi}{3}$ **d** $\dfrac{\pi}{18}$
   e $\dfrac{\sqrt{5}\pi}{20}$ **f** $\dfrac{\pi}{3}$

4 a i $\sin^{-1}\dfrac{x}{6} + c$
    ii $-\cos^{-1}\dfrac{x}{6} + c$
   b i $\dfrac{\pi}{3}$ **ii** $\dfrac{\pi}{3}$

Exercise 6D (Review) (p. 172)

1 a $-\frac{1}{3}\cos(x^3) + c$ **b** $-\frac{1}{7}\cos^7 x + c$
   c $\frac{1}{7}\tan^7 x + c$

2 a $\frac{1}{3}\sin 3x + c$ **b** $\frac{1}{2}\tan 2x + c$
   c $\sin^2 2x + c$

4 a $B\left(\frac{1}{2}\pi, 0\right)$ **b** $P\left(\dfrac{\pi}{4}, \dfrac{\sqrt{2}}{2}\right)$

6 a 0.144 (3 s.f.)   **b** 0.169 (3 s.f.)

7 $\dfrac{\pi^2}{2}$

8 a $\frac{3}{2}x\sin 2x + \frac{3}{4}\cos 2x + c$
   b $-\frac{3}{16}e^{-4x}(4x + 1) + c$
   c $\frac{1}{49}x^7(7\ln x - 1) + c$
   d $x^3\sin x + 3x^2\cos x - 6x\sin x$
     $- 6\cos x + c$

9 a $\sin^{-1}\dfrac{x}{7} + c$ **b** $\sin^{-1}\dfrac{\sqrt{2}x}{10} + c$

10 a $\dfrac{1}{9}\tan^{-1}\dfrac{x}{9} + c$ **b** $\dfrac{\sqrt{5}}{20}\tan^{-1}\dfrac{\sqrt{5}x}{20} + c$

11 a $\dfrac{\sqrt{3}\pi}{36}$ **b** $\dfrac{\pi}{2}$

7 Numerical methods

Exercise 7A (p. 181)

1 a $-43, -11, 3, 5, 1, -3, -1, 13, 45$
   b $(-3, -2), (0, 1), (2, 3)$

2 b $-2.2, 0.7, 1.7$

3 c 1.86

4 a 2 roots   **c** $1.0, -2.0$

5 c 2.88

6 c 0.5

7 b $1.4, 9.9$

Exercise 7B (p. 186)

1 b 0.11 (2 d.p.)

2 b i $x_2 = 0.208$, $x_3 = 0.208\,65$,
     $x_4 = 0.208\,71$
    ii $x_2 = 4.75$, $x_3 = 4.789$, $x_4 = 4.791$
   c $0.209, 4.79$

3 a $x_2 = 4$, $x_3 = 3.5$, $x_4 = 3.464\,286$,
     $x_5 = 3.464\,101\,62$, $x_6 = 3.464\,101\,615$

4 b 11.26943

5 b 0.1001

6 b $1.934\,5632$

7 c 1.97

8 c i $-0.339\,8769$
    ii $2.601\,6791$
    iii $-2.261\,8022$

9 c 0.26
   d 2.3

10 a $x^2 - 4x + 1 = 0$
   b $x^3 - 10 = 0$
   c $x^5 - 50 = 0$
   d $x^3 + x + 5 = 0$
   e $(x - 1)e^x - 4x^2 = 0$
   f $x^2 + 4x - \sin x - 2 = 0$

11 a 0.2679
   b 2.1544
   c 2.1867
   d -1.5160
   e 2.8688
   f 0.5547

12 a 2.414 **b** 1.27

13 c $1.749\,031$ **d** 0.5064

14 c $(2.0867, 11.2602)$

15 b $2.857\,391$

Exercise 7C (p. 194)

1 a $1.45;\ 1.46$ **b** $0.748;\ 0.759$
  c $0.775;\ 0.773$ **d** $2.68;\ 2.65$

2 $0.2585;\ 0.2585$

3 0.3161

4 1.48

5 108

6 a $4.01;\ 3.98;\ 3.98$
  b I is 4.0 to 2 s.f.

Exercise 7D (Review) (p. 195)

1 b $0.347\,2964$

2 a 0.54
  b $x + x\cos x - 1 = 0$
  c $1, 0, 1, 0, 1$

3 b $-1.841\,41,\ 1.146\,19$

4 c $x_{n+1} = \frac{1}{6}(x_n^3 + 1),\ 0.167\,449$

5 a $c = 1$
  b $(0, 1),\ (2, 3)$
  c $0.36,\ 2.15$

6 b $1.213\,\mathrm{m}^3;\ 1.218\,\mathrm{m}^3$

7 a 1.49 **b** 2.62

8 0.4069

8 Proof

Exercise 8B (p. 205)

2 a Not true   **b** Not true
  c True   **d** True

3 a Not true
  b True
  c True
  d Not true
  e True
  f Not true, e.g. when $x = 41$

Examination questions 2 (p. 210)

1 2.83

2 a i $-2e^{-2x} - \dfrac{3}{x^2}$
  b $f(x) > 3$

3 b $3e^{3x} - 24,\ 9e^{3x}$
  c $x = \ln 2,\ 72,$ minimum
  d $\frac{1}{3}(e^9 - e^6) - 60$

4 a -2 **b** $0.5\ln 3$
  c i $2e^{2x}$ **d** $-1 + \frac{3}{2}\ln 3$

5 a $6x(x + 1)$ **b** $k = \frac{1}{6}\ln 5$

6 a $6\ln\sec\theta - \tan\theta + c$
  b $1.1,\ 1.3,\ 4.2,\ 4.5$

7 a 3
  b i $-2e^{2x}$ **ii** -8
  c i $\ln 2$ **ii** $4x - \frac{1}{2}e^{2x} + c$
  iii $2\ln 2 - 1$

8 b i $\frac{1}{2}e^{2x} - 2x + c$

9 a $x = \frac{\pi}{8}$ **c** $\frac{1}{4}\ln 2 - \frac{\pi^2}{32}$

10 a $\frac{6}{5}\pi$
  b $y = \frac{1}{2}(5x - 1)$
  c i $3x\sqrt{x^2 + 3}$ **ii** $\frac{1}{3}\left(8 - 3\sqrt{3}\right)$

11 b $0 \leqslant f(x) \leqslant \sqrt{3}$
  c i $\frac{1}{5}\ln 2$

12 a $\cos 2x + 2x\sin 2x$

13 $\dfrac{1}{2}\left[-\dfrac{1}{1 + 2x} + \dfrac{1}{2(1 + 2x)^2}\right] + c$

14 c $k = 8$

15 a 0.737
  b i $0.660,\ 0.633,\ 0.645,\ 0.640$
  ii $x = 2^{-x}$

16 c $4.88,\ 4.95$

17 1.434

18 a 0.38182 **b** 0.38177

C3 sample exam paper (p. 217)

1 a i $-\sin x$
  ii $-2\ln(1 + \cos x) + c$

2 b $x_1 = 0.39602\ldots,\ x_2 = 0.39541\ldots,$
  $x_3 = 0.39532\ldots,\ x_4 = 0.39530\ldots;$
  0.395

3 a i $f(x) > -1$
  ii $(0,\ e^3 - 1),\ (-\frac{3}{2},\ 0),\ y = -1$
  b $a = 2,\ b = -1.5$

4 b $x = 26.6°,\ 135°,\ 206.6°,\ 315°$

5 a $x\ln x - x + c$
  b Translation (in x-direction) of $\begin{bmatrix} -2 \\ 0 \end{bmatrix}$
  one way stretch in y-direction scaling factor 3
  d i 4.871 **ii** 138

6 a $3\sin^2 x + 6x\sin x\cos x$
  b $x + 3y - 5\pi = 0$
  c $\dfrac{25\pi^2}{6}$

7 b ii $2 \pm \sqrt{19}$

9 Algebra

Exercise 9A (p. 223)

1 a $\dfrac{y - x}{xy}$ **b** $\dfrac{x^2 + y^2}{xy}$

c $\dfrac{1 + a}{a^2}$ **d** $\dfrac{a + b}{a^2 b^2}$

e $\dfrac{2x}{(x - h)(x + h)}$ **f** $\dfrac{-h(2x + h)}{x^2(x + h)^2}$

g $\dfrac{3x}{(1 - x)(2 + x)}$ **h** $\dfrac{-(x^2 - 2x + 4)}{(x^2 + 2)(2 + x)}$

i $\dfrac{n + 1}{n + 2}$ **j** $\dfrac{x^2 + 3x + 3}{(x + 1)^2}$

2 a $\dfrac{x + 3}{x}$ **b** $\dfrac{2(x^2 - 2)}{x^2}$

c $\dfrac{6x + 11}{4(2x - 3)}$ **d** $\dfrac{5x^2 - 7}{x^2 - 1}$

e $\dfrac{22 - 13x}{2(3x - 4)}$ **f** $\dfrac{3 - 2x}{x^2 + 2x - 3}$

g $\dfrac{2(2x^2 + 2x + 1)}{2x + 1}$ **h** $\dfrac{3(x - 1)^2}{x - 2}$

i $\dfrac{18xyz^2}{3z^2 - 2}$

3 a $\frac{1}{4}$ **b** $\dfrac{x}{a}$ **c** $\dfrac{1}{a - b}$

d -1 **e** $\dfrac{1}{x + 3}$

4 a $\dfrac{xy}{2}$ **b** $\dfrac{c}{a + b}$ **c** $\dfrac{x^2}{y^2}$

d $\dfrac{y}{x}$ **e** -5 **f** $\dfrac{30by}{a^2}$

g $\dfrac{3t(t - 5)}{2(t - 1)(t + 5)}$

5 a $\dfrac{b}{a}$ **b** $\dfrac{x}{a + b}$ **c** $\dfrac{x^2}{x - 1}$

d $x + y$ **e** $\dfrac{3x + 1}{3x - 1}$ **f** $-y$

g $\dfrac{1 - x}{1 + x}$ **h** $\pm\left(\dfrac{1 - t}{1 + t}\right)$

6 a $\dfrac{7 - 3x - 5x^2}{(x + 2)^2(3x - 1)}$

b $\dfrac{2 - 4x - 3x^2}{(2 + 3x^2)(1 - x)}$

c $\dfrac{2x + 3}{(x + 3)(x - 2)}$

d $\dfrac{1 - x}{(x + 2)(x + 1)}$

e $\dfrac{-x^3 + 6x^2 - 7x + 6}{(x^2 + 1)(x - 1)^2}$

f $\dfrac{2x^2 + 6x + 1}{(x - 1)(x + 2)(x + 3)}$

g $\dfrac{8x^2 - x + 23}{(x - 1)(x^2 + 4)}$

h $\dfrac{3x(4x^2 + 3)}{(2x - 1)(2x^2 + 1)}$

i $\dfrac{x^3 + 3x^2 - 5x - 9}{(x + 4)(x - 1)}$

j $\dfrac{2x^3 - 3x^2 - x + 8}{x^2 - 1}$

k $\dfrac{2x+9}{(x-5)(2-x)}$

l $\dfrac{3x^2+5x+4}{x^2-1}$

m $\sqrt{a+b}$

n $(1+x^3)^{-\frac{3}{2}}$

7 a $\dfrac{x^2+50}{x(x-5)}$ b $x=-4$ or $x=15$

8 a $\dfrac{5x^2+1}{(x-1)(x+2)}$ b $x=\dfrac{3}{2}$ or $x=4$

Exercise 9B (p. 230)

1 a $x^2-3x+4,\ 5$ b $5x^2+7x-3,\ -7$
c $x^2-7,\ 3$ d $x^3-7x+4,\ -1$
e $2x^2+x-1,\ 3$ f $x^2-3x-2,\ -5$

2 a $1+\dfrac{4}{x+3}$ b $1-\dfrac{8}{x+5}$

c $2-\dfrac{5}{x+3}$ d $1-\dfrac{6}{x^2+4}$

e $1-\dfrac{2x-4}{x^2+2x}$ f $1-\dfrac{4}{x+3}$

g $-1+\dfrac{10}{x+7}$ h $-2-\dfrac{3}{x-2}$

3 a x^2-3x+1 b x^2+3x-1
c $3x^2-2x+1$ d $2a-1$

4 a $x+3$ b $x-2$
c x^2-2x+3 d x^2+3x-4
e $2x^2-3x-1$ f x^2+4x-5
g $4x^2+2x-5$ h $2x^2+3x-4$
i x^2-x+3

5 a x^2+3, remainder -4
b x^2+4x+4, remainder -8
c x^2-5x+3, remainder -15
d $2x^2+5x-3$, remainder $20x-12$
e $4x^2+8$, remainder $-4x+16$
f x^2-4x-2, remainder 6
g x^2-3x+4, remainder 12
h $4x^2+2x-7$, remainder -2
i x^2+2x+3, remainder $-9x+3$

6 a $A=5, B=3$
b $A=2, B=-5$
c $A=3, B=4$
d $A=4, B=-2, C=1$
e $A=1, B=2, C=-6$
f $A=2, B=3, C=4$

7 $\dfrac{3x-5}{4-3x}$

8 a $2x-1,\ \dfrac{2x+1}{2x+7}$

b $2x-5,\ \dfrac{x+3}{x-3}$

c $2x+1,\ \dfrac{2x-1}{x(x-5)}$

d $x+2,\ \dfrac{x(x-9)}{2x-3}$

e $3x-1,\ \dfrac{x+1}{x^2(x-5)}$

f $2x-1,\ \dfrac{3x+1}{x^2-9}$

Exercise 9C (p. 239)

1 a $\dfrac{1}{x}-\dfrac{3}{x+1}$ b $\dfrac{2}{4x-1}-\dfrac{2}{x}$

c $\dfrac{1}{x-2}-\dfrac{1}{x+2}$ d $\dfrac{1}{2-x}-\dfrac{1}{2+x}$

e $\dfrac{1}{x-1}+\dfrac{1}{x+2}$ f $\dfrac{1}{x-2}-\dfrac{2}{3x-5}$

g $\dfrac{3}{x-3}-\dfrac{4}{x-1}$ h $\dfrac{2}{1+2x}+\dfrac{1}{x-2}$

2 a $\dfrac{1}{2(x+1)}+\dfrac{1}{2(x-3)}-\dfrac{1}{x+2}$

b $\dfrac{1}{x+1}-\dfrac{2}{x-1}+\dfrac{3}{x+2}$

c $\dfrac{1}{x+1}-\dfrac{2}{x+2}+\dfrac{1}{x-2}$

d $\dfrac{1}{x}-\dfrac{3}{2x+1}+\dfrac{2}{x-1}$

e $\dfrac{1}{2(x+2)}+\dfrac{1}{x+1}-\dfrac{1}{2(x-2)}$

f $\dfrac{2}{1-x}+\dfrac{1}{2-x}-\dfrac{3}{3-x}$

3 a $\dfrac{1}{x+3}-\dfrac{2}{(x+3)^2}$

b $\dfrac{5}{x-2}-\dfrac{3}{x-1}-\dfrac{4}{(x-1)^2}$

c $\dfrac{4}{x+1}-\dfrac{2}{(x+1)^2}-\dfrac{3}{x-4}$

d $\dfrac{1}{x+1}+\dfrac{3}{(x+1)^2}-\dfrac{4}{x-1}$

e $\dfrac{1}{2(2x-1)}-\dfrac{1}{(2x-1)^2}-\dfrac{3}{2(x+4)}$

f $\dfrac{3}{x-4}-\dfrac{2}{x+1}+\dfrac{1}{(x+1)^2}$

4 a $1+\dfrac{2}{x-2}$

b $1+\dfrac{1}{x-1}-\dfrac{1}{x+1}$

c $2+\dfrac{5}{2(x-1)}-\dfrac{5}{2(x+1)}$

d $1+\dfrac{3}{x}-\dfrac{4}{x+1}$

e $1+\dfrac{2}{x+1}-\dfrac{1}{x-2}$

f $x+\dfrac{3}{4(x-1)}+\dfrac{1}{4(x+3)}$

g $x-1-\dfrac{3}{4(x-2)}+\dfrac{3}{4(x+2)}$

h $x^2+x+1+\dfrac{2}{x-2}-\dfrac{1}{x+3}$

5 a $\dfrac{4}{x+1}+\dfrac{2}{x-2}-\dfrac{3}{x-3}$

b $\dfrac{3}{x-1}-\dfrac{1}{x}+\dfrac{2}{x+1}$

c $\dfrac{23}{4(3x+1)}-\dfrac{1}{4(x+1)}-\dfrac{7}{2(x+1)^2}$

d $\dfrac{1}{2(5-x)}-\dfrac{1}{2(5+x)}$

e $1+\dfrac{5}{3(x-2)}-\dfrac{2}{3(x+1)}$

6 a $\dfrac{1}{6(x+2)}+\dfrac{10}{3(x-1)}-\dfrac{7}{2x}$

b $\dfrac{3}{2x^2}-\dfrac{3}{4x}+\dfrac{3}{4(x+2)}$

c $\dfrac{5}{3+x}+\dfrac{2}{4-x}-\dfrac{3}{4+x}$

d $\dfrac{2}{(2x+1)^2}-\dfrac{5}{2x+1}+\dfrac{3}{x-3}$

e $x+2-\dfrac{1}{2x+1}+\dfrac{3}{x-2}$

Exercise 9D (p. 242)

1 $\ln\dfrac{2x-1}{3x+2}+c$

2 a $A=2, B=-3$
b $2\ln(x+1)+3(x+1)^{-1}+c$

3 a $\frac{1}{6}\ln\dfrac{x-3}{x+3}+c$ b $\frac{1}{12}\ln\dfrac{2x-3}{2x+3}+c$

4 a and b $-\frac{1}{2}\ln(4-x^2)+c$

5 a $\ln(2x+1)^2$
$-\frac{1}{2}\ln((x-3)(x+3)^3)+c$

b $\frac{1}{3}\ln\dfrac{x-2}{x+1}-\dfrac{4}{x-2}+c$

Exercise 9E (Review) (p. 243)

1 a $\dfrac{x-12}{(x+3)(x-2)}$

b $\dfrac{3-x}{1-x^2}$

c $\dfrac{(x+2)(x-1)}{(x^2+1)(x+1)}$

d $\dfrac{3x^2-x+4}{(x-1)^2(x+1)}$

e $\dfrac{7x^2+11x+1}{(x+2)^2(1-3x)}$

f $\dfrac{4x}{(2x-1)(2x+1)^2}$

2 a $\dfrac{1}{3}$ b $\dfrac{1}{x+y}$ c -1

d $\dfrac{1}{x+4}$ e $\dfrac{x}{x-1}$ f $\dfrac{x-7}{x+2}$

3 a $2b^2c$ b yz c $\dfrac{3y}{2}$

d $\dfrac{p}{p+q}$ e 2 f 10

4 a $x^2-7,\ -1$ b $x^2-2x+1,\ 11$
c $x^2+x+1,\ 0$ d $x^2+x,\ 1$
e $x^2+2x+3,\ x-1$
f $x^2+2x+7,\ 0$

5 $x(3x+2)(4x-1),\ (3x+2)(x+2)(x-2);$
$\dfrac{x(4x-1)}{x^2-4}$

6 a $\dfrac{3}{x+2}+\dfrac{1}{x-4}$

b $\dfrac{1}{x-4}-\dfrac{2}{2x+1}$

c $\dfrac{3}{5(x-1)}+\dfrac{2}{5(x-6)}$

d $\dfrac{1}{x+1}+\dfrac{2}{x-1}-\dfrac{3}{2x+1}$

e $\dfrac{2}{x+3}-\dfrac{1}{(x+3)^2}+\dfrac{1}{x-1}$

7 a $2+\dfrac{1}{x}-\dfrac{3}{x-1}$

b $x-\dfrac{2}{x-1}+\dfrac{3}{x+4}$

c $2x+1-\dfrac{4}{x+2}-\dfrac{3}{2x+1}$

8 a $A=1,\ B=1,\ C=-1$

b $x+\ln\dfrac{x-1}{x+1}+c$

10 Binomial expansion

Exercise 10A (p. 254)

1 a $1-2x+3x^2-4x^3,\ -1<x<1$

b $1+\frac{1}{3}x-\frac{1}{9}x^2+\frac{5}{81}x^3,\ -1<x<1$

c $1+\frac{3}{2}x+\frac{3}{8}x^2-\frac{1}{16}x^3,\ -1<x<1$

d $1-x-\frac{1}{2}x^2-\frac{1}{2}x^3,\ -\frac{1}{2}<x<\frac{1}{2}$

e $1-\frac{3}{2}x+\frac{3}{2}x^2-\frac{5}{4}x^3,\ -2<x<2$

f $1+\frac{3}{2}x+\frac{27}{8}x^2+\frac{135}{16}x^3,\ -\frac{1}{3}<x<\frac{1}{3}$

g $1-3x+9x^2-27x^3,\ -\frac{1}{3}<x<\frac{1}{3}$

h $1-\frac{1}{2}x^2,\ -1<x<1$

i $1-\frac{1}{3}x-\frac{1}{9}x^2-\frac{5}{81}x^3,\ -1<x<1$

j $1-x+\frac{3}{2}x^2-\frac{5}{2}x^3,\ -\frac{1}{2}<x<\frac{1}{2}$

k $1-x+\frac{3}{4}x^2-\frac{1}{2}x^3,\ -2<x<2$

l $1-3x+\frac{3}{2}x^2+\frac{1}{2}x^3,\ -\frac{1}{2}<x<\frac{1}{2}$

m $\frac{1}{4}-\frac{1}{16}x+\frac{1}{64}x^2-\frac{1}{256}x^3,\ -4<x<4$

n $\frac{1}{9}+\frac{2}{27}x+\frac{1}{27}x^2+\frac{4}{243}x^3,\ -3<x<3$

o $2+\frac{5}{4}x-\frac{25}{64}x^2+\frac{125}{512}x^3,\ -\frac{4}{5}<x<\frac{4}{5}$

p $2+\frac{1}{4}x-\frac{1}{32}x^2+\frac{5}{768}x^3,\ -\frac{8}{3}<x<\frac{8}{3}$

2 a $1-2x+4x^2,\ |x|<\frac{1}{2}$

b $1+3x+9x^2,\ |x|<\frac{1}{3}$

c $1-\frac{2}{3}x+\frac{4}{9}x^2,\ |x|<\frac{3}{2}$

d $1-x^2+x^4,\ |x|<1$

e $\frac{1}{2}-\frac{x}{4}+\frac{x^2}{8},\ |x|<2$

f $3+\frac{9}{4}x+\frac{27}{16}x^2,\ |x|<\frac{4}{3}$

3 a $1-4x+12x^2,\ |x|<\frac{1}{2}$

b $1+6x+27x^2,\ |x|<\frac{1}{3}$

c $1-\frac{4}{3}x+\frac{4}{3}x^2,\ |x|<\frac{3}{2}$

d $1+2x^3+3x^6,\ |x|<1$

e $\frac{1}{4}-\frac{x}{4}+\frac{3x^2}{16},\ |x|<2$

f $\frac{1}{2}+\frac{3x}{4}+\frac{27x^2}{16},\ |x|<\frac{4}{3}$

4 a $1+\frac{x}{2}-\frac{x^2}{8}+\frac{x^3}{16}$

b $1.000\,500$

5 a $1-4x+12x^2-32x^3$

b 0.9612

6 a $0.998\,999$ **b** 1.0099

c 1.0102

7 a $1+2x+2x^2+2x^3$

b $2-3x+4x^2-5x^3$

c $1-\frac{3}{2}x+\frac{7}{8}x^2-\frac{11}{16}x^3$

8 a $1-3x+6x^2-10x^3$

b $0.970\,59$

c $9.970\,0599\times10^{-10}$

9 a $1+x-\dfrac{x^2}{2}$

b $1.009\,95,\ 10.0995$

c $1000.000\,999\,999\,5$

10 $1-4x-8x^2-32x^3,\ 4.7958$

11 $1+x-\frac{3}{2}x^2+\frac{7}{2}x^3,\ 10.000\,999\,850$

12 $n=4,\ 1+8x+24x^2+32x^3$
$n=-3,\ 1-6x+24x^2-80x^3$

13 $k=-3,\ n=-2;\ 108x^3$

14 a $a=-3,\ n=-4$

b $|x|<\frac{1}{3}$

15 a $a=5,\ b=9,\ c=13$

b $|x|<1$

16 a $\dfrac{1}{4(3+x)}+\dfrac{1}{4(1-x)}$

17 $-\frac{1}{4}+\frac{3}{16}x-\frac{13}{64}x^2$

18 $4+4x+8x^2,\ |x|<1$

19 $\frac{1}{2}+\frac{3}{4}x+\frac{9}{8}x^2,\ |x|<1$

20 a $\dfrac{2}{7(1+2x)}+\dfrac{1}{7(3-x)}$

b $\frac{1}{3}-\frac{5}{9}x+\frac{31}{27}x^2$

21 a $1-\dfrac{x^3}{2}-\dfrac{x^6}{8}$

11 Further differentiation

Exercise 11A (p. 264)

1 a $2y\dfrac{dy}{dx}$ **b** $3y^2\dfrac{dy}{dx}$

c $12y^3\dfrac{dy}{dx}$ **d** $x\dfrac{dy}{dx}+y$

e $x^2\dfrac{dy}{dx}+2xy$ **f** $2xy\dfrac{dy}{dx}+y^2$

g $\dfrac{1}{y}\dfrac{dy}{dx}$ **h** $\dfrac{5}{y}\dfrac{dy}{dx}$

i $\dfrac{2}{x}+\dfrac{3}{y}\dfrac{dy}{dx}$ **j** $\cos y\dfrac{dy}{dx}$

k $\cos y-x\sin y\dfrac{dy}{dx}$

l $2xe^{2y}\left(x\dfrac{dy}{dx}+1\right)$

m $e^x y\left(2\dfrac{dy}{dx}+y\right)$

n $-\dfrac{1}{y^2}\dfrac{dy}{dx}$

o $\dfrac{y-3x\dfrac{dy}{dx}}{y^4}$

p $\cos(x+y)\left[1+\dfrac{dy}{dx}\right]$

2 a $-\dfrac{x}{y}$ **b** $-\dfrac{x^2}{2y^3}$

c $\dfrac{2x+3y}{4y-3x}$ **d** $\dfrac{3x^2-2y^2+7}{4xy}$

e $\dfrac{x(4-3y^2)}{3y(x^2-2)}$

f $-\dfrac{9x^2+4xy+5y^2}{2x^2+10xy+12y^2}$

g $-\dfrac{y^2}{x^2}$ **h** $\dfrac{x-2}{10y}$

3 a $\dfrac{3e^y-4e^x y}{4e^x-3xe^y}$

b $-\dfrac{1}{x}\sin y\cos y$

c $-\dfrac{\sin y+y\cos x}{x\cos y+\sin x}$

d $\dfrac{3y\ln y}{4y^2-3x}$

e $\frac{2}{3}\cot 2x\cot 3y$

f $\dfrac{e^x y\ln y}{y-e^x}$

g $\dfrac{y(4y-3\ln y)}{3x-4xy}$

h $\dfrac{3(x+y)-\sin^2 y}{x\sin 2y-3(x+y)}$

4 a 1 **b** $\frac{1}{9}$

c -1 **d** -1

e $-\frac{1}{6}$ **f** $-\dfrac{\sqrt{3}}{3}$

5 a $-\frac{1}{5}$ **b** $-\frac{8}{11}$ **c** 1

d -1 **e** -2 **f** 0

g $-\dfrac{e}{2}$ **h** $\frac{24}{45}$

7 $4x+3y-20=0$

8 $x+8y+6=0,\ y=8x-17$

9 $16x+21y-90=0$ at $(3,2)$,
$5x+21y+90=0$ at $(3,-5)$

10 $y=4-x,\ y=x$

11 $(e^2+e)x+y=e^2+2e$

12 $x+3y-\dfrac{5\pi}{6}=0,\ y=3x-\dfrac{5\pi}{6}$

13 $7x-6y-10=0$,
Area $=\frac{25}{21}$ square units

14 a $(0, -\sqrt{3}), (0, \sqrt{3})$
  b $(0, \sqrt{5}), (0, -\sqrt{5})$
  c $(-9, -3), (9, 3)$

Exercise 11B (p. 268)

1 a dt **b** dr **c** dx **d** $\dfrac{dA}{dc}$

2 $\dfrac{1}{18}$

3 a $6\pi\,\text{cm s}^{-1}$ **b** $60\pi\,\text{cm}^2\,\text{s}^{-1}$

4 $80\pi\,\text{cm}^3\,\text{s}^{-1}$

5 $72\,\text{cm}^3\,\text{s}^{-1}$

6 $16\pi\,\text{cm}^2\,\text{s}^{-1}$

7 24

8 $\dfrac{2}{27}\,\text{cm s}^{-1}$

9 $0.25\,\text{cm s}^{-1}$

10 Decreasing at $8\pi\,\text{cm}^2\,\text{s}^{-1}$

11 $\dfrac{4}{15}$

12 $\dfrac{1}{8\pi}\,\text{cm s}^{-1}$

Exercise 11C (p. 275)

1 a 5440 **b** $14\,800$
  c 896 per year

2 $m_0 = 30$, 0.15 grams per hour (2 s.f.)

3 a $A = 10, k = \dfrac{\ln 2}{1600}$
  c $0.002\,17$ grams per year (3 s.f.)
  d Using this model, the life of radium is infinitely long. The half-life however can be calculated. In reality, the number of atoms eventually reduces to zero, so the model is imperfect.

4 a $A = 40$
  b $T = 20 + 40e^{-\frac{1}{4}t}$
  c $35°$ (nearest degree)
  d 5.5 mins (2 s.f.)
  e $6.1°$C/min (2 s.f.)

5 a $(\ln 5)5^x$ **b** $(3\ln 2)2^x$
  c $(3\ln 2)2^{3x-4}$ **d** $\left(\frac{1}{2}\ln 10\right)10^x$
  e $-(\ln 6)6^{-x}$ **f** $(3\ln 2)2^{3x}$

6 a $\dfrac{1}{x\ln 3}$ **b** $\dfrac{1}{(x+2)\ln 10}$
  c $\dfrac{1}{x\ln 2}$ **d** $\dfrac{1}{x}$

7 a $4\ln 4$
  b $y = (16\ln 4)x + 16(1 - 2\ln 4)$
  c $(\ln 4)y + x = \ln 4$

8 $y = \dfrac{10}{\ln 10}x - \left(\dfrac{1}{\ln 10} + 1\right)$

Exercise 11D (Review) (p. 277)

1 $\pm\frac{1}{3}$

2 $-1, \frac{11}{3}$

3 $\frac{3}{7}$

4 $-\frac{3}{2}$

5 $\dfrac{2(x - y - 1)}{2x - 2y - 3}$

6 $(9, 3), (-1, 3)$

7 a $-\dfrac{2y}{3x}$ **b** $\dfrac{y(2x - y)}{x(2y - x)}$

8 $\dfrac{4y - 3x}{3y - 4x}$

9 -1

10 $\dfrac{3y - 2x - 4}{2y - 3x - 2}$

11 $\dfrac{3x - 2y}{2x}$

12 $5x - 4y = 9$

13 $4x - 7y - 6 = 0,\ 7x + 4y - 43 = 0$

14 $3x + 2y \pm 2\sqrt{10} = 0$

15 $y = \dfrac{\pi}{4} - x,\ y = x + \dfrac{\pi}{4}$

16 a y_0 **b** $k = \ln 2$
  c 3.19 p.m. **d** $c = 32\ln 2$

17 $\dfrac{1}{2\pi}\,\text{cm s}^{-1}$

18 a $240\,\text{cm}^2\,\text{s}^{-1}$ **b** $600\,\text{cm}^3\,\text{s}^{-1}$
  c $24\,\text{cm s}^{-1}$ **d** $2\sqrt{2}\,\text{cm s}^{-1}$
  e $2\sqrt{3}\,\text{cm s}^{-1}$

12 Further trigonometry

Exercise 12A (p. 285)

1 a $\frac{1}{4}(\sqrt{6} + \sqrt{2})$ **b** $\frac{1}{4}(\sqrt{6} + \sqrt{2})$
  c $\frac{1}{4}(\sqrt{2} - \sqrt{6})$ **d** $\frac{1}{4}(\sqrt{6} - \sqrt{2})$
  e $2 - \sqrt{3}$ **f** $2 - \sqrt{3}$

2 a $\frac{1}{4}(\sqrt{6} + \sqrt{2})$ **b** $-\frac{1}{4}(\sqrt{6} + \sqrt{2})$
  c $\frac{1}{4}(\sqrt{6} - \sqrt{2})$ **d** $\frac{1}{4}(\sqrt{6} + \sqrt{2})$
  e $\sqrt{6} - \sqrt{2}$ **f** $2 - \sqrt{3}$

3 a $\frac{56}{65}$ **b** $\frac{33}{65}$ **c** $\frac{33}{56}$

4 a $\frac{63}{65}$ **b** $-\frac{63}{16}$ **c** $-\frac{56}{33}$

5 a $\dfrac{2}{\sqrt{5}}$ **b** $\dfrac{1}{5\sqrt{2}}$ **c** 1

6 a $\dfrac{1}{a}$ **b** $\dfrac{ab - 1}{\sqrt{(1 + a^2)(1 + b^2)}}$
  c $\dfrac{ab + 1}{b - a}$

7 a $\frac{56}{65}$ **b** $\frac{56}{33}$ **c** $-\frac{65}{63}$

8 a $x = 9.9°, 189.9°$
  b $x = 157\frac{1}{2}°, 337\frac{1}{2}°$
  c $x = 49.1°, 229.1°$
  d $x = 56.5°, 236.5°$

9 $\tan B = -2$

10 $A + B = 45°$

11 $A - B = 135°$

12 a $\cos(x + 60°) = \sin(30° - x)$
  b $\cos(45° - x) = \sin(45° + x)$
  c $\tan(x + 60°)$
  d $\sin 26°$
  e $\sec 39°$
  f $\cos 15° = \sin 105° = \sin 75°$

13 a $\frac{1}{2}$ **b** $\frac{1}{2}$ **c** $\dfrac{\sqrt{3}}{3}$ **d** 0
  e $\frac{1}{2}$ **f** $\dfrac{\sqrt{2}}{2}$ **g** $\dfrac{\sqrt{3}}{3}$ **h** $\dfrac{\sqrt{6}}{2}$

14 $\tan A = 2$

15 $\cot B = \frac{12}{31}$

16 a $\frac{1}{3}$ **b** 1 **c** $-\frac{7}{4}$ **d** $2 - \sqrt{3}$

Exercise 12B (p. 291)

1 a $\sin 34°$ **b** $\tan 60°$ **c** $\cos 84°$
  d $\sin\theta$ **e** $\cos\dfrac{\pi}{4}$ **f** $\tan\theta$
  g $\cos\dfrac{\pi}{6}$ **h** $\sin 4A$ **i** $\cos\theta$
  j $\cos 6\theta$ **k** $\frac{1}{2}\tan 4\theta$ **l** $\frac{1}{2}\sin 2x$
  m $2\cot 40°$ **n** $2\operatorname{cosec} 2\theta$ **o** $\cos\theta$

2 a $\frac{1}{2}$ **b** 1 **c** $-\dfrac{\sqrt{3}}{2}$ **d** $-\dfrac{\sqrt{2}}{2}$
  e $\dfrac{\sqrt{2}}{2}$ **f** $2\sqrt{3}$ **g** 1 **h** $2\sqrt{2}$

3 a $\pm\frac{24}{25}, \frac{7}{25}$ **b** $\pm\frac{120}{169}, \frac{119}{169}$
  c $\pm\frac{1}{2}\sqrt{3}, -\frac{1}{2}$

4 a $-\frac{24}{7}$ **b** $\frac{240}{161}$ **c** $\pm\frac{120}{119}$

5 a $\pm\frac{3}{4}, \pm\frac{1}{4}\sqrt{7}$ **b** $\pm\frac{4}{5}, \pm\frac{3}{5}$
  c $\pm\frac{5}{13}, \pm\frac{12}{13}$

6 a $\frac{1}{3}, -3$ **b** $\frac{1}{2}, -2$ **c** $-\frac{2}{3}, \frac{3}{2}$

7 $\sqrt{2} - 1$

8 a $\theta = 90°, 120°, 240°, 270°$
  b $\theta = 60°, 300°$
  c $\theta = 30°, 150°, 90°, 270°$
  d $\theta = 0°, 180°, 360°; \theta = 120°, 240°;$
    $\theta = 36.9°, 323.1°$
  e $\theta = 45°, 225°; \theta = 121.0°, 301.0°$

9 a $\theta = -\pi, 0, \pi; \theta = -\dfrac{\pi}{3}, \dfrac{\pi}{3}$
  b $\theta = -\dfrac{\pi}{2}, 0.99, 2.16$
  c $\theta = 0; \theta = \pm 2.97$
  d $\theta = -\pi, 0, \pi; \theta = -\dfrac{5\pi}{6}, \dfrac{\pi}{6};$
    $\theta = -\dfrac{\pi}{6}, \dfrac{5\pi}{6}$
  e $\theta = -2.82, 0.32; \theta = -0.32, 2.82$

10 a $y = 2x^2 - 1$
  b $2y = 3(2 - x^2)$
  c $y(1 - x^2) = 2x$
  d $x^2 y = 8 - x^2$

13 $\dfrac{5\sqrt{26}}{26}$, $\dfrac{\sqrt{26}}{26}$, 5

14 b $\theta = 0°, 112.6°, 360°$

Extension Exercise 12C (p. 295)

1 a $\sin(x+y) + \sin(x-y)$
 b $\sin(x+y) - \sin(x-y)$
 c $\sin 4\theta + \sin 2\theta$
 d $\sin 2S + \sin 2T$
 e $\sin 8x - \sin 2x$
 f $\sin 2x - \sin 2y$

2 a $\cos(x+y) + \cos(x-y)$
 b $\cos(x+y) - \cos(x-y)$
 c $\cos 4\theta + \cos 2\theta$
 d $\cos 2S - \cos 2T$
 e $\cos 2x - \cos 8x$
 f $\cos 2x + \cos 2y$

3 a $2\cos\frac12(x+y)\cos\frac12(x-y)$
 b $2\sin 4x\cos x$
 c $2\cos(y+z)\sin(y-z)$
 d $2\cos 6x\cos x$
 e $-2\sin\frac32 A\sin\frac12 A$
 f $2\cos 3x\sin x$
 g $2\sin 4A\sin A$
 h $2\sin 6\theta\cos\theta$
 i $\sqrt3\sin x$
 j $\sqrt2\cos(y-35°)$
 k $-2\cos 4\theta\sin\theta$
 l $-\sin x$
 m $-2\sin x\sin\frac12 x$
 n $2\sin 2x\cos 80°$

4 a $2\cos(45°-\frac12 x+\frac12 y)\times$
  $\cos(45°-\frac12 x-\frac12 y)$
 b $2\cos(45°-\frac12 A+\frac12 B)\times$
  $\cos(45°-\frac12 A-\frac12 B)$
 c $2\sin(\frac32 x+45°)\cos(\frac32 x-45°)$
 d $2\sin(x+45°)\cos(x-45°)$
 e $2\cos(45°-\frac12 A+\frac12 B)\times$
  $\sin(45°-\frac12 A-\frac12 B)$
 f $2\cos(30°+\theta)\cos(30°-\theta)$

5 a $1+\dfrac{\sqrt3}{2}$ **b** $-\dfrac{\sqrt6}{2}$
 c $-\dfrac{\sqrt6}{2}$ **d** $-\frac14(\sqrt3+1)$

6 a $x=\dfrac{\pi}{6},\dfrac{\pi}{2},\dfrac{5\pi}{6},\dfrac{7\pi}{6},\dfrac{3\pi}{2},\dfrac{11\pi}{6};$
  $\dfrac{\pi}{4},\dfrac{3\pi}{4},\dfrac{5\pi}{4},\dfrac{7\pi}{4}$
 b $x=0,\dfrac{2\pi}{3},\dfrac{4\pi}{3},2\pi;\dfrac{2\pi}{5},\dfrac{4\pi}{5},\dfrac{6\pi}{5},\dfrac{8\pi}{5}$
 c $x=0,\pi,2\pi;\dfrac{\pi}{4},\dfrac{3\pi}{4},\dfrac{5\pi}{4},\dfrac{7\pi}{4}$
 d $x=0,\dfrac{2\pi}{5},\dfrac{4\pi}{5},\pi,\dfrac{6\pi}{5},\dfrac{8\pi}{5},2\pi$

7 a $x=0°,60°,120°,180°,240°,300°,$
  $360°;90°,270°$
 b $x=175°,355°$
 c $45°,135°,225°,315°$
 d $25°,205°$

9 a $2\sin 2x\cos x$
 c $x=0,\dfrac{\pi}{2},\pi;\dfrac{2\pi}{3}$

10 a $\theta=\dfrac{\pi}{3},\dfrac{2\pi}{3};\dfrac{\pi}{6},\dfrac{\pi}{2},\dfrac{5\pi}{6}$
 b $\theta=0,\dfrac{\pi}{3},\dfrac{\pi}{2},\pi$

Exercise 12D (p. 300)

1 a $5,\frac34$ **b** $\sqrt{13},\frac32$
 c $5\sqrt2,7$ **d** $5,\frac43$

2 a $\theta=90°,330°$
 b $\theta=94.9°,219.9°$
 c $\theta=114.3°,335.7°$
 d $\theta=45°,225°;161.6°,341.6°$

3 a $\theta=-\dfrac{\pi}{2},\dfrac{\pi}{6}$
 b $\theta=-0.71,1.27$
 c $\theta=-2.71,-0.15$
 d $\theta=-\pi,0,\pi;-2.19,0.96$

6 $13\sin(\theta-337.4°)$

7 b i $5,53.1°;-5,233.1°$
 ii $11,53.1°;1,233.1°$
 iii $15,233.1°;5,53.1°$
 iv $-\frac15,233.1°;\frac15,53.1°$
 v $-\frac13,233.1°;\frac13,53.1°$
 vi $1,233.1°;\frac{7}{17},53.1°$
 vii $25,53.1°;0,143.1°$
 viii $1,143.1°;\frac{1}{26},53.1°$

8 a $17,298.1°;-17,118.1°$
 b $6,22.5°$ and $202.5°;2,112.5°$ and $292.5°$
 c $\frac12,60°;\frac16,240°$
 d $5,305.3°;\frac57,125.3°$

9 $\angle\mathrm{BAC}=63°$

Exercise 12E (p. 309)

2 a $3\sin 2x$ **b** $-\frac32\sin x\cos^2 x$
 c $8\sin x\cos^3 x$ **d** $8\sin 4x$
 e $\dfrac{\cos x}{2\sqrt{\sin x}}$

3 $\frac16\pi,\frac12\pi,\frac56\pi,\frac76\pi,\frac32\pi,\frac{11}{6}\pi$

5 a $\frac12 x-\frac14\sin 2x+c$
 b $\frac12 x+\frac{1}{12}\sin 6x+c$
 c $\frac38 x-\frac18\sin 4x-\frac{1}{64}\sin 8x+c$
 d $x+\frac12\sin 2x+c$

6 a $\frac14\pi$ **b** $\frac{3}{16}\pi$

7 a $A(-\frac14\pi,\frac12), B(\frac14\pi,\frac12)$
 b 1

8 a $\frac13\sin^3 x+c$ **b** $-\frac16\cos^6 x+c$
 c $-\frac{1}{16}\cos^4 4x+c$ **d** $\frac15\sin^5 x+c$
 e $-\frac13\cos^3 x+c$ **f** $\frac14\sin^4 x+c$

9 a $\cos x=2\cos^2(\frac12 x)-1$
 c $2\sqrt2\sin(\frac12 x)+c$

10 a $2\sin x\cos x$
 b i $-\frac25\cos^5 x+c$
 ii $-\frac43\cos^3(\frac12 x)+c$
 iii $\frac12\sin^4 x+c$

11 $-\frac14 x\cos 2x+\frac18\sin 2x+c$

12 a $\frac12 x+\frac14\sin 2x+c$
 b $\frac14 x^2+\frac14 x\sin 2x+\frac18\cos 2x+c$

13 $\dfrac{\pi^2}{2}$

14 2

15 $1\frac14$

16 a $-\frac15\cos^5 x+\frac17\cos^7 x+c$
 b $\frac13\sin^3 x-\frac15\sin^5 x+c$
 c $\frac14\sin^4 x-\frac16\sin^6 x+c$ or
  $-\frac14\cos^4 x+\frac16\cos^6 x+c$

17 a $-\cos x+\frac13\cos^3 x+c$
 b $\frac12\sin 2x-\frac16\sin^3 2x+c$

18 $\frac{8}{15}$

19 a $2\sin 2x\cos x$
 b $\sin 5x+\sin x$
 c $-\frac{1}{10}\cos 5x-\frac12\cos x+c$

20 a $\frac14\cos 2x-\frac18\cos 4x+c$
 b $\frac12\sin 2x+\sin x+c$
 c $\frac16\sin 3x-\frac{1}{10}\sin 5x+c$

21 a $\frac14\sec^4 x+c$
 b $\frac15\tan^5 x+c$
 c $\frac13\tan^3 x+c$
 d $\frac13\sec^3 x-\sec x+c$
 e $\tan x+\frac13\tan^3 x+c$
 f $\frac13\tan^3 x-\tan x+x+c$

Exercise 12F (Review) (p. 311)

1 a $\frac{140}{221}$ **b** $-\frac{21}{221}$ **c** $\frac{171}{140}$
2 a $\frac{468}{493}$ **b** $-\frac{475}{493}$ **c** $\frac{475}{132}$
3 a $\frac12$ **b** 1 **c** 1
4 a $\pm\frac23,\pm\frac13\sqrt5$ **b** $\pm\frac49,\pm\frac19\sqrt{65}$
5 a $\frac52,-\frac25$ **b** $\frac29,-\frac92$
6 a $\frac{840}{1369}$ **b** $-\frac{1369}{1081}$
7 a $\theta=39.3°,129.3°,219.3°,309.3°$
 b $\theta=0°,180°,360°;41.4°,318.6°$
 c $\theta=60°,300°$
 d $\theta=90°;210°,330°$
 e $\theta=45°,225°;76.0°,256.0°$
 f $\theta=15°,75°,195°,255°$
 g No solution
 h $\theta=0°,360°;126.9°,233.1°$
8 a $\theta=0,\pi,2\pi;\dfrac{\pi}{3},\dfrac{2\pi}{3},\dfrac{4\pi}{3},\dfrac{5\pi}{3}$
 b $\theta=\dfrac{\pi}{4},\dfrac{5\pi}{4}$

9 a $5\cos(\theta + 36.9°)$
  b $\theta = 41.6°, 244.7°$

10 $\theta = -0.22, 1.39$

11 a $13, 292.6°; -13, 112.6°$
  b $37, 288.9°; -37, 108.9°$

15 a $9x = 4y^2 - 18$
  b $y(4 - x^2) = 4x$

16 a $\theta = 0, \dfrac{\pi}{4}, \dfrac{\pi}{2}, \dfrac{3\pi}{4}, \pi; \dfrac{\pi}{3}, \dfrac{2\pi}{3}$
  b $\theta = \dfrac{\pi}{3}, \pi$
  c $\theta = \dfrac{\pi}{4}, \dfrac{3\pi}{4}; \dfrac{\pi}{6}, \dfrac{5\pi}{6}$

21 $y = -2x + \frac{1}{2}(\pi + 1)$

23 a $\frac{1}{2}\sin 2x - \frac{1}{6}\sin^3 2x + c$
  b $\frac{3}{8}x - \frac{1}{8}\sin 4x + \frac{1}{64}\sin 8x + c$

24 a $\frac{1}{2}(\sin 4x + \sin 2x)$

25 a $\frac{1}{2}(\cos 5x + \cos x)$
  b $\frac{1}{5}\sqrt{2}$

26 $\sqrt{3} - 1$

27 a $2\sqrt{2}$ **b** π^2

28 $-2\ln\cos x + c$

Examination questions 3 (p. 316)

1 $\dfrac{2}{x - 3} - \dfrac{4}{2x + 1}$

2 $\dfrac{1}{x - 3} + \dfrac{4}{(x + 2)^2} - \dfrac{3}{x + 2}$

3 b $\dfrac{2x + 1}{x - 2}$

4 a 30 **b** $\dfrac{2}{x + 3} - \dfrac{7}{x - 2} + \dfrac{5}{x - 4}$
  c $N = 11, M = 15$

5 a $\dfrac{1}{x - 1} + \dfrac{4}{x + 2}$ **b** $4\ln 5 - 7\ln 2$

6 a $1 + \dfrac{4}{2x - 3} - \dfrac{3}{x + 2}$
  b $4 - 3\ln 2 + 4\ln 3$

7 a ii $A = 6, B = -3, C = 8$
  b $3\ln 2 + \frac{4}{3}$

8 a $1 - 4x + 12x^2 - 32x^3 + \dots$
  b $|x| < \frac{1}{2}$

9 a 0.8467
  b $1 + \frac{1}{2}x^3 - \frac{1}{8}x^6$; 0.8475 (4 d.p.)

10 a $1 + \dfrac{3x}{10} + \dfrac{3x^2}{50} + \dfrac{x^3}{100} + \dots$
  b $K = \frac{1}{2}$
  c $1.030\,610\,152\,128\,36$

11 b $4x + 3y = 7$

12 a $\dfrac{dy}{dx} = \dfrac{2(x + y)}{3y^2 - 2x}$ **b** $y = \frac{1}{2}x - 2$

13 b 4.54

14 b 14.3

15 a $\cos 3\theta$ **b** $25°, 95°, 145°$

16 No

17 a $R = 13, \alpha = 67.4°$
  b $105°, 209°$

18 c $\dfrac{5\pi}{6}, \dfrac{11\pi}{6}$

19 a $3 - 3\cos 2\theta$ **b** $\dfrac{\pi - 3}{4}$
  c 0.766, 2.375

20 a $6\cos 2x - 2\sin 2x$
  b $y + 2x = 3 + \dfrac{\pi}{2}$
  c i $A = 3, B = -4, C = 5$
    ii 17.0

13 Coordinate geometry

Exercise 13A (p. 325)

2 a $(y + 2)^2 = x - 3$
  b $x^2 = y^3$
  c $xy = 25$
  d $2x + y = 10$
  e $y^2 = 4ax$
  f $xy^2 = 100$
  g $y^2 = x^2(x - 1)$
  h $9x^2 + 4y^2 = 36$
  i $y^2 = (1 - x^2)^3$
  j $25x^2 - 9y^2 = 225$
  k $a^2y^2 = 4(a^2 - x^2)x^2$
  l $9x = 4y^2 - 18$
  m $y = x^2 + 3x + 2$
  n $5x + y - 13 = 0$

3 a $t = \frac{4}{3}, x = \frac{4}{3}$
  b $t = \pm\frac{3}{2}, y = \pm 3a$
  c $t = -2, x = -\frac{1}{3}$
  d $\theta = \dfrac{\pi}{3}, y = \dfrac{\sqrt{3}}{2}b$

4 $3x - 2y + 1 = 0$

5 $-\frac{4}{13}, -\frac{4}{3}$

Exercise 13B (p. 331)

1 a t **b** $\dfrac{2(1 - t)}{2t + 1}$
  c $-2\tan t$ **d** $\dfrac{t^2\cos t + 2t\sin t}{\cos t - t\sin t}$
  e $-\dfrac{e^t}{t}$ **f** $\frac{3}{2}(t + 1)$
  g $5t^{\frac{3}{2}}$ **h** $\dfrac{2 + 2t}{1 + \ln t}$

2 a 8   **b** $2\sqrt{2}$ **c** 24
  d 0   **e** 2   **f** 8
  g -2 **h** $e^6 - 2e^{-2}$

3 a $\dfrac{1}{t}$ **b** $-\dfrac{1}{t^2}$
  c $-\dfrac{b}{a}\cot t$ **d** $\dfrac{b}{a}\operatorname{cosec} t$
  e $-\tan t$

4 a $t^2 - 2t$ **b** $\dfrac{9(t + 2)^2}{4(t + 3)^2}$
  c $\dfrac{2t}{t^2 - 1}$ **d** $-\dfrac{1}{2t}$
  e -1

5 a $\frac{3}{2}t$ **b** $\frac{3}{2}\sqrt{x}$

6 a $3x + 2y - 1 = 0, 2x - 3y - 5 = 0$
  b $4x + y - 3 = 0, 4x - 16y + 31 = 0$
  c $x + y + a = 0, x - y - 3a = 0$
  d $x + y + 2c = 0, x - y = 0$
  e $x + 2y + 9 = 0, 2x - y + 3 = 0$
  f $2x + 3\sqrt{3}y - 12 = 0,$
    $6\sqrt{3}x - 4y - 5\sqrt{3} = 0$
  g $4x - 5y + 4a = 0,$
    $15x + 12y - 26a = 0$
  h $(3 + e)x - y = 11 + 3e,$
    $x + (3 + e)y = e^2 + 4e + 7$

7 $y = 12, (16, 12)$

8 $y = 2x - 2, \left(\frac{3}{4}, -\frac{1}{2}\right)$

9 a Maximum at $(1, 2)$,
    minimum at $(1, -2)$
  b Maximum at $(0, 2)$,
    minimum at $(0, -2)$

Exercise 13C (Review) (p. 333)

1 a $y = x^{\frac{2}{5}} - 4$ **b** $xy = 9$
  c $y = 3 - \frac{3}{16}x^2$ **d** $2x + y^2 = 2$

3 a $2(t^2 + t + 1)$
  b $t\cot t - 4\operatorname{cosec} t + 1$
  c $2(t^3 - 2)e^{-2t}$

4 $8\sqrt{3}x - 18y - 21\sqrt{3} = 0$

5 $y = 4x - 20$

14 Differential equations

Exercise 14A (p. 344)

1 a $y = 4x^2 - 2x + c$
  b $y^2 = -2e^{-x} + c$
  c $x^2 = t^2 + 2t + c$
  d $y^2 = 2\sin x + c$
  e $\tan y = \frac{1}{2}x^2 + c$
  f $v = \dfrac{1}{t + c}$
  g $\sqrt{y} = -\dfrac{1}{6x^3} + c$
  h $e^{-y} = -\frac{1}{2}x^2 - 2x + c$
  i $x^2 - x = y^2 + c$
  j $4y^3 = 3x^4 + c$

k $e^{2y} = x^2 + c$

l $\sin y = e^x + c$

2 a $y = Ae^x$

b $y = A(x + 3)$

c $y^2 = 2\ln x + c$

d $\sin y = Ax$

e $x\cos y = A$

f $y = Ax - 2$

g $y = Ae^{x^2 + x}$

h $\theta = Ae^{-\frac{1}{t}}$

i $A = Ce^{y^2 + 8y} - 1.5$

3 a $\dfrac{3}{x+2} - \dfrac{1}{x+1}$ **b** $y = A\dfrac{(x+2)^3}{x+1}$

4 b $y = A(x - 3)^2$

c $y^2 - 2y = 2x^2 + c$

d $y = Axe^x$

5 $3y^2 + 18y = 2x^3 + 5$

6 $y = \frac{3}{10}x - \frac{7}{2}$

7 $y = x^2 - 3x + 1$

8 a $s = -\dfrac{1}{6}\ln t + c$ **b** $s = \dfrac{1}{6}\ln\dfrac{2}{t}$

9 a $y = 5e^{2x^2}$

b $x^2 + y^2 = 16$
Circle, centre $(0, 0)$, radius 4

10 $y = \ln(2e - 1)$

11 $a = 3,\ b = 12$

12 b $\sin y = \frac{1}{2}x + \frac{1}{4}\sin 2x + c$

c $\sin y = \frac{1}{2}x + \frac{1}{4}\sin 2x + 1 - \frac{1}{4}\pi$

13 a $s = \frac{1}{2}t^2 + 4\ln t + c$

b $a = 6,\ b = 8$

14 a $x\sin x + \cos x + c$

b $y^2 = 2x\sin x + 2\cos x + c$

c $y^2 = 2x\sin x + 2\cos x - 1$

15 a $-1, 0$ **c** $xy = 1$

d $y = \dfrac{A}{x+1} + 2$

Exercise 14B (p. 353)

1 a $12\,532$ **b** 17 hours 47 mins

2 a $2.43\,\text{g}$

b 0.693 (3 s.f.)

c half life is 0.693 seconds

3 a $\dfrac{1}{1-x} - \dfrac{1}{2-x}$

b $\ln\dfrac{2-x}{1-x} + c$

c i 0.56 (2 s.f.)

ii $\dfrac{2(e^t - 1)}{2e^t - 1}$

iii 1

4 a $3.01\,\text{mm}$ (3 s.f.)

b Approx $10\frac{1}{4}$ minutes

5 a $\dfrac{1}{4}\ln\left(\dfrac{2+x}{2-x}\right) + c$

b $x = \dfrac{2(e^{40t} - 1)}{e^{40t} + 1}$

6 b £149.33

7 a $147.39\,\text{m}^2$ **b** 12.0 days (3 s.f.)

9 83.4 minutes

10 £6040

11 a $\dfrac{dx}{dt} = -kx$ **c** 10 minutes

d i $x = x_0 e^{-(\frac{1}{3}\ln 2)t}$

12 a $h = -\frac{1}{8}t + 10$

b $\dfrac{dh}{dt} = -kh,\ h = 10e^{-(\frac{1}{40}\ln 2)t}$

c $A{:}h = 2.5,\ B{:}h = 3.5$

d B's estimate is better

Exercise 14C (Review) (p. 356)

1 a $\frac{1}{2}y^2 = \frac{1}{3}e^{3x} + 2x + c$

b $y = A(x - 1)^3$

c $x = Ae^{-\frac{1}{t}} - 1$

d $y = x - \sin x + c$

e $\sqrt{y} = \frac{1}{2}\ln x + c$

f $\theta = Ae^{\frac{1}{2}t^2 + t}$

2 a $y = \left(\frac{1}{2}(3x^2 - 1)\right)^{\frac{1}{3}}$

b $y = -\dfrac{2}{x^2 - 3}$

c $y = \dfrac{1}{1 - \ln x}$

d $y = (3\ln x + 1)^{\frac{1}{3}}$

3 $y = 10\sqrt{x} + 3$

4 a $y = \ln(e^x + e - 1)$

b $y = x + 1 - \ln(e - e^{x+1} + e^x)$

5 $y = 3 - \dfrac{4}{x^2}$

6 $\dfrac{1}{2(1+x)} - \dfrac{1}{2(3+x)};\ y = 2\sqrt{\dfrac{2(1+x)}{3+x}}$

7 $\ln y = \frac{1}{3}xe^{3x} - \frac{1}{9}e^{3x} + \frac{1}{9}$

8 a $T = -6x + 470,\ 360°\text{C}$

b $\dfrac{dT}{dx} = -kx,\ T = -\frac{1}{10}x^2 + 380$

9 $n = 450e^{2t} + 50$; the number increases infinitely

15 Vectors

Exercise 15A (p. 365)

1 a 5 **b** 13

c $\sqrt{2}$ **d** 6

e 4 **f** $\sqrt{13}$

g $\sqrt{41}$ **h** $2\sqrt{13}$

i 3 **j** $5\sqrt{2}$

k $\sqrt{3}$ **l** 7

2 a $13,\ 22.6°$

b $2\sqrt{2},\ -45°$

c $2,\ -60°$

d $17,\ 118.1°$

e $8,\ 180°$

f $16,\ 90°$

3 $\begin{pmatrix} -5\sqrt{3} \\ 5 \end{pmatrix}$

4 $8\mathbf{i} - 8\sqrt{3}\mathbf{j}$

5 $-6\mathbf{i} - 6\mathbf{j}$

6 a $\begin{pmatrix} \frac{5}{13} \\ -\frac{12}{13} \end{pmatrix}$

b $\begin{pmatrix} \frac{3}{13} \\ \frac{4}{13} \\ -\frac{12}{13} \end{pmatrix}$

c $\dfrac{1}{\sqrt{21}}\mathbf{i} - \dfrac{2}{\sqrt{21}}\mathbf{j} + \dfrac{4}{\sqrt{21}}\mathbf{k}$

d $\dfrac{1}{\sqrt{3}}\mathbf{i} + \dfrac{1}{\sqrt{3}}\mathbf{j} + \dfrac{1}{\sqrt{3}}\mathbf{k}$

e $\frac{3}{5}\mathbf{i} + \frac{4}{5}\mathbf{j}$

f $\dfrac{1}{\sqrt{30}}\mathbf{i} - \dfrac{2}{\sqrt{30}}\mathbf{j} + \dfrac{5}{\sqrt{30}}\mathbf{k}$

7 $8\mathbf{i} - 4\mathbf{j} + 8\mathbf{k}$

8 $\frac{3}{5}\mathbf{i} - \frac{4}{5}\mathbf{j} + \mathbf{k}$

9 $\overrightarrow{AC} = 7\mathbf{i} + 9\mathbf{j} + 4\mathbf{k}$

10 $\overrightarrow{CE} = 9\mathbf{j}$

11 $\overrightarrow{PR} = \begin{pmatrix} -2 \\ -3 \\ 9 \end{pmatrix}$

12 $\overrightarrow{FG} = 11\mathbf{i} - 4\mathbf{j} + 6\mathbf{k}$

13 $\overrightarrow{NM} = 16\mathbf{i} - 9\mathbf{j} + 6\mathbf{k}$

14 $\overrightarrow{YZ} = \begin{pmatrix} 8 \\ 4 \\ -12 \end{pmatrix}$

15 a $6\mathbf{i} + 8\mathbf{j} - 14\mathbf{k}$

b $6\mathbf{i} + 6\mathbf{j} - 15\mathbf{k}$

c $\mathbf{i} + 2\mathbf{j} - 2\mathbf{k}$

d $4\mathbf{i} + 6\mathbf{j} - 9\mathbf{k}$

e $-\mathbf{i} + 3\mathbf{k}$

f $2\mathbf{j} + \mathbf{k}$

g $5\mathbf{i} + 6\mathbf{j} - 12\mathbf{k}$

h $12\mathbf{i} + 14\mathbf{j} - 29\mathbf{k}$

i $-23\mathbf{i} - 28\mathbf{j} + 55\mathbf{k}$

j $-\mathbf{i} - 2\mathbf{j} + 2\mathbf{k}$

k $-23\mathbf{i} - 28\mathbf{j} + 55\mathbf{k}$

l $-21\mathbf{i} - 24\mathbf{j} + 51\mathbf{k}$

16 $\alpha = -5,\ \beta = 8$

17 $\lambda = \pm 4$

18 a $2\sqrt{3}$ **b** $2\sqrt{89}$

c $\sqrt{170}$ **d** $2\sqrt{170}$

$$19 \quad \begin{pmatrix} -\dfrac{3}{\sqrt{14}} \\ \dfrac{2}{\sqrt{14}} \\ -\dfrac{1}{\sqrt{14}} \end{pmatrix}$$

20 a 2**a**   **b** 3**b**
   c **a** + **b**   **d** −**a** − **b**
   e 2**a** + **b**   **f** 5**a** + 2**b**
   g 3**a** − 2**b**   **h** 2**b** − 3**a**
   i −2**a** − 2**b**   **j** −2**b**
   k 2**b** − **a**   **l** 0

Exercise 15B (p. 370)

1 a **q** − **p**   **b** $\frac{1}{4}$(**q** − **p**)
   c $\frac{3}{4}$**p** + $\frac{1}{4}$**q**

2 i $\overrightarrow{BD} = \frac{1}{5}$(**q** − **p**)
     $\overrightarrow{AD} = \frac{4}{5}$**p** + $\frac{1}{5}$**q**
  ii $\overrightarrow{BD} = \frac{2}{5}$(**q** − **p**)
     $\overrightarrow{AD} = \frac{3}{5}$**p** + $\frac{2}{5}$**q**
  iii $\overrightarrow{BD} = \frac{4}{5}$(**q** − **p**)
     $\overrightarrow{AD} = \frac{1}{5}$**p** + $\frac{4}{5}$**q**

  iv b $\dfrac{l}{l+m}$(**q** − **p**)

     c $\dfrac{m}{l+m}$**p** + $\dfrac{l}{l+m}$**q**

3 a i $\overrightarrow{AC}$ = **p** + **q**

   ii $\overrightarrow{FD}$ = **p** + **q**

   iii $\overrightarrow{FC}$ = 2**p**

  b i AC and FD are parallel and of
     equal length
   ii FC and AB are parallel and
     FC = 2AB

4 a i **a** + **b**
   ii $\frac{1}{2}$**a** + $\frac{1}{2}$**b**
   iii $\frac{1}{2}$**a** + $\frac{1}{2}$**b**
  b The diagonals of a parallelogram
    bisect each other.

5 a i $\overrightarrow{OM} = \frac{1}{4}$**a**
   ii $\overrightarrow{AC}$ = **b**
   iii $\overrightarrow{AN} = \frac{3}{4}$**b**
   iv $\overrightarrow{ON}$ = **a** + $\frac{3}{4}$**b**
   v $\overrightarrow{MN} = \frac{3}{4}$**a** + $\frac{3}{4}$**b**
   vi $\overrightarrow{OC}$ = **a** + **b**
  b MN is parallel to OC and three
    quarters of its length.

6 $\overrightarrow{MN} = \frac{1}{2}$**b** − $\frac{1}{2}$**a**; MN is parallel to AB
   and half its length.

7 $\overrightarrow{MN} = \dfrac{1}{1+\lambda}$(**b** − **a**); MN is parallel

   to AB and MN = $\dfrac{1}{1+\lambda}$ AB

8 Collinear: a, b, d, e. Not collinear: c.

9 a **d** = **a** + **b**, **e** = **a** + **c**, **f** = **b** + **c**,
    g = **a** + **b** + **c**
  b All are $\frac{1}{2}$(**a** + **b** + **c**). Diagonals of a
    parallelepiped bisect each other.

10 **d** = $\frac{1}{2}$**a** + $\frac{1}{2}$**b**, **e** = $\frac{1}{4}$**b** + $\frac{3}{4}$**c**,
    f = $\frac{3}{2}$**c** − $\frac{1}{4}$**a**. E is the mid-point of DF.

Exercise 15C (p. 374)

1 a 22   **b** 0   **c** −44   **d** 73
   e 12   **f** 40   **g** 3   **h** 50
   i −4   **j** 40

2 a 10   **b** 3   **c** 13   **d** 13
   e 7   **f** −10

3 a 58   **b** 26   **c** 26   **d** 0
   e 84   **f** 6   **g** 26

4 $-14\sqrt{3}$

5 $q = 4$

6 a Perpendicular   **b** Parallel
   c Neither   **d** Perpendicular

7 a 70.5°   **b** 46.8°   **c** 90°
   d 88.6°   **e** 60°   **f** 109.4°
   g 98.6°   **h** 48.2°   **i** 42.8°
   j 112.4°

8 $\lambda = 1.5$

9 $\lambda = -2.5$ or $\lambda = 3$

10 $\lambda = \pm 4$

11 a $\mu = -58$ **b** $\mu = 4$

12 2**a**.**c**

13 $\cos \theta = -\dfrac{7\sqrt{2}}{26}$

14 a 120°   **b** 82.0°   **c** 110.9°
   d 90°   **e** 45°   **f** 150°

15 ∠BAC = 6.5°

16 ∠DCE

17 b For example $\begin{pmatrix} 2 \\ -2 \\ -1 \end{pmatrix}$

18 a $m = 3.5$ **b** 86.7°

Exercise 15D (p. 381)

1 a **r** = 2**i** + 3**j** − 4**k** + t(**i** + **j** + **k**)
   b **r** = 2**i** − 4**j** + 6**k** + t(7**i** + **j**)
   c **r** = 5**j** + 3**k** + t(2**i** − 5**j** + 7**k**)
   d **r** = t(**i** + 6**j** − **k**)

   e **r** = $\begin{pmatrix} 7 \\ 0 \\ 1 \end{pmatrix} + t \begin{pmatrix} 0 \\ 1 \\ 0 \end{pmatrix}$

   f **r** = $\begin{pmatrix} 5 \\ 3 \\ -4 \end{pmatrix} + t \begin{pmatrix} 1 \\ 3 \\ -2 \end{pmatrix}$

2 a **r** = 3**i** + 2**j** − **k** + t(2**i** + 5**j** + 4**k**)
   b **r** = t(4**i** − 2**j** + 7**k**)
   c **r** = 3**i** + 8**j** − 5**k** + t(2**i** + 5**j** − 5**k**)

   d **r** = 6**i** + 4**j** + 2**k** + t(**i** − **k**)
   e **r** = 10**i** + 20**j** − 15**k** + t(**i** + **j** + **k**)
   f **r** = −**i** + **j** + 6**k** + t(3**i** − 2**j** + 2**k**)

3 **r** = 3**i** + 5**j** + t(5**i** + 2**j**)

4 **r** = 2**i** + 10**j** + t(3**i** − 5**j**)

5 b 53.1°

6 a Intersect at (4, −1, −5), 38.3°
   b Intersect at (0, 0, 0), 48.6°
   c Skew, 85.9°
   d Parallel
   e Skew, 29.2°

7 a **r** = 2**i** + **j** + **k** + t(−2**i** + 4**j** + 2**k**)
   d 18.7°

8 a 4**i** − **j** − 3**k**   **b** 71.4°

9 a $\sqrt{105}$ **b** $3\sqrt{5}$
   c $2\sqrt{11}$ **d** 0

Exercise 15E (Review) (p. 382)

1 a $\overrightarrow{OB}$ = **a** + **c**
   b $\overrightarrow{AC}$ = **c** − **a**
   c $\overrightarrow{OQ} = \frac{2}{3}$**a** + $\frac{1}{2}$**c**
   d $\overrightarrow{PR} = \frac{1}{2}$**c** − $\frac{2}{3}$**a**
   e $\overrightarrow{RC} = \frac{1}{2}$**c**
   f $\overrightarrow{AQ} = \frac{1}{2}$**c** − $\frac{1}{3}$**a**
   g $\overrightarrow{QC} = \frac{1}{2}$**c** − $\frac{2}{3}$**a**
   h $\overrightarrow{PB} = \frac{1}{3}$**a** + **c**
   i $\overrightarrow{PC}$ = **c** − $\frac{2}{3}$**a**
   j $\overrightarrow{BQ} = -\frac{1}{3}$**a** − $\frac{1}{2}$**c**

2 a $\sqrt{58}$ **b** $\sqrt{182}$ **c** $2\sqrt{58}$

3 9**i** − 12**j** + 36**k**

4 $m = \frac{3}{5}$, $n = \frac{6}{5}$, 3:2

5 a $m = 10$, $n = -9$

   b $n = \dfrac{5m + 8}{2}$

6 75.7°

8 $\dfrac{1}{\sqrt{6}}$(**i** − **j** − 2**k**)

9 $\frac{4}{5}$; 16

10 **r** = 4**i** + 3**j** + t(2**i** − **j**)

11 (17, −5, 21), 18.2°

13 (1, 2, 1)

14 a $s = 6$
   b AC: **r** = 2**i** − 3**j** + 5**k**
     + s(2**i** + 9**j** − 3**k**)
     OB: **r** = t(6**i** + 3**j** + 7**k**)
   c $\frac{9}{47}$

15 a P(−2, 2, 1)   **b** 3

Examination questions 4 (p. 387)

1 b $3y + 5x = 36$

   c $x = \left(\dfrac{x-y}{2}\right)^2 + \left(\dfrac{x-y}{2}\right)$

or $y = \left(\dfrac{x-y}{2}\right)^2 - \left(\dfrac{x-y}{2}\right)$

or $x^2 + y^2 = 2(x + xy + y)$

2 i $\dfrac{dy}{dx} = \dfrac{-1}{4\cos t}$ **ii** $t = \dfrac{\pi}{3}$

3 a $\dfrac{dy}{dx} = t^3 + 2t$ **b** $y = 7 - \frac{1}{3}x$

5 a i $x + y = 2t,\ x - y = \dfrac{6}{t}$

    b i $\dfrac{t^2 + 3}{t^2 - 3}$ **ii** $\dfrac{x}{y}$

6 a $xe^x - e^x + c$ **b** $\ln y = xe^x - e^x + 1$

7 b $t = \sqrt{A} - \frac{1}{2}$ **c** 17 days

8 a $\dfrac{1}{3 + \cos x} + c$

    b $\ln y = \dfrac{1}{3 + \cos x} + \dfrac{3}{4}$

9 $y = 2e^{\tan 2x}$

10 a $\dfrac{3}{x - 2} - \dfrac{2}{x - 1},\ \ln\dfrac{(x-2)^3}{(x-1)^2} + c$

    b $y = \dfrac{32(x-2)^3}{(x-1)^2} - 4$

11 a $-2xe^{-2x} - e^{-2x} + c$

    b $y = \frac{1}{4}\left(7 - e^{-2x}(1 + 2x)\right)^2$

12 a i $\dfrac{dt}{dN} = \dfrac{1}{4N}$

    ii $t = \frac{1}{4}\left(\ln N - \ln 200\right)$

    b 19 minutes

13 a $\dfrac{dy}{dx} = x + y$

    b i $\dfrac{dy}{dx} = \dfrac{dz}{dx} - 1$

    c i $\ln(1 + z) = x + c$

    d $y - (e - 2) = -\dfrac{1}{e - 1}(x - 1)$

14 a $\mathbf{r} = 2\mathbf{i} + \mathbf{j} + 3\mathbf{k} + t(2\mathbf{i} + \mathbf{j})$

    b (4, 2, 3)

    c $56.8°$

15 $a = 12,\ \mathbf{p} = 4\mathbf{i} + 3\mathbf{j} + 4\mathbf{k}$

16 a $p = -9,\ \mathbf{x} = 15\mathbf{i} + 2\mathbf{j} - 3\mathbf{k}$

    b $42°$

17 b (10, 1, 2)   **c** $2\sqrt{3}$

C4 sample exam paper (p. 392)

1 $\dfrac{2}{x + 2}$

2 a 2   **b** $x + 2y + 1 = 0$

3 a $\frac{1}{2} + \frac{1}{16}x + \frac{3}{256}x^2 + \frac{5}{2048}x^3 + \ldots$

    b -3

4 a 9000

    c i £6364   **ii** 84 months

5 b $R = 13.34,\ \alpha = 12.99$

    c $x = 26.0$

6 a $\mathbf{r} = 4\mathbf{i} + 3\mathbf{j} - \mathbf{k} + \lambda(-3\mathbf{i} + \mathbf{j} + 5\mathbf{k})$

    b i $a = -3,\ b = 3$

    ii $7\mathbf{i} + 2\mathbf{j} - 6\mathbf{k}$

7 a $1 - \dfrac{1}{x - 1} + \dfrac{5}{x - 3}$

8 b $\ln y = -3\cos x + \frac{1}{3}\cos 3x + \frac{11}{6}$

    c $4e^{\frac{11}{6}}$

Glossary

abscissa

The x-coordinate

algorithm

A systematic procedure for solving a problem

altitude of a triangle

The perpendicular distance from a vertex to the base

angle of elevation or depression

The angle between the line of sight and the horizontal

arithmetic series

A series whose consecutive terms have a common difference

ascending powers of x

In order, smallest power of x first, e.g.
$a_0 + a_1 x + a_2 x^2 + \cdots$

asymptote

A line to which a curve approaches

axiom

A statement which is self-evidently true or accepted as true

base of a log

b as in $\log_b x$

base of a power

b as in b^x; *see also* exponent, index

base vector

In 3D space, the (usual) base vectors are

$$\mathbf{i} = \begin{pmatrix} 1 \\ 0 \\ 0 \end{pmatrix}, \mathbf{j} = \begin{pmatrix} 0 \\ 1 \\ 0 \end{pmatrix}, \mathbf{k} = \begin{pmatrix} 0 \\ 0 \\ 1 \end{pmatrix}$$

in terms of which any other vector can be written

bearing

A direction measured from the North clockwise

binomial

An expression consisting of two terms, e.g. $x + y$

bisect

Cut into two equal parts

Cartesian equation

Relationship connecting x and y coordinates

chain rule

A rule used to differentiate a composite function

chord

A straight line joining two points on a curve

circumcentre of a triangle

The centre of the circle which passes through all three vertices of the triangle

coefficient

The numerical factor in a term containing variables, e.g. -5 in $-5x^2 y$

collinear

Lying on the same straight line

commutative

An operation which is independent of order, e.g. addition is commutative: $a + b = b + a$

complex number

A number of the form $a + ib$ where $i = \sqrt{-1}$

composite function

A function which results from combining two functions so that the output of the first becomes the input of the second

composite number

A positive integer with factors other than 1 and itself

concurrent

Meeting at a point (of three or more lines)

congruent

Identical in shape and size

constant

A quantity whose value is fixed

constant term

The term in an expression which has no variable component , e.g. -3 in $x^2 - 5x - 3$

continuous curve or function

One whose graph has no break in it

continuous variable

A variable that can take all real values

convergent

Approaching closer and closer to a limit

converse

The converse of $a \Rightarrow b$ is $b \Rightarrow a$

coordinate

A magnitude used to specify a position

coprime numbers

Two positive integers whose HCF is 1

corollary

A result which follows directly from one proved

counter-example

An example which disproves a hypothesis

cubic equation

An equation of the form $ax^3 + bx^2 + cx + d = 0$, the highest power of x being 3

definite integral

An integral with limits

degree of a polynomial

The highest power of the variable, e.g. 2 in $x^2 - 5x - 3$

denominator

'Bottom' of a fraction – the divisor – remember D for Down

descending powers of x

In order, largest power of x first, e.g.
$a_4 x^4 + a_3 x^3 + a_2 x^2 + \cdots$

difference of squares

$$x^2 - y^2 = (x + y)(x - y)$$

differential coefficient

$$\frac{dy}{dx}, \frac{d^2y}{dx^2}, \text{ etc.}$$

differential equation

An equation containing at least one differential coefficient

direction vector (of a line)

A vector in the direction of the given line

discontinuous curve or function

One whose graph has a break in it

discriminant of a quadratic equation

The value of $b^2 - 4ac$ for the equation $ax^2 + bx + c = 0$

displacement

Change in position from a given point

displacement vector

A vector representing the translation (movement) from one point to another point

distributive

An operation which allows brackets to be removed, e.g. multiplication is distributive over addition: $a(b + c) = ab + ac$

divergent

Not convergent

dividend

A number (or expression) which is divided by a divisor to produce a quotient and possibly a remainder

divisibility tests

A number is divisible by

2 if the last digit is even

3 if the digit sum is divisible by 3

4 if the number formed by the last two digits is divisible by 4

5 if the last digit is 0 or 5

8 if the number formed by the last three digits is divisible by 8

9 if the digit sum is divisible by 9

11 if the sum of the digits in the odd positions differs from the sum of the digits in the even positions by 0 or any multiple of 11

For a composite number, such as 6 ($= 2 \times 3$) use tests for 2 and 3

divisor

A number (or expression) by which another is divided to produce a quotient and possibly a remainder

domain

The set of 'inputs' to a function

even function

A function where $f(x) = f(-x)$; the function is symmetrical about the y-axis

explicit function

A function expressed in the form $y = f(x)$, e.g. $y = x^3 - \ln x$

exponent

In the power 3^4, 4 is the exponent; also called index

exponential function

A function of the form $a^{f(x)}$ where a is constant, e.g. 2^x; and e^x, *the* exponential function

foot of a perpendicular

The point where the perpendicular meets a specified line

frustum of a cone or pyramid

The part remaining when the top is cut off by a plane parallel to the base

function

A one–one or many–one relationship between the elements of two sets; for any value in the domain, the value in the range is uniquely determined

general solution

A solution, given in terms of a variable, which generates all required solutions

general solution (of a differential equation)

A solution containing an arbitrary integration constant

geometric series

A series whose consecutive terms have a common ratio

HCF

Highest common factor

heptagon

A seven-sided 2D figure

hypotenuse

The side of a right-angled triangle opposite the right angle

identity

An equation which is true for all values of the variable(s)

implicit function

A function not expressed in the form $y = f(x)$, e.g. $x^2 + 3xy - \sin y = 0$

improper fraction (algebraic)

Fraction where the degree of the numerator is greater than or equal to the degree of the denominator

improper fraction (numerical)

$\frac{p}{q}$ where $p > q$; p, q are positive integers; *see also* proper fractions

incentre of a triangle

The centre of the circle which touches all three sides of the triangle

included angle

The angle between two given sides

increment

A small change in the value of a quantity

indefinite integral

An integral without limits

index (pl. indices)

In 3^4, 4 is the index; also called exponent

infinity (∞)

The concept of 'without end'

integer

A whole number, +ve or −ve or zero

integrand

The function to be integrated

integration by parts

A rule used to integrate an expression consisting of the product of two functions

inverse function

The function $f^{-1}(x)$ which 'undoes' the function $f(x)$

irrational number

A real number which is not rational, e.g $\sqrt{2}$, π, e

isosceles trapezium

A trapezium with an axis of symmetry through the mid-points of the parallel sides

kite

A quadrilateral with one diagonal an axis of symmetry

LCM

Lowest (or least) common multiple

LHS

Left-hand side, for example, of an equation

limit

The value to which a sequence converges

line segment

A finite part of an infinite line

linear function

A function whose highest power of x is x^1, e.g. $3x + 2$, $4x$

ln

Napierian or natural log, to base e

locus

A set of points satisfying some specified conditions

logarithm (log) of a number

The power to which a base must be raised to obtain the number

lowest terms

In its lowest terms, a fraction which cannot be cancelled, the numerator and denominator having no common factor

major

The larger arc, sector or segment

mapping

A relationship between two sets

median of a triangle

A line joining a vertex to the mid-point of the opposite side

minor

The smaller arc, sector or segment

modulus (of a vector)

An alternative name for the magnitude of a vector

monomial

An expression consisting of one term

Napierian or natural log

log to the base e, ln

normal at a point

A line which passes through the point and is perpendicular to the curve at that point

numerator

'Top' of a fraction; the dividend

odd function

A function where $f(x) = -f(-x)$; the function has 180° rotational symmetry about the origin

order (of a differential equation)

The order of the highest differential coefficient

e.g. $\dfrac{d^2y}{dx^2} + 2\dfrac{dy}{dx} = 5$ (second order),

$\dfrac{dy}{dx} = 3x$ (first order)

ordinate

The y-coordinate

oscillating sequence

A sequence which neither converges to a limit, nor diverges to $+\infty$ or $-\infty$

parallelogram

A quadrilateral with both pairs of opposite sides parallel

parametric equation of a curve

An equation in which x and y are each expressed in terms of a third variable

partial fractions

An expression decomposed into the sum of two or more separate fractions

particular solution

A specific member of the general solution of a differential equation

period

The smallest interval (or number of terms) after which a function (or sequence) regularly repeats

periodic function or sequence

One which repeats at regular intervals

perpendicular

At right angles

perpendicular bisector of AB

The line which bisects AB at right angles; the set of points equidistant from A and B

Platonic solid

A solid all of whose faces are identical regular polygons

point of contact

The point at which a tangent touches a curve

polygon

A plane figure with many sides

polynomial (of degree n)

A sum of terms of the form $a_0 + a_1 x + a_2 x^2 + \cdots + a_n x^n$

position vector

A displacement vector from the origin to a point, e.g. $\overrightarrow{OP}$, written as **p**

power

For example, $81 = 3^4$ is the fourth power of 3; *see also* base, exponent, index

prime number

A positive integer which is divisible only by itself and 1
NB: 1 is not included in the set of prime numbers

prism

A solid with uniform cross-section

produce

Extend, as of a line

product rule

A rule used to differentiate an expression consisting of the product of two functions

proper fraction (algebraic)

Fraction where the degree of the numerator is less than the degree of the denominator

proper fraction (numerical)

$\frac{p}{q}$ where , $p < q$; p, q are positive integers; *see also* improper fraction

quadrant

One of the four parts into which the plane is divided by the coordinate axes

quadratic equation

An equation of the form $ax^2 + bx + c = 0$, the highest power of x being 2

quartic equation

An equation of the form $ax^4 + bx^3 + cx^2 + dx + e = 0$, the highest power of x being 4

quotient

The result of dividing one number or expression (dividend) by another (divisor) – there may be a remainder

quotient rule

A rule used to differentiate an expression consisting of one function divided by another

radian

Measure of an angle; 1 radian = angle subtended at the centre of a circle radius r by an arc of length r; 1 radian $\approx 57°$

range

The set of 'outputs' of a function

rational number

A number which can be expressed as $\frac{p}{q}$ where p and q are integers, $q \neq 0$

real number

A number corresponding to some point on the number line

reciprocal of $\frac{a}{b}$

$\frac{b}{a}$; and vice versa, reciprocal of $\frac{b}{a}$ is $\frac{a}{b}$

reductio ad absurdum

Proof by assuming the result is not true and arriving at a contradiction

reflex angle

An angle between 180° and 360°

regular polygon

A polygon with all sides and all angles equal

respectively

In the order mentioned

rhombus

A parallelogram with four equal sides; the diagonals bisect each other at 90°

RHS

Right-hand side, for example, of an equation

right cone or pyramid

The vertex being vertically above the centre of the base

root of a number

$\sqrt[n]{a}$ (nth root of a); if nth root of a is b, then $b^n = a$

root of an equation

A solution of the equation

scalar

A quantity that has magnitude but no direction

scalar product (of two vectors)

$\mathbf{a}.\mathbf{b} = |\mathbf{a}||\mathbf{b}| \cos \theta$, where θ is the angle between the vectors **a** and **b**

scale factor

The number by which corresponding lengths are multiplied in similar figures or in a transformation

scalene

A triangle with three unequal length sides

sector of a circle

Part of a circle bounded by an arc of the circle and two radii

segment of a circle

Part of a circle bounded by an arc of the circle and a chord

separating the variables

A technique used to solve a type of first-order differential equation

sequence

An ordered list of numbers or terms, e.g. 1, 2, 4, 8, ...

series

The sum of a sequence, e.g. $1 + 2 + 4 + 8 + \cdots$

sigma (Σ)

Symbol indicating summation, e.g. $\sum_{1}^{n} r = 1 + 2 + \cdots + n$

similar

Having the same shape (all corresponding lengths being multiplied by the same scale factor)

skew lines

A pair of lines in 3D space which are not parallel and do not meet

slant height of a cone

The distance from the vertex to a point on the circumference of the base

solid of revolution

The solid formed when a curve or area is rotated about a line

solution

A value (or values) which satisfies the given problem

standard form

A number in the form $a \times 10^n$ where $1 \leqslant a < 10$ and n is an integer

subtended angle

Angle subtended by the line segment AB at C is the angle ACB

surd

An irrational root, e.g. $\sqrt{2}$, $\sqrt{7}$, $\sqrt[3]{11}$

tangent at a point

A line which passes through the point and touches the curve at that point

term (of a sequence)

One of a sequence, e.g. 4 in 1, 2, 4, 8, ...

term (of an expression)

Part of an expression. e.g. x^2, $-5x$ or -3 in $x^2 - 5x - 3$

theorem

A proposition proved by logical reasoning

trapezium

A quadrilateral with one pair of sides parallel

trinomial

An expression consisting of three terms

unit vector

A vector whose magnitude is one

unknown

A letter which represents a specific value or values

variable

A letter which represents various values

vector

A quantity which has both magnitude and direction

vertex (pl. vertices) of a parabola

The turning point of a parabola

vertex (pl. vertices) of a polygon

The point where two sides meet

vertex (pl. vertices) of a solid

The point of a cone or the point where faces of the solid meet

volume of revolution

The volume of a solid of revolution

zero vector

A vector of magnitude zero (the zero vector has no direction)

SINGLE USER LICENCE AGREEMENT FOR A2 CORE FOR AQA CD-ROM
IMPORTANT: READ CAREFULLY

WARNING: BY OPENING THE PACKAGE YOU AGREE TO BE BOUND BY THE TERMS OF THE LICENCE AGREEMENT BELOW.

This is a legally binding agreement between You (the user or purchaser) and Pearson Education Limited. By retaining this licence, any software media or accompanying written materials or carrying out any of the permitted activities You agree to be bound by the terms of the licence agreement below.

If You do not agree to these terms then promptly return the entire publication (this licence and all software, written materials, packaging and any other components received with it) with Your sales receipt to Your supplier for a full refund.

YOU ARE PERMITTED TO:

| | Use (load into temporary memory or permanent storage) a single copy of the *Live Player* software on only one computer at a time. If this computer is linked to a network then the *Live Player* software may only be installed in a manner such that it is not accessible to other machines on the network.

| | Transfer the *Live Player* software from one computer to another provided that you only use it on one computer at a time.

| | Print a single copy of any PDF file from the CD-ROM for the sole use of the user.

YOU MAY NOT:

| | Rent or lease the software or any part of the publication.

| | Copy any part of the documentation, except where specifically indicated otherwise.

| | Make copies of the software, other than for backup purposes.

| | Reverse engineer, decompile or disassemble the software.

| | Use the software on more than one computer at a time.

| | Install the software on any networked computer in a way that could allow access to it from more than one machine on the network.

| | Use the software in any way not specified above without the prior written consent of Pearson Education Limited.

| | Print off multiple copies of any PDF file.

ONE COPY ONLY

This licence is for a single user copy of the software

PEARSON EDUCATION LIMITED RESERVES THE RIGHT TO TERMINATE THIS LICENCE BY WRITTEN NOTICE AND TO TAKE ACTION TO RECOVER ANY DAMAGES SUFFERED BY PEARSON EDUCATION LIMITED IF YOU BREACH ANY PROVISION OF THIS AGREEMENT.

You only own the disk on which the software is supplied.

Pearson Education Limited warrants that the diskette or CD-ROM on which the software is supplied are free from defects in materials and workmanship under normal use for ninety (90) days from the date You receive them. This warranty is limited to You and is not transferable. Pearson Education Limited does not warrant that the functions of the software meet Your requirements or that the media is compatible with any computer system on which it is used or that the operation of the software will be unlimited or error free.

You assume responsibility for selecting the software to achieve Your intended results and for the installation of, the use of and the results obtained from the software. The entire liability of Pearson Education Limited and its suppliers and your only remedy shall be replacement of the components that do not meet this warranty free of charge.

This limited warranty is void if any damage has resulted from accident, abuse, misapplication, service or modification by someone other than Pearson Education Limited. In no event shall Pearson Education Limited or its suppliers be liable for any damages whatsoever arising out of installation of the software, even if advised of the possibility of such damages. Pearson Education Limited will not be liable for any loss or damage of any nature suffered by any party as a result of reliance upon or reproduction of or any errors in the content of the publication.

Pearson Education Limited does not limit its liability for death or personal injury caused by its negligence.

This licence agreement shall be governed by and interpreted and construed in accordance with English law.